Biotechnology in the Developing World and Countries in Economic Transition

Biotechnology in the Developing World and Countries in Economic Transition

G.T. Tzotzos
United Nations Industrial Development Organization
Vienna
Austria

K.G. Skryabin
Russian Academy of Sciences
Moscow
Russia

CABI *Publishing*

CABI *Publishing* is a division of CAB *International*

CABI Publishing
CAB International
Wallingford
Oxon OX10 8DE
UK

Tel: +44 (0)1491 832111
Fax: +44 (0)1491 833508
Email: cabi@cabi.org

CABI Publishing
10 E 40th Street
Suite 3203
New York, NY 10016
USA

Tel: +1 212 481 7018
Fax: +1 212 686 7993
Email: cabi-nao@cabi.org

A catalogue record for this book is available from the British Library, London, UK.
A catalogue record for this book is available from the Library of Congress, Washington DC.

ISBN 0 85199 331 1

Printed and bound in the UK by Cromwell Press, Trowbridge, Wiltshire, from copy supplied by the author.

Contents

List of Contributors

Authors of Chapters

Tom Crompton

Liaison Office of the International Centre for Genetic Engineering and Biotechnology/ United Nations Industrial Development Organization, Vienna, and Visiting Fellow at the Department of Environmental Science and Health, University of East London, Romford Road, London, E15 4LZ. E-mail: crompton@uel.ac.uk

John Komen

John Komen is Research Analyst, Intermediary Biotechnology Service (IBS), International Service for National Agricultural Research (ISNAR), The Hague, The Netherlands.

George Tzotzos

George Tzotzos is Chief, Biodiversity Unit at the United Nations Industrial Development Organization (UNIDO), Sectoral Support and Environmental Sustainability Division, Cleaner Production and Environmental Management Branch, PO Box 300, A-1400, Vienna, Austria. Tel: +43 1 26226 4336; Fax: +43 1 26026 6810; E-mail: george@binas.unido.org

Authors of Country Profiles

Belarus

N.A. Kartel, A.G. Lobanok, V.M. Dorogoush and O.A. Kazuchets
Prof N.A. Kartel is Director of the Institute of Genetics and Cytology of the National Academy of Sciences of Belarus, ul.F.Skoriniy 27, 220734 Minsk 72, Belarus. Tel: +375 17 2684848; Fax: +375 17 2684917.
Prof A.G. Lobanok is Director of the Institute of Microbiology of the National Academy of Sciences of Belarus, ul.Zhodinskaya 2, Academgorodok, 220141 Minsk, Belarus. Tel./ Fax: +375 17 2644766; E-mail: lobanok@mbio.basnet.minsk.by
Dr V.M. Dorogoush is Head of Department Innovation, Concern 'Belbiopharm', ul. Varvasheni 17, 220029 Minsk, Belarus. Tel: +375 17 2342336; Fax: +375 17 2769402.
Dr O.A. Kazuchets, Department Innovation of the Concern 'Belbiopharm'.

Brazil

Leila Macedo Oda, Bernardo Elias Correa Soares and Maria Cleria Valadares-Inglis
Leila Macedo Oda is Representative of the Ministry of Health and Head of the Health Working Group in the National Technical Biosafety Committee of Brazil and Senior Research Scientist at the Oswaldo Cruz Foundation (FIOCRUZ), Av. Brasil, 4036/sl. 716 Manguinhos 21040-361, Rio de Janeiro, Brazil. Tel: +55 21 2605979; Fax: +55 21 2609944; E-mail: oda@dcc001.cict.fiocruz.br
Bernardo Elias Correa Soares is Technical Adviser of the Brazilian National Capacity Building Programme on Biosafety and Senior Technologist at the Oswaldo Cruz Foundation (FIOCRUZ). E-mail: bernardo@dcc001.cict.fiocruz.br
Maria Cleria Valadares-Inglis is Executive Secretary for Biotechnology EMBRAPA, Parque Rural, Final W3 Norte, CP02372, CEP 70770-900, Norte Brasilia – DF, Brazil. Tel: +55 61 3484603, 3483604; Fax: +55 61 3403658; E-mail: cleria@cenargen.embrapa.br

Bulgaria

Pavlin Petrinsky and Atanas Atanassov
Prof A. Atanassov is Director of the Institute of Genetic Engineering, Plant Biotechnology Research Centre, Via Sofia, 2232 Kostinbrod-2, Bulgaria. Tel: +359 7212552; Fax: +359 7214985; E-mail: atanas@icgeb.trieste.it and also dmuij@bgcict.acad.bg
Pavlin Petrinsky is management consultant at Complete Business Solutions (CBS) Ltd., Office 335, 127 Rakovski Str., Sofia 1000, Bulgaria. Tel: +359 2 8602472; Fax: +359 2 8602483; E-mail: CBS@BGnet.bg

Chile

Lionel Gil and Carlos Irarrázabal
Dr Lionel Gil and Dr Carlos Irarrázabal, University of Chile, Facultad de Medicina, Departmento de Bioquimica, Laboratorio de Bioquimica y Toxicologia Ambiental, Independencia 1027, Casilla 70086, Santiago 7, Chile. Tel: +56 2 6786068, 6786061; Fax: +56 2 7356373; E-mail: lgil@machi.med.uchile.cl

China

Wei Jiewei
China National Centre for Biotechnology Development (CNCBD), 7B Zao Jun Miao, Haidan Qu, P.O. Box 8118, Beijing 100081, China. Tel: +86 10 62251955; Fax: +86 10 62254106.

Colombia

Rafael H. Aramendis and Elizabeth Hodson de Jaramillo
Instituto Colombiano para el Desarrollo y la Tecnología, Francisco José de Caldas – COLCIENCIAS (Colombian Institute for Science and Technology Development), Programa Nacional de Biotecnología, Transversal 9A 133-28, Santa Fé de Bogotá, Colombia. Tel: +57 1 2169800; Fax: +57 1 6251788; E-mail: raramend@colciencias.gov.co; simbiosis@colciencias.gov.co; ehodson@colciencias.gov.co; http://www.colciencias.gov.co/simbiosis/

Cuba

Pedro A. López-Saura
Dr Pedro López-Saura is Vice-director of Regulatory Affairs at the Centre for Genetic Engineering and Biotechnology (CGEB), Apartado Postal 6162, Ave. 31e/158 y 190, Cubanacán, Playa, Ciudad de la Habana, Cuba. Tel: +53 7 217377-79; Fax: +53 7 218070; E-mail: clintr@cigbdec.cigb.edu.cu

Egypt

Hassan Moawad and Magdy Madkour
Prof Magdy Madkour is Director of the Agricultural Genetic Engineering Research Institute (AGERI), A.R.C., 9 Gamaa Street, Giza 12619, Egypt. Tel: +20 2 5727831; Fax: +20 2 5689519; E-mail: nagel@frcu.eun.eg; madkour@ageri.sci.eg
Dr H. Moawad is Dean of the Genetic Engineering and Biotechnology Research Institute (GEBRI).

Hungary

Ferenc Rudan
Dr F. Rudan is Director of the National Committee for Technological Development, Szervita tér 3, Budapest, H-1052 Hungary. Tel: +36 1 1187076; Fax: +36 1 1187998; E-mail: ferenc.rudan@omfb.x400gw.itb.hu

India

Manju Sharma and Renu Swarup
Dr Manju Sharma is Secretary, and Dr. Renu Swarup, Principal Scientific Officer at the Department of Biotechnology, Ministry of Science and Technology, New Delhi, India.

Indonesia

Sugiono Moeljopawiro
Dr S. Moeljopawiro is Head of the Molecular Biology Division, Research Institute for Food Crop Biotechnology (RIFCB), Jalan Tentara Pelajar 3A, Bogor 16114, Indonesia. Tel: +62 251316897; Fax: +62 251338893; E-mail: sugionom@indo.net.id

Kenya

Ahmed Hassanali
Prof Ahmed Hassanali is Deputy Director General of the International Centre of Insect Physiology and Ecology (ICIPE), Nairobi, Kenya.

Korea

Cherl-Ho Lee
Prof Cherl-Ho Lee, Centre for Advanced Food Science and Technology, Graduate School of Biotechnology, Korea University, 1 Anam-Dong, Sungbuk-ku, Seoul, 136-701 Korea. Tel: +82 2 3293414; Fax: +82 2 9275201; E-mail: chlee@kuccnx.korea.ac.kr

Kuwait

Sabah AlMomin
Dr S. AlMomin is Director of the Biotechnology Department, Kuwait Institute for Scientific Research, P.O. Box 24885, Safat 13109, Kuwait.

Latvia

Martin Beker, Alexander Rapoport and Uldis Viesturs
Dr M. Beker, Institute of Microbiology and Biotechnology, University of Latvia, Kronvald Blvd. 4, LV-1586 Riga, Latvia. Tel: +371 7325558; Fax: +371 7323065; E-mail: muter@acad.latnet.lv
Dr A. Rapoport, Institute of Microbiology and Biotechnology, University of Latvia, Kronvald Blvd. 4, LV-1586 Riga, Latvia. Tel: +371 7323019; Fax: +371 7323065; E-mail: rapoport@latnet.lv
Dr U. Viesturs is Director of the Institute of Microbiology and Biotechnology, University of Latvia, Kronvald Blvd. 4, LV-1586 Riga, Latvia. Tel: +371 7325123; Fax: +371 7323065, E-mail: koks@edzi.lza.lv

Lithuania

Arvydas Janulaitis
Prof A. Janulaitis is Director of the Institute of Biotechnology, V. Graiciuno 8, 2028 Vilnius, Lithuania.

Malaysia

A.H. Zakri
Prof A.H. Zakri is Deputy Vice Chancellor of Academic Affairs, Universiti Kebangsaan Malaysia, 43600 UKM Bangi, Selangor Darul Ehsan, Malaysia. Tel: +60 3 8250001; Fax: +60 3 8256484; E-mail: zakri@pkrisc.cc.ukm.my

Mexico

José Luis Solleiro
Dr J.L. Solleiro is Coordinator of CamBioTec (Canada-Latin America Initiative on Biotechnology for Sustainable Development), P.O. Box 20-103, 01000 México, D.F., Mexico. E-mail: solleiro@servidor.unam.mx

Morocco

Arifi Abdelaziz
Dr A. Abdelaziz is Director of the National Institute of Agronomic Research (INRA), Avenue de la Victoire, B.P. 415 RP, Rabat, Morocco. Tel: +212 7 772639; Fax: +212 7 770049.

Nigeria

Nduka Okafor
Prof N. Okafor is President of the Foundation for African Development through

International Biotechnology (FADIB), P.O. Box 1457, Enugu, Nigeria. Tel: +234 42 459360; Fax: +234 42 453202; E-mail: 107705.3607@compuserve.com

Pakistan

S. Riazuddin

Prof S. Riazuddin is Director of the National Centre of Excellence in Molecular Biology, University of the Punjab, Canal Bank Road, Thokar Niaz Baig, Lahore 53700, Pakistan. Tel: +92 42 5421235; Fax: +92 42 5164155; E-mail: riaz@lhr.comsats.net.pk

Philippines

Virginia G. Novenario-Enriquez

Virginia G. Novenario-Enriquez is Chief Science Research Specialist, Research Information and Utilisation Division, Philippine Council for Advanced Science and Technology Research and Development, Department of Science and Technology, Republic of the Philippines. Tel: +63 2 8373168; E-mail: vgn@sun1.dost.gov.ph

Poland

Tomasz Twardowski, Andrzej B. Legocki and Andrzej Zabza

Prof Tomasz Twardowski, Institute of Bioorganic Chemistry, Polish Academy of Sciences, ul. Noskowskiego 12/14, 61-704 Poznañ, Poland. Tel: +48 61 8528503; Fax: +48 61 8520532; E-mail: twardows@ibch.poznan.pl

Prof Andrzej B. Legocki is Director of the Institute of Bioorganic Chemistry of the Polish Academy of Sciences.

Prof Andrzej Zabza, Institute of Organic Chemistry, Biochemistry and Biotechnology, Technical University of Wroclaw, ul. Wybrzeze Wyspianskiego 27, 50-370 Wroclaw, Poland. Tel: +48 71 3203446; Fax: +48 71 3284064; E-mail: zabza@kchf.ch.pwr.wroc.pl

Russian Federation

K.G. Skryabin and L.P. Matyash

Prof K.G. Skryabin is Chairman of the Scientific Council for Biotechnology, Russian Academy of Sciences, Prosp. 60let Oktyabrya 7 kor.1., 117312 Moscow, Russian Federation. Tel: +7 095 1357319; Fax: +7 095 1350571; E-mail: Skryabin@biengi.ac.ru

Dr L.P. Matyash is Executive Secretary of the Scientific Council for Biotechnology, Russian Academy of Sciences.

South Africa

Muffy Koch and Jocelyn Webster

Mrs Muffy Koch, Director, Innovation Biotechnology, Box 309, Noordwyk, 1687 South Africa. Tel/Fax: +27 11 3181397; E-mail: mk000005@pixie.co.za

Dr Jocelyn R. Webster, Director, Probio, Box 873, Irene 0062, South Africa. Tel: +27 12 6672689; Fax: +27 12 6671920.

Thailand

Sutat Sriwatanapongse, Morakot Tanticharoen and Sakarindr Bhumiratana
Dr S. Bhumiratana is Director and Drs S. Sriwatanapongse and M. Tanticharoen are Deputy Directors of the National Centre for Genetic Engineering and Biotechnology (BIOTEC) of the National Science and Technology Development Agency (NSTDA), Yothi Road, Phyathai, Bangkok 10400, Thailand. Tel: +66 2 6448150-4; Fax: +66 2 6448107; E-mail: sakarind@biotec.or.th

Ukraine

L.N. Shynkarenko, O.V. Novozhilov and Ya.B. Blume
Dr L.N. Shynkarenko is Head of the Department of Biotechnology and Medicine, Ministry of Science and Technology of Ukraine, T. Shevchenko blvrd.16, Kyiv MSP-30, 252601, Ukraine. Tel: +380 44 2463894; Fax: +380 44 2463884.
Dr O.V. Novozhilov, Senior Officer, Presidium of National Academy of Sciences of Ukraine, Department of Science Organisation, Volodymyrs'ka St. 54, Kyiv MSP-30, 252601, Ukraine. Tel: +380 44 2216508; Fax: +380 44 2961063.
Prof Ya.B. Blume, Deputy Director, National Academy of Sciences of Ukraine, Institute of Cell Biology and Genetic Engineering, acad. Zabolotny St. 148, Kyiv MSP-22, 252650, Ukraine. Tel: +380 44 2667104; Fax: +380 44 2521786; E-mail: yablume@cellbio.freenet.viaduk.net

Venezuela

Edgar Otaiza-Vásquez and Asdrúbal Arcia
Edgar Otaiza-Vásquez, Professor of Biochemistry and Genetics, School of Veterinary Sciences, Central University of Venezuela. Address: Avda. Portugal No. 4 - 41, Urb. Santa Elena, Barquisimeto 3001, Venezuela. Tel: +58 51 549152; E-mail: eotaiza@reacciun.ve
Dr Asdrúbal Arcia, Professor of Plant Breeding, School of Agronomy, Central University of Venezuela. Address: Avda. Circunvalación 1, No. 19, Urb. El Castaño, Maracay, Venezuela. Tel: +58 43 411768; E-mail: aarcia@reacciun.ve

Zimbabwe

Joseph Gopo
Dr J.M. Gopo is Director of the Biotechnology Research Institute, Scientific and Industrial Research and Development Centre (SIRDC), P. Bag 7726, Causeway, Harare, Zimbabwe. Tel: +263 4 860320-9; Fax: +263 4 860351.

Preface

The prevailing image of biotechnology is now that of an all-pervasive profit-generating technology playing a strategic role in maintaining and enhancing national competitiveness in an environment of global economic interaction. While the beneficial impacts of the technology are now becoming evident in the industrialized world, the developing world has yet to share any of these. Only a few developing countries - mostly top-tier - have an incipient biotechnology sector, or, in some cases, the research and development (R&D) component thereof; and even in many of these instances, the future holds a great deal of uncertainty. As for the 'ground-level' social benefits of biotechnology, they have yet to make themselves felt.

The present volume is essentially the second edition of an earlier volume under the title 'Biotechnology Worldwide' published by the Committee on Biotechnology (COBIOTECH) of the International Council of Scientific Unions. In view of the availability of a good number of books and reports surveying biotechnology developments in the industrialized world we have concentrated our effort in reviewing recent developments in selected countries of the developing world and of Central and Eastern Europe; a fact that is reflected in the title of the volume in hand.

The scope of coverage has been determined by the adopted definition of biotechnology as 'the use of contemporary technologies, specifically recombinant DNA and related techniques, to make or modify products or processes for specific use'. Although this definition is restrictive we felt that the value of this work is in presenting national efforts in cutting-edge fields of biotechnology rather than in the more conventional ones. This in no way implies that the latter are of no relevance to developing countries.

The fact that the volume has been written by a large number of authors from different parts of the world has inevitably resulted in different approaches to the subject matter as well as to style. Every effort has been made to keep heterogeneity of content to a minimum without compromising the work of the authors. Where we failed to do so, we offer our sincere apologies to both authors and readers. In any case, without the contribution of these authors, the book would not have materialized. Also, we would like to acknowledge with sincere thanks the contribution of Susan Tzotzos, who did the layout and extensive editorial research. We have made every possible effort to ensure that the work contained in this volume is both factual and informative.

George T. Tzotzos
Konstantin G. Skryabin

Chapter One
Industrial Biotechnology: Challenges and Opportunities

George T. Tzotzos

In the industrialized world biotechnology is now beginning to revolutionize all major sectors of industrial activity. Indeed, innovation capability in biotechnology is seen as a key ingredient in maintaining and/or expanding the national share in a global economy.

The driving force of advanced biotechnology is the exceptional dynamism of the science-technology interface; science and technology are intertwined in such a way that technology is simultaneously highly dependent upon its scientific underpinnings and constantly pushing the knowledge base forward, thus creating the grounds for further technological breakthroughs. The pace of scientific and technological advancement has been such that many powerful and sophisticated techniques, technologies and equipment are used routinely not only in industrial countries but also in an increasing number of laboratories in the developing world. Unfortunately, this does not mean that the innovation capacity in the developing world has been enhanced accordingly.

In the top-tier industrialized countries, innovation in biotechnology has been accompanied by a series of dynamic adjustments at the institutional and social levels. These include innovative approaches to technology transfer and to financing; structural and strategic reorganization at the enterprise level; skill upgrading through continuous education schemes; and important shifts in science and industrial policy. On balance, these mutually complementary adjustments have created a positive institutional climate for biotechnology innovation. Where such institutional adjustments failed to occur because of systemic rigidities, technological innovation and diffusion has been uneven and slower than expected.

Conditions for and barriers to commercial entry into biotechnology are subject to contextual variability. Requirements tend to combine differently from one country or region to another and, similarly, a wide range of factors including resource endowment, previous level of technology capacity building and macroeconomic environment, to mention but a few, come into play. Furthermore, it is acknowledged that these not only vary from one country (or region) to another but some are dependent on domestic conditions; others lie utterly beyond national control. In this regard, it is deceptive to talk of developing countries as a single homogeneous group.

Whereas, most developing countries lack many if not all of the critical ingredients that go into building a biotechnology capability there is another group of countries (e.g. the Asian newly industrialized countries, China, India, Brazil, Mexico) that have a working understanding of the complex and multiple requirements that must be met in order to develop an economically viable biotechnology industry. For the least developed countries, the vast majority of which are in Africa, biotechnology, like the other advanced generic technologies, may contribute to widening the growth gap between themselves and the industrialized world. It may also accentuate the difference between them and those developing countries that are effectively exploiting knowledge-intensive technologies.

The asymmetry of problems, opportunities and priorities between the industrial and developing world on one hand and amongst developing countries themselves on the other, confounds generalizations regarding biotechnology's potential impacts. In the following, an attempt is made to assess biotechnology's potential economic and social impact on developing countries in the areas of health care, agriculture and the environment/industry in the broadest possible terms. What does the technology offer in these industrial sectors? How does this relate to developing countries' needs and demand? What are the prospects for domestic technological innovation?

Mapping the Landscape

The so-called industrial biotechnology sector 'embraces all companies which use modern biological, rather than conventional, techniques to develop commercial products for human or animal healthcare, agricultural productivity, food processing, and the environmental services sector' (Ernst and Young, 1997).

These companies engage in commercial activities in the health care (diagnostic and therapeutic), agri-food, chemical, environmental and supplier (instrumentation, lab supplies, reagents) sectors as well as in the provision of specialized services. The unique characteristic of the biotechnology industry (as defined above) is its research-intensive nature. Research and development investment reached US $9.9 billion in 1998 in USA alone. 'The top five biotech companies spent an average of US $121,400 per employee on R&D. This compares with an average of US $31,200 per employee for the top pharmaceutical companies' (The Biotechnology Industry Organization, 1999). Sales of the US biotechnology industry reached US $13 billion in 1997, a 20% increase over 1996 (The Biotechnology Industry Organization, 1999).

Healthcare

Most of the R&D investment (estimated to have reached US $8 billion in 1996 (James, 1997)) has been channelled into the development of novel drugs and vaccines intended for the treatment and prevention of degenerative diseases (e.g. cancer, cardiovascular, etc.) affecting large numbers of people. Table 1.1 shows the number of biotech companies working on treatments and cures for these diseases.

Yet, more than 90% of the world's preventable mortality occurs in the developing world with least developed countries carrying the lion's share of the burden. An estimated 14 million children under the age of five die annually from acute respiratory infections, diarrhoeal disease and a combination of other diseases such as parasitic diseases, typhoid, tetanus and meningitis. The global distribution of illness and mortality from preventable causes is presented in Table 1.2.

Non-infectious/degenerative diseases are the main cause of death in the industrialized world. In the developing world, because of improving living standards and increasing life expectancy, healthcare systems will be challenged to cope with both infectious and degenerative diseases. Only 5% of the global investment in health research, which amounted to US $30 billion in 1986, is directed to health problems affecting the developing world. Although there is some evidence that the actual figure may be currently higher it is still small to deal with the magnitude of the problems faced.

Table 1.1. Number of US biotechnology companies currently working on treatments and cures for common diseases.

Disease or therapy	Number of biotech companies working on therapies/cures
AIDS/HIV	151
Cancer:	
Breast cancer	68
Colon cancer	25
Leukaemia	35
Liver cancer	17
Lung cancer	45
Ovarian cancer	27
Prostate cancer	56
Stomach cancer	4
Kidney cancer	16
Skin cancer	16
Diabetes	54
Heart/circulatory system diseases	16
Infectious diseases	70
Osteoporosis	44
Neurological disorders	14
Vaccines	98
Gene therapy	53
Skin and tissue replacement	10

Source: The Biotechnology Industry Organization (1999).

Conventional drugs, improvements in sanitation, food quality and water supply and control of infectiveness will continue to play a significant role in the prevention and treatment of infective diseases and constitute the primary approaches of health care systems in the developing world. Yet, some of the major killer diseases (e.g. malaria) still elude effective treatment and the most promising strategies to combat them are through the development of recombinant drugs and vaccines. Correlation of the data of Tables 1.1 and 1.2 shows little evidence that R&D investment is being directed for the development of drugs and vaccines for infectious diseases endemic in the developing world, obviously due to lack of commercial interest.

For the vast majority of developing countries there is no obvious commercial entry to healthcare biotechnology as development costs for new drugs and vaccines are prohibitive. ‘The approval process for biotechnology medicines is both lengthy and expensive. According to a study by the US Office of Technology Assessment, it costs US $200 million to US $350 million and takes from seven to 12 years for a product to move through the drug development and the FDA (Food and Drug Administration) approval process’ (The Biotechnology Industry Organization, 1999).

Some countries (e.g. Brazil, China, India, Mexico) with large domestic markets and sizable pharmaceutical industries have little investment incentives due to, *inter alia,* adverse intellectual property and pricing policies and shrinking drug lists for public reimbursement. For example, while India is, by volume, the second largest pharmaceutical market in the world, it ‘comes a lowly 13th by value (approximately US $4 billion). [But] because of India’s patent laws, which were all but scrapped in the early 1970s, margins have been pared wafer thin by competition from dozens of locally-produced generic drugs. Add to

Table 1.2. Estimated annual number of deaths by cause and in industrialized and developing countries, 1990 (thousands of persons).

Health problem	Industrialized countries	Developing countries
Pre-transition		
Infectious and parasitic diseases:	**(478)**	**(17,000)**
Acute respiratory infections	367	6,500
Diarrhoeal diseases		4,180
Tuberculosis	38	3,300
Malaria		1,000
Other infections and parasitic diseases	73	1,700
Measles		220
Pertussis		100
Post-transition		
Circulatory and degenerative diseases:	**(12,174)**	**(19,952)**
Malignant neoplasma	2,700	2,421
Chronic lung disease	388	2,500
Maternal causes		500
Perinatal conditions	86	3,030
Injury and poisoning	570	2,400
Other and unknown causes	1,930	3,670
Diseases of the circulatory system	6,500	5,431

Source: Modified from *Global Health Situation and Projections: Estimates*, WHO (1992).

this the low income of most Indians and a plethora of government price controls, and drug prices are often less than a tenth of European levels' (Financial Times, 1999). Notwithstanding this, the innovation potential of the top-tier developing countries and economies in transition is not inconsiderable.

A second avenue of technology acquisition through licensing does not appear to be very attractive. Royalties range from 1-5% for research reagents, 5-10% for therapeutic products and vaccines and may go as high as 15% for bulk product sales (Correa, 1991).

Agriculture

Population expansion at the projected levels and increasing land erosion pose a real threat to food security in most of the developing world. The annual growth of world agricultural output would need to be commensurate with that achieved over the last 40 years (2.43% from 1950 to 1985) to meet the demands of a population which some expect to reach ten billion by the year 2050. This population growth is predicted to occur mainly in the least developed countries according to a recent United Nations report (United Nations, 1995). In Africa in particular, population size over the period 1950 to 2015 is predicted to show a sixfold increase (United Nations, 1995). As illustrated in Figure 1.1, population size is levelling off in the developed world, whereas in developing countries it is increasing rapidly. It is precisely in the developing world where needs are most urgent to ensure food security. Although food security should not be equated with hunger alleviation, as the latter is intimately linked to issues such as management of land and water resources, distribution networks and pricing policy to mention but a few, plant biotechnology can expand the ecological limits of population growth by bringing more land into production and improving harvests. Current agri-biotechnology applications do not increase plant productivity *per se* but can be extremely effective in pest control lowering pre- and post-harvest losses by up to 25% (James, 1997).

Much the same way as in medicine, the private sector plays the dominant role in setting priority agri-biotechnology R&D targets. Research and development effort has focused on the major commercial crop varieties and a few selected traits (Figures 1.2 and 1.3). Transformation of crops of importance for the developing world, such as sweet potato or cassava, perennials, like cocoa and coffee, and legumes are conspicuously absent from the list of early commercialized products. The same applies for varieties of the major crops (e.g. rice, potato, maize) grown in tropical or subtropical climates where different biotic and abiotic stresses apply. The transformation of crops of little commercial interest for the North American and European markets is left in the domain of national and international research programmes (see Chapter 2).

In the livestock sector, veterinary growth promoters such as bovine and porcine somatotropin have been known to increase dairy productivity considerably. The development of recombinant animal diagnostics and vaccines has received a new impetus from human health care technological 'spin-offs'. A number of diagnostic systems based on monoclonal antibodies are already available. In the short term, new vaccines for cattle and poultry will have a major impact by increasing productivity, though lesser but still important improvements in animal performance are likely to be derived from the development of more flexible feed formulations. The creation of transgenic animals for

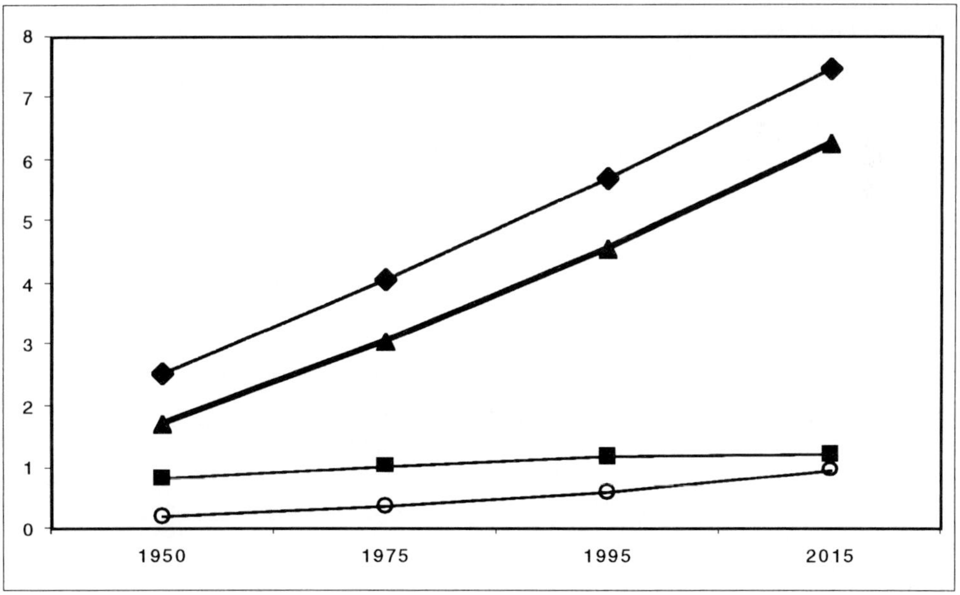

Figure 1.1. World population trends 1950-2015 (population size in billions for the world ◆, developing countries ▲, developed countries ■, and least developed countries O). Source: United Nations (1995).

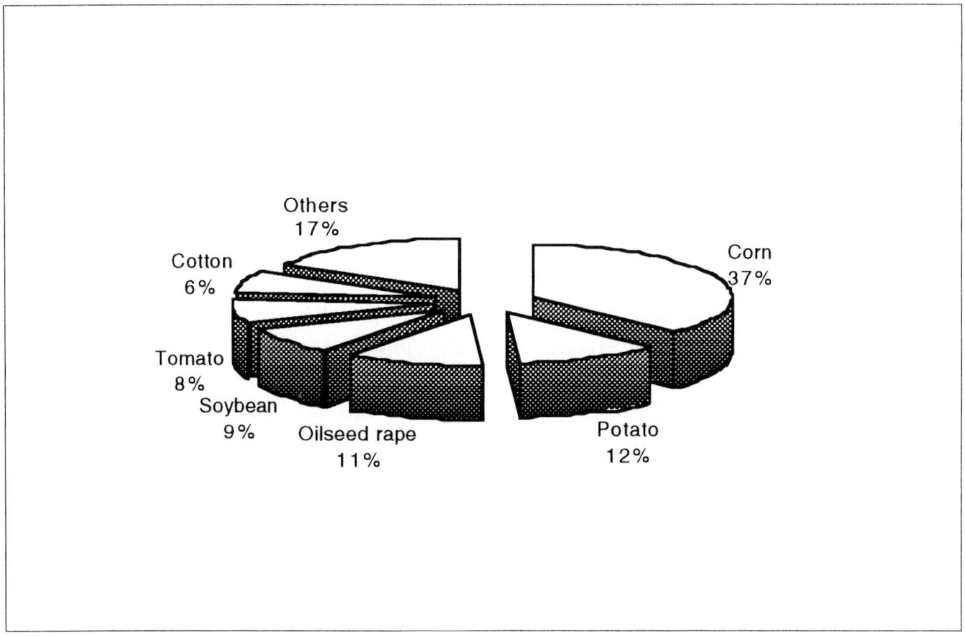

Figure 1.2. Global distribution of genetically modified crop species.
Source: BIOTRACK (1999).

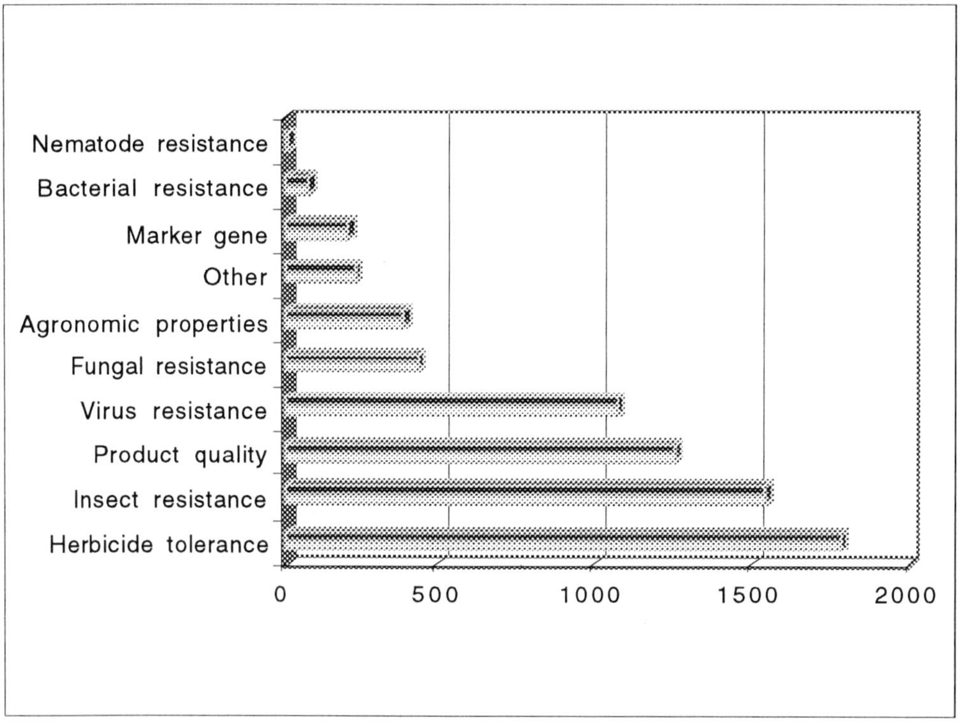

Figure 1.3. Introduced traits in genetically modified crop species (figures given are the number of field trials carried out in the USA for crops carrying the respective trait).
Source: National Biological Impact Assessment Program (1999).

drug development and food represents, in technological terms, the greatest leap beyond conventional agriculture.

However, unlike medical biotechnology, for many developing countries entry into commercial agri-biotechnology is not out of bounds. The wide range of technologies of varying levels of sophistication means that they can be utilized according to needs and priorities of the highly diversified agricultural systems of the developing world. Furthermore, market forces are not necessarily prohibitive to investments in agri-biotechnology. Nonetheless, entry barriers are expected to rise with the strengthening of intellectual property protection, the domination of the major marketing and distribution networks of the multinational companies and the limited ability of developing countries to integrate biotechnology with information and engineering systems. There is evidence of the involvement of international seed companies in plant breeding activities predominately in Latin America and South East Asia but the same evidence suggests that transgenic varieties are produced exclusively for export (FAO, 1998).

Environment/Industry

The scope of biotechnology for cleaner industrial production and pollution reduction and/ or abatement is quite considerable. Biotechnology has generated inroads in using old or hitherto unknown biomaterials for industrial purposes. Novel biocatalysts, recombinant organisms and extremophiles are starting to be introduced into a number of industrial processes. Furthermore, the range of new materials such as biodegradable plastics, biopolymers, fibres and biopesticides (Tables 1.3 and 1.4) developed by means of biotechnology is rapidly increasing. The same applies for the energy, metals and minerals sectors and process industries (e.g. chemicals and food industry (Table 1.5), pulp and paper production, textiles, leather). 'In developed countries, these sectors account for between 30 and 50% of all manufacturing. Biotechnological processes have helped them to improve their sometimes poor environmental image and, in many cases, increased their efficiency.' (OECD, 1999).

Yet despite its potential, industrial and environmental biotechnology has remained a 'backwater' (OECD, 1998). Industrial applications of modern biotechnology have been to a large extent 'spin-offs' of medical biotechnology and are only now beginning to penetrate the energy, environment and manufacturing industries. Furthermore, national and international regulations both in the industrialized and developing world are not conducive to the use of recombinant organisms for pollution control and remediation in unconfined environments.

In terms of international initiatives little has been done to date despite the fact that environmental pollution is a major scourge in the developing world not only in terms of irreversible ecosystem degradation but also as a cause of chronic disease.

The United Nations Conference on Environment and Development (UNCED) recognizes that biotechnology can contribute to, halt and reverse environmental degradation. Agenda 21 (Chapter 16) calls for the prevention of environmental degradation at source, and proposes 'the inauguration as soon as possible of specific programmes with specific targets:

- To adopt production processes making optimal use of natural resources, by recycling biomass, recovering energy and minimizing waste generation.
- To promote the use of biotechnologies, with emphasis on bioremediation of land and water, waste treatment, soil conservation, reforestation, afforestation and land rehabilitation.

Table 1.3. Biopesticide products on the global market.

Product	Producer	Use/mode of action
Laginex™ Bioinsecticide	AgraQuest, Inc.	Controls mosquito larvae in rice, wetlands and other bodies of water.
Disease Free Kleentek™	Crop Genetics International	Increases yield of sugar per acre.
Spod-X™	Crop Genetics International	Uses a naturally occurring insect virus to control the beet armyworm, which is becoming resistant to many chemical insecticides.
Aspire™ Biofungicide	Ecogen	Used on citrus, pome fruits, berries and grapes to protect from post-harvest rot. The active ingredient is a naturally-occurring yeast that is harmless to all non-targeted organisms.
Mattch®	Mycogen	Controls caterpillars in vegetables, field crops, nuts, grapes, turf, stored grain and ornamental and nursery crops.
MVP®	Mycogen	Used on tree fruits and nuts, vines, cotton and vegetables to control leaf-eating caterpillar pests.
M-Peril™	Mycogen	Used on corn to combat the European corn borer.
M-Pede™	Mycogen	Used on fruits, vegetables, grapes and ornamentals to resist soft-body insects and powdery mildew.
DeMoss™	Mycogen	Used to resist moss, algae and lichens on buildings, etc.
Thinex®	Mycogen	Used for blossom thinning in apples, pears and fruits.
M/C®	Mycogen	Used on vegetables to control various caterpillar pests.
Scythe® Herbicide	Mycogen	Used to control weeds in horticulture and landscape management.

Source: The Biotechnology Industry Organization (1999).

- To apply biotechnologies and their products to protect environmental integrity with a view to long-term ecological security' (UNCED, 1992).

Table 1.4. Biopesticide products expected on the global market within three years.

Product	Producer	Use/mode of action
Sonata™ Biofungicide	AgraQuest, Inc.	For diseases of grapes, vegetables and fruit (downy mildews, bunch rot, grey mould and brown rot).
Serenade™ Biofungicide	AgraQuest, Inc.	For diseases of grapes, other fruits and vegetables, turf and ornamentals (powdery mildew, grey mould, scab, damping off and late blight).
QST 776 Insect Repellent	AgraQuest, Inc.	Natural product for repelling mosquitoes, ticks and other biting insects.
Ecologix™ Cockroach Bait	Dominion Biosciences	First commercial product developed from a unique insect growth-regulator technology. It is highly effective in eliminating insect populations, yet completely non-toxic to users, pets and the environment.
Leone™ Biofungicide	Dominion Biosciences	Controls a number of plant diseases by relying on highly active, antimicrobial 'predator' bacteria.
Crymax™	Ecogen	Genetically engineered bioinsecticide effective against a broad range of pests. It will be used on vegetables, trees, nuts and vines.
EG7826™	Ecogen	Genetically modified insecticide that will control the fall armyworm, a major insect pest affecting sweet corn.

Source: The Biotechnology Industry Organization (1999).

UNIDO is the only agency within the UN system with a clearly defined biotechnology programme for cleaner industrial production and bioremediation.

Bioremediation has been on the agenda of the OECD since 1991, when the committee for Scientific and Technological Policy set up an 'Ad Hoc Group of Government Experts on Biotechnology for a Clean Environment', leading to a report in 1994 (OECD, 1994). Most recently, the group held a workshop on Biotechnology for Water Use and Conservation in Mexico during October 1996, jointly with the Mexican National Science and Technology Council (CONACYT) and sponsored by the Mexican Government. This focused both on biotechnological approaches to controlling water-borne diseases and on the use of biotechnology to cut water consumption (for example, the use of non-water-based sewage treatment for municipal effluents).

Table 1.5. Enzymes for food applications.

Product	Producer/distributor	Mode of action/use
Optiren™	Gist-Brocades International B.V.	An aspartic proteinase enzyme preparation produced by pure culture fermentation of *Aspergillus oryzae*, genetically modified to contain the aspartic proteinase gene from *Rhizomucor miehei*. The enzyme is used as a milk-clotting enzyme in the production of cheese.
Maxiren®	Gist-Brocades International B.V.	The biotechnology-produced version of calf rennet (chymosin) and is used as a milk-clotting enzyme in the production of cheese. It is produced by pure culture fermentation of the yeast *Kluyveromyces lactis*, modified to contain the calf gene encoding for chymosin.
Natuphos®	Produced by Gist-Brocades International B.V. and distributed by BASF Corp.	A phytase enzyme preparation produced in pure culture fermentation by a genetically modified strain of *Aspergillus niger*. It is used as a supplement in animal diets resulting in a marked improvement of the bioavailability of phosphorus in pigs, poultry and fish. As a consequence, phosporus excretion in manure is decreased by about 30%.
Natugrain® Blend	Produced by Gist-Brocades International B.V. and distributed by BASF Corp.	An enzyme blend, optimized in endoxylanase activity. Has a marked favourable effect on the feed conversion ratio in poultry grown on diets with a high inclusion rate of wheat, barley and rye. The endoxylanase enzyme is produced by pure culture fermentation of a genetically modified strain of *Aspergillus niger*.

Source: The Biotechnology Industry Organization (1999).

Both the Swedish Agency for Research Collaboration with Developing Countries (SAREC) and the Netherlands Directorate-General for Industrial Cooperation (DGIS) specifically address the development of biotechnologies for environmental protection in their policy statements. SAREC, for example, provided US $461,000 to support research on the use of recombinant proteins as biosensors for environmental applications in India and Ethiopia over the period 1986-1994. In addition, the International Centre for Biotechnology, at Osaka University, Japan, lent support to projects for pollution control and bioremediation of enclosed water bodies, in collaboration with universities in the Philippines, Thailand, Malaysia and Indonesia.

Conclusion

Disease, food scarcity and environmental degradation are closely intertwined and constitute major scourges of the developing world. Gressel states that '...there are two HIV problems in the Third World: one is the AIDS virus, and the other stands for "highly invasive vegetation", an acronym coined by P.C. Quimby of the United States Department of Agriculture Rangeland Weeds Laboratory in Bozeman, Montana. In fact, one type of HIV is related to the other, as the decreased crop yields cause the men to leave their wives and families in their native villages and look for work in the cities, where AIDS is rampant. Low yields facilitate disease epidemics by causing malnutrition, which exacerbates illnesses. There is all too often a tendency to say the major scourge of Africa is disease, while in fact its medical problems are inseparable from that of lack of food' (Gressel, 1996).

It is quite clear that modern biotechnology could play an extremely important role in alleviating disease, securing food supply and reducing environmental pollution. In a recent report, the prestigious Nuffield Council for Bioethics recognizes that biotechnology can 'produce more food, or more employment or income for those who need it most urgently' (Nuffield Council for Bioethics, 1999). The same report also acknowledges that patenting of new technologies presents 'potentially serious difficulties for developing countries' and stresses the need to direct research effort more evenly. In today's global set-up of shrinking public budgets and proprietary nature of innovations, it is not at all clear how R&D investment could be purposely directed into commercially unattractive projects.

Sachs states that 'The imbalance of global science reflects several forces. First, of course, science follows the market. This is especially true in an age when technological leaps require expensive scientific equipment and well-provisioned research laboratories. Second, scientific advance tends to have increasing returns to scale: adding more scientists to a community does not diminish individual marginal productivity but tends to increase it.' He notes that the 'inequalities of income across the globe are actually exceeded by the inequalities of scientific output and technological innovation' and demonstrates the massive dominance of industrialized countries in scientific publications and more notably patents (Sachs, 1999)(Table 1.6).

Notwithstanding the above, it should also be recognized that technological innovation and diffusion are not solely determined by commercial considerations alone. The role of mediators such as venture capital institutions and other financial brokers also influence the capital flow to public and private institutions and firms, particularly the new technology-based ones. Government policies likewise. These have often been instrumental in guiding the pathways of innovation through funding of research, regulation, technology transfer legislation, and appropriate patenting policies. For example, 'the newly industrializing Asian countries have been particularly skilful in deploying industrial policy to promote commercial biotechnology. They use grants, direct capital investment, fiscal stimulants

Table 1.6. Indicators of global science.

Resource	Advanced economies %[a]	Developing countries %[a]
Population	17	83
GDP	58	42
Publications	85	15
Patents (EU)	96	4
Patents (US)	98	2

[a]Figures are given as percentages of world totals for 1995.
Source: Sachs (1999).

and support skill-building and downstream development. Korea, Singapore and Taiwan have made systematic use of foreign technology licensing and the repatriation of scientists trained abroad to strengthen their relatively weak research bases. In Latin America, most governments have tended to confine their activities to R&D support, treating science and technology policy as if it were industrial policy. Although some of the larger economies are beginning to develop a more comprehensive approach to biopolicy, it will be some time before this shift begins to make itself felt in the market place.' (Tzotzos and Leopold, 1995).

In this regard, the effectiveness with which biotechnology can meet domestic needs is crucially dependent upon actions that strengthen the capacity of public and private research systems. Such actions include both the absorption of technological spin-offs that can be adapted to serve domestic needs, and the introduction of policy and institutional reforms that are conducive to investment in biotechnology. Several developing countries are moving in this direction and are beginning to gear an important part of their biotechnology research capacity building towards the effective exploitation of existing knowledge, rather than the generation of new knowledge. The use, assimilation and adaptation of new knowledge is becoming, in some cases, an integral part in the cumulative learning process increasing a country's potential for upgrading its R&D capability. Clearly, biotechnology R&D alone is not the be all and end all. It is, nonetheless, critical in overcoming some severe bottlenecks of conventional health care, agricultural and environment-related programmes and/or in enhancing their delivery prospects.

References

BIOTRACK (1999) http://www.olis.oecd.org/biotrack.nsf

Correa, C. M. (1991) Developing private biopharmaceutical capacity in developing countries. *Biotechnology and Development Monitor,* 9, 7-8.

Ernst and Young (1997) *European Biotech 97: A New Economy.* Ernst and Young International Ltd., London.

FAO (1998) *The State of the World's Plant Genetic Resources for Food and Agriculture.* FAO, Rome, p. 169.

Financial Times (1999) Survey: Life Sciences in Asia, *Financial Times*, 15 July, p. 4.

Gressel, J. (1996) *The Scientist*, 10, No.19, 30 September, p. 10.

James, C. (1997) *Global Status of Transgenic Crops in 1997*, ISAAA Briefs, p. 26.

National Biological Impact Assessment Program (1999) http://www.nbiap.vt.edu/cfu/cfdocs/fieldtests1.cfm

Nuffield Council for Bioethics (1999) *Report, May 1999.* Nuffield Council for Bioethics, London.

OECD (1994) *Biotechnology for a Clean Environment: Prevention, Detection, Remediation*, http://www.oecd.org/dsti/biotech/biorem.html

OECD (1998) *Biotechnology for Clean Industrial Products and Processes: Towards Industrial Sustainability*. OECD, Paris.

OECD (1999) OECD Policy Brief, Modern Biotechnology and OECD. *OECD Observer 1999*, OECD, Paris.

Sachs, J. (1999) *The Economist*, 14 August, p. 17.

The Biotechnology Industry Organization (1999) http://www.bio.org/whatis/guide1.html#stats (accessed 10 September, 1999).

Tzotzos G.T. and Leopold, M. (1995) Commercial Biotechnology: Developing World Prospects. In: Brauer, D. (Ed.) *Biotechnology, Vol. 12: Legal, Economic and Ethical Dimensions*. VCH, Weinheim, Germany, pp. 339-367.

UNCED (1992) United Nations Conference on Environment and Development, Agenda 21, 16.22. http://www.igc.apc.org/habitat/agenda21

United Nations (1995) *Review and Appraisal of the World Population Plan of Action. 1994 Report*. United Nations, New York.

WHO (1992) *Global Health Situation and Projections: Estimates*, WHO, HST/92.1, Geneva.

Chapter Two
International Initiatives in Agri-food Biotechnology

John Komen

Introduction

In 1994, large-scale field trials of transgenic potatoes (variety Alpha) with resistance to potato viruses X and Y were conducted by the Centre for Advanced Studies (CINVESTAV) in Mexico. In the autumn of 1996, field tests of transgenic potatoes with tuber moth resistance were carried out in Egypt by the Agricultural Genetic Engineering Research Institute (AGERI), and trials for virus-resistant tomatoes and cucurbits occurred in 1997. These are just two examples of field trials organized by public institutions, as part of international collaborative projects. While the application of agricultural biotechnology in industrialized countries is dominated by the private sector, and new products are launched by large international companies, the situation is quite different in developing countries. Here, agricultural research is still predominantly carried out by the public sector.

A wide range of international collaborative opportunities is available for agricultural research organizations in developing countries who plan or implement research programmes and national capacity in agricultural biotechnology. From around 1985 onwards, a number of international initiatives in agricultural biotechnology have been established, and they provide an important source of information or assistance.

For the purpose of this chapter, international initiatives in agri-food biotechnology are defined as *those organizations or programmes that conduct, support, or coordinate collaborative biotechnology research that addresses developing-country agriculture.*

A discussion of international initiatives will be presented below, while a full overview of the various initiatives is provided in Appendix III, organized by the following five categories:

- International research programmes for plant biotechnology.
- International research programmes for livestock biotechnology.
- International or regional biotechnology networks for specific crops or regions.
- Special programmes initiated and coordinated by bilateral or multilateral donor organizations.
- Organizations providing advisory services on policy and research management issues.

This chapter gives an introduction to, and overview of the different international initiatives in agricultural biotechnology that have been established. This is followed by an analysis of the research and training opportunities they provide to partner institutions in the developing countries. Finally, the chapter discusses the tasks ahead and challenges that the organizations and countries involved are facing, including: local capacity development, biosafety, private-sector involvement, and intellectual property rights.

International Initiatives in Agri-food Biotechnology

The origins of most international initiatives in agri-food biotechnology are found in the 1980s, when representatives of multilateral and bilateral donor organizations, and representatives of national and international agricultural research institutes met on several occasions, to discuss the potential benefits and challenges of agricultural biotechnology for developing countries (see, for example, IRRI, 1985; Cohen, 1989; Sasson and Costarini, 1989; NRC, 1990; Persley 1990). It was recognized that applying biotechnology for tropical agriculture and transferring these new techniques to developing countries posed some specific challenges, such as the necessity for safety review of rDNA products and the need to involve the private sector which had become more involved in agricultural research in general, and agricultural biotechnology in particular. A common message from those meetings was that special initiatives and new ways of doing business were needed to realize the potential benefits for developing-country agriculture, and to prevent a growing 'technology gap' between industrialized and developing countries. The international donor community responded in a variety of ways:

- By setting up a specialized international institute for biotechnology, the International Centre for Genetic Engineering and Biotechnology (ICGEB), and supporting the establishment of specialized national facilities for biotechnology in several developing countries.
- By creating special biotechnology initiatives and networks aimed at the transfer of advanced technology to developing countries.
- By supporting the development of biotechnology capacity in established international agricultural research programmes.
- By strengthening the international component and outreach activities of advanced national research institutes in industrialized countries.
- By supporting special advisory programmes on the policy and management aspects of agricultural biotechnology.

These actions have resulted in a wide range of initiatives which are summarized in Appendix III. They vary widely in scope and content. Some examples include the Rockefeller Foundation's International Rice Biotechnology Programme, ICGEB's plant biology programme, the Cassava Biotechnology Network (CBN), and the Special Programme Biotechnology and Development Cooperation of the Government of The Netherlands. Such initiatives provide linkages with a number of public and private advanced research institutes worldwide. In addition, they provide information, services and expertise on important issues related to biotechnology research management, such as priority setting, infrastructure development, and biosafety and intellectual property rights.

The Intermediary Biotechnology Service[a] (IBS) set up a database on international agricultural biotechnology, which currently contains information on 44 organizations

[a] The Intermediary Biotechnology Service was established in 1992 by a group of donor agencies to act as an independent advisor to organizations in developing countries on matters of biotechnology research, management and policy. IBS is a special project executed by the International Service for National Agricultural Research (ISNAR), and is supported by the governments of Switzerland and The Netherlands. IBS has the following overall objectives:

- To support policy and strategy development that enhances capacity in agricultural biotechnology;
- To assist the development of leadership skills among policy-makers and research managers for integrating biotechnology into agricultural research programmes;
- To harness the internationally available expertise to respond to identified technical, policy, and management needs.

The survey and database discussed in this chapter respond to the third objective defined for the IBS project.

supporting agricultural biotechnology for developing countries. The information was initially collected through a survey conducted in June-December 1993, while analysis, updating and expanding of the information was undertaken in 1994-1997.[b] The survey requested information on:

- Overall goals and priorities.
- Agricultural research focus in terms of crops, livestock species, livestock diseases, or other.
- Regional focus and collaborating institutes from developed and developing countries.
- Type, number and location of training opportunities provided.
- Activities and methods developed in programme planning, policy and management.
- Information products and services.
- Support for infrastructure development in developing-country programmes.
- Research and development projects, their progress, and technology transfer routes.
- Funding sources and expenditures.

Donor Agencies Supporting Agricultural Biotechnology: Some Examples

This section gives an overview of the main donor agencies supporting the international initiatives as described above. It will also provide representative examples of the international initiatives, rather than introducing each one individually.

Private Foundations

The overwhelming share of funding for international biotechnology by foundations is provided by the Rockefeller Foundation and, while the Rockefeller Foundation does support a number of smaller biotechnology research programmes, the major share of its effort is directed to its International Rice Biotechnology Programme. This programme involves an international network of researchers in universities and public research institutions in both developed and developing countries, as well as at international institutions such as the International Rice Research Institute (IRRI), ICGEB, the West Africa Rice Development Association and the International Centre for Tropical Agriculture (CIAT). The programme was initiated in 1985, and to date around US $77 million have been invested. By 1994, there were 76, mainly Asian, developing-country research institutions participating in the rice programme.

As one of the early international initiatives in agricultural biotechnology, it is relevant to look at the origins of the Rockefeller rice programme. In the early 1980s, the Foundation commissioned an external review of its Agricultural Sciences Programme as it was looking for new directions in its work in international agriculture. Until then, the Foundation had primarily worked on developing the network of International Agricultural Research Centres (IARCs), and on national agricultural research programmes in individual countries. The

---

[b] It is appropriate, at the outset, to point to the limitations and strengths of the survey. Firstly, it does not pretend to be an exhaustive review of international activities in agricultural biotechnology. Undoubtedly some of the growing number of institutions, networks or donor agencies active in supporting biotechnology initiatives were not identified in the course of the survey and were therefore not included, or simply did not respond to the questionnaire. Still, the survey illustrates the wide range of activities and organizations involved in international agricultural biotechnology. Secondly, the survey form was designed to be as comprehensive as possible and therefore has its limitations regarding the level of detail of information requested. Finally, it is not possible to verify all information submitted. This implies that although they do provide a clear indication of the relative importance of various funding organizations and programme elements, the figures on funding and expenditures should be regarded with caution.

external review recommended to support a more narrow range of projects, and to target research grants on the molecular biology of plants. The Foundation then decided to concentrate on rice, on the basis of four criteria: the importance of the crop in developing countries; the maturity of breeding programme in international centres and developing countries; the limited extent of biotechnology research efforts on rice; and, the probability of near-term success. This last factor was considered tentative, but was outweighed by the first three. It was realized that a considerable amount of time and resources would be required before significant results would be seen in the farmer's field. Consequently, the programme started with a ten to 15 year time horizon, and an initial yearly allocation of around US $3 million (Toenniessen, 1995).

The major share of grant funding under the rice programme initially went to 'first world' laboratories in order to build a knowledge base on rice biotechnology. Currently, the lion's share goes to the international centres, and to developing-country institutes with the strongest components found in India, Thailand and China. These together work on the following priorities:

- Transformation for disease resistance: tungro virus, stripe virus, yellow stemborers.
- Marker-assisted selection for resistance to blast and blight.
- Hybrid rice production.

The UN System

One of the most recent developments arising from the United Nations system is the International Centre for Genetic Engineering and Biotechnology (ICGEB). Initiated as a project executed by the United Nations Industrial Development Organization (UNIDO), ICGEB became an autonomous intergovernmental organization in 1995, with research facilities in Trieste (Italy) and New Delhi (India). Its research and training activities began in 1988, with programmes in human health and plant biology. The plant biology programme has a fundamental research component which develops crop transformation methods and studies the mechanisms of DNA replication and antiviral protein production in plants. A more applied component of the plant biology programme aims at developing rice material with gall-midge resistance, and studies different *Bacillus thuringiensis* isolates for their activity against rice insect pests. In addition to the research conducted at the Trieste and New Delhi sites, ICGEB supports research projects in its 'Affiliated Centres'.

Other UN organizations also support regional and international collaboration in agricultural biotechnology. The United Nations Development Programme (UNDP) has been a major supporter of biotechnology-based agricultural research at IARCs such as the International Potato Centre (CIP), the International Centre for Maize and Wheat Improvement (CIMMYT) and the International Livestock Research Institute (ILRI). In addition, UNDP was instrumental in setting up national institutes such as the Agricultural Genetic Engineering Research Institute (AGERI) in Egypt. The UNDP, UNIDO and the Food and Agriculture Organization of the United Nations (FAO) recently established a joint project 'Farmer Centred Agricultural Resource Management' (FARM), involving a number of Asian countries. FARM has a biotechnology-biodiversity component managed by India. Member countries collaborate in research and development on topics such as biofertilizers, biocontrol agents, commercial micropropagation, diagnostics and vaccines, biotechnology in agroforestry. UNDP has been the main sponsor of the Regional Programme of Biotechnology (RPB) for Latin America and the Caribbean, which also involved UNIDO, as the implementing organization. The programme ended in 1996, and has played an important role through setting up national biotechnology committees, and stimulating regional research collaboration. Some of the RPB regional research activities involved agricultural biotechnology, for instance on potato transformation for virus resistance. The FAO also promotes regional collaboration in Latin America and the Caribbean, through the Technical Cooperation Network on Plant Biotechnology (REDBIO by its Spanish acronym). REDBIO was established in 1990, and currently links over 400 research units located in different Latin American and Caribbean countries. The network encourages regional collaboration through an active information exchange programme and regular regional and sub-regional conferences.

Bilateral Donors

A number of bilateral donor agencies have become involved in agricultural biotechnology, with very different approaches and programmes. Examples will be discussed in this section.

United States of America

The US Agency for International Development (USAID) supports a number of recent initiatives that utilize recombinant DNA or cell culture techniques to enhance crop resistance to pests and pathogens. The current efforts in support of plant biotechnology originated from a panel convened in 1990 by the National Research Council. It brought together the views from national agricultural research organizations in developing countries, universities, private industry, government agencies and donor agencies to suggest the future orientation for USAID involvement. The panel recommended a product-oriented, collaborative programme which would integrate research, human resource development, biosafety and intellectual property rights issues, and involve the public as well as the private sector. These recommendations are incorporated in the Agricultural Biotechnology for Sustainable Productivity (ABSP) project, which started in 1992 with a six-year budget of US $6.7 million. The ABSP works with public and private organizations in the USA, and in Egypt, Indonesia and Kenya. Research activities involve:

- Potato transformation for resistance to tuber moth.
- Cucurbit transformation for resistance to potyvirus.
- Tomato transformation for resistance to gemini virus.
- Maize transformation for resistance to stemborers.
- Micropropagation of banana, pineapple, coffee, ornamental palms.

Another, smaller USAID project in plant biotechnology involves the transfer of proprietary technology from an international private company, Monsanto, to the Kenya Agricultural Research Institute (KARI). Monsanto's transformation technologies are applied for the development of genetically engineered sweet potato resistant to feathery mottle virus.

Other programmes supported by USAID focus on the generation of recombinant livestock vaccines and disease diagnostics to improve livestock production. The main international programme here is the International Programme on Vectors and Vector-borne Diseases, based at the University of Florida. The programme was initiated in 1985 and has been conducting research on improved vaccines and diagnostic tests for tick-borne disease in livestock: cowdriosis (heartwater), anaplasmosis and babesiosis. Currently, the programme entirely focuses on heartwater, and has concentrated its activities in sub-Saharan Africa, working in collaboration with the Veterinary Research Laboratory in Zimbabwe. During the period 1985-1996, USAID has contributed more than US $17 million. USAID also supports the International Laboratory of Molecular Biology for

Tropical Disease Agents, based at the University of California, for its work on developing recombinant vaccines for rinderpest and foot-and-mouth disease.

United Kingdom

The UK's Department for International Development (DFID) supports international agricultural biotechnology research through two programmes. The first and largest is the Plant Sciences Research Programme (PSRP), which is managed by the Centre for Arid Zones Studies at the University of Wales. The programme is composed of several advanced plant breeding and crop physiology projects at various British public and private institutions, and international research centres. Since its inception in 1990, the programme has expanded gradually and DFID now contributes approximately US $3.5 million annually to this programme. The PSRP has a plant breeding component which entirely consists of biotechnology research. It involves the use of molecular markers for 'tagging' genes of economic importance, and the use of genetic engineering to produce virus-, pest-, and disease-resistance plants. The programme constitutes some 30 projects in 20 universities and research institutes in the UK, in collaboration with institutes in developing countries. Examples of the larger projects are:

- Development of RFLP maps in pearl millet.
- Transformation of groundnut for virus resistance.
- Transfer of salt tolerance from wild relatives to cultivated wheat by chromosome engineering.
- Incorporation of the CpTi (cowpea trypsin inhibitor) gene into sweet potato for insect resistance.

The second DFID programme, with funds totalling around US $600,000 per year, is the Biotechnology Programme, managed directly by the Natural Resources Research Department in DFID. Funds are distributed over some 12 research projects, mostly at British public institutions.

France

The French government is one of the major bilateral supporters for agricultural biotechnology, through the Centre de Coopération International en Recherche Agronomique pour le Développement (CIRAD). CIRAD has a total annual budget of approximately US $200 million and works in some 50 different countries. Plant biotechnology is being conducted in the Plant Breeding Division and it is estimated that the annual expenditure on plant biotechnology amounts to US $7.5 million. Table 2.1 gives an impression of the wide range of research activities and crops involved.

Switzerland

The longest-established bilateral initiative in biotechnology is the Indo-Swiss Collaboration in Biotechnology (ISCB) programme, based on a bilateral agreement between the Indian and Swiss governments. The Institute of Biotechnology at the Federal Institute of Technology (Zürich) initiated the agreement in 1974. The programme was restructured in 1988 and currently involves six Swiss research institutes and six Indian research institutes. It is funded by the Swiss Agency for Development and Cooperation which has committed around US $6.5 million to the programme since 1988. The ISCB is not directly related to agricultural issues, but rather focuses on methodology development, capacity building and technology transfer. Programme activities range from human health, to animal diseases, to biopesticides production.

Table 2.1. Crop biotechnology activities at CIRAD's plant breeding division.

Crop	R&D activity
Banana / plantain	Mass propagation of banana through somatic embryogenesis. Mass propagation of banana through meristem proliferation. Cellular regeneration system (somatic embryogenesis) for bananas and plantains for genetic engineering. Molecular markers, probes, genome mapping, field evaluation for bananas and plantains.
Citrus	Shoot-tip grafting for cleaning citrus cultivars. Somatic hybridization of citrus for rootstock breeding. Genetic diversity analysis and genome mapping of citrus.
Cocoa	Cocoa genome mapping. Mass propagation of cocoa through somatic embryogenesis.
Coconut	Coconut somatic embryogenesis. Coconut genetic diversity with molecular markers.
Coffee	Genetic transformation of coffee with Bt toxin genes. Mass propagation of coffee through microcuttings. Mass propagation of coffee through somatic embryogenesis. Coffee genome mapping.
Cotton	Insect-resistant cotton through the transfer of Bt toxin genes or protease inhibitor genes.
Oilpalm	Oilpalm genetic diversity with molecular markers. Oilpalm somatic embryogenesis.
Rice	Stemborer-resistant engineered rice through Bt toxin gene transfer. Molecular tagging of Rice Yellow Mottle Virus resistance genes. Production of improved double haploid lines of rice through anther culture. Conversion of F1 hybrid rice parentals to male sterile lines.
Rubber tree	Physiology of latex production. Genome mapping of rubber. *In vitro* propagation of rubber.
Sorghum	Molecular tagging of sorghum genes controlling grain qualities corresponding to consumer demand.
Sugarcane	RFLP studies in sugarcane genetics. Long-term storage of sugarcane. *In situ* hybridization of sugarcane. Rapid micropropagation of sugarcane.

Table 2.1. Continued.

Crop	R&D activity
Forestry species	Regeneration of acacia. Isolation and characterization of *Bradyrhizobium* strains of acacia (ELISA, PCR, RFLP). Evaluation of genetic diversity in acacia. Genetic transformation of acacia through particle gun and *Agrobacterium.* Effect of soil nutrients (oligoelements) on establishment and efficiency of symbiosis. Vegetative propagation of rattan by somatic embryogenesis. Genetic diversity studies of rattan. Micropropagation of acacia. Vegetative propagation of teak by meristem culture. Eucalyptus mapping with RAPD. Micropropagation of selected eucalyptus clones.

Source: IBS *BioServe* Database (1997).

The Netherlands

A different approach to advance agricultural biotechnology in developing countries has been developed under the Special Programme Biotechnology and Development Cooperation, launched in 1992 by the Directorate-General for International Cooperation (DGIS, by its Dutch acronym) of the Ministry of Foreign Affairs of The Netherlands. A total of US $27 million has been allocated to the programme, for the period 1992-1997. The programme has three unique features. Firstly, it is specifically directed towards small-scale producers and women in developing countries. Secondly, while focusing on a limited number of countries - Kenya, Zimbabwe, Colombia and India - the programme is to a large extent being decentralized, with local steering committees taking responsibility for project implementation and management. Thirdly, these country programmes should address farmers' needs, identified through a participatory, 'bottom-up' approach. Local steering committees include representatives from farmers' organizations and individual farmers. The process of needs assessments and priority setting has been completed in each focal country, and the following projects have been initiated:

- Marker-assisted selection of drought-tolerant and insect-resistant maize (Zimbabwe and Kenya, in collaboration with CIMMYT).
- Micropropagation of disease-free planting material (all countries).
- Production of biofertilizers and biopesticides (Kenya, India).

Policy issues such as biosafety and intellectual property rights have been addressed from the outset through regional and national meetings, and commissioned studies. The Special Programme has an international component through supporting the Intermediary Biotechnology Service (IBS) at the International Service for National Agricultural Research (ISNAR), and the Cassava Biotechnology Network (CBN) at CIAT. The review of social and economic aspects of biotechnology and information dissemination is encouraged through the journal 'Biotechnology and Development Monitor'.

Analysis: Realizing the Promises

The majority of international initiatives discussed in this chapter focus on tropical food crop improvement. As shown in Appendix III, the database developed by IBS includes 20 international programmes on crop biotechnology, six livestock programmes, and two research programmes with a mixed focus on crop-related as well as livestock research. It should be noted that most networks and donor-agency programmes also primarily focus on crop biotechnology.

In a ten-year period, 1985-1994, the various organizations covered by the IBS survey contributed an estimated US $260 million in *grant funds* to the international biotechnology initiatives. In the same period, the total biotechnology component of World Bank loans and credits for national agricultural research projects in developing countries has been estimated at around US $150 million. Consequently, total international investment (grants plus loans) since 1985 can be estimated at US $400 million. The IBS survey assessed the different categories of expenditure within the research programmes, and these are shown in Table 2.2.

Table 2.2. Research programmes: categories of expenditures (percentages).

Programme element	Share (%)
Research and development	50.2
Human resource development	18.3
National programme participation	9.6
Infrastructure support	8.1
Information products and services	5.6
Policy and programme management	5.3
Other	2.9
ALL	100

The figures indicate a strong emphasis on research and development (R&D), followed at a distance by human resource development. These two components are discussed in more detail below. In this section, the analysis focuses on the 28 identified research programmes (crop and livestock), in order to keep the information comparable.

Research and Development

What then are the type of opportunities available in collaborative research? Regarding crop biotechnology, as shown in Table 2.3, current research activities primarily focus on developing varieties with enhanced pest resistance and on pest control (diseases, viruses, insects). Of the 126 projects recorded in crop biotechnology, around 60% deal with pest resistance and control.

Cereal grains, including rice, maize and sorghum, are the main crop category in terms of research activities. It is followed by root crops (potato, cassava, yam, sweet potato) and tropical perennials. By far the most important crop is rice, which accounts for 17% of all crop research projects recorded. This indicates the relative importance of the Rockefeller Foundation's Rice Biotechnology Programme, which supports advanced rice research in US-based and European research organizations, Asian national agricultural research organizations and international programmes at IRRI, CIAT, ICGEB and the International Laboratory for Tropical Agricultural Biotechnology (ILTAB). In addition, CIRAD has a firm rice research programme.

Table 2.3. Crop research objectives (by number of projects).[a]

	Objectives					
Crops	Disease resistance	Insect resistance	Virus resistance	Quality traits	Micropropagation	All
Cereals:	**9**	**13**	**8**	**12**		**42**
Rice	5	4	6	6		21
Maize	1	6	2	3		12
Sorghum	1	3		2		6
Other	2			1		3
Root crops:	**4**	**5**	**7**	**2**	**1**	**19**
Potato	1	3	2			6
Cassava	1		3	2		6
Yam	2		1		1	4
Sweet potato		2	1			3
Legumes:	**4**	**6**	**4**	**6**		**20**
Bean	1	2	1	2		6
Groundnut	1	1	3	1		6
Chickpea	1	1		2		4
Other	1	2		1		4
Horticulture	**2**		**3**		**1**	**6**
Perennial:	**2**	**1**	**2**	**2**	**15**	**22**
Banana/plantain	2		1		5	8
Industrial crops				1	4	5
Coffee		1			4	5
Sugarcane			1	1	1	3
Cocoa					1	1
Forestry species				**2**	**5**	**7**
Miscellaneous	**3**	**3**		**2**	**2**	**10**
ALL	**24**	**28**	**24**	**26**	**24**	**126**

[a] Figures are based on information gathered from the 22 international research programmes that include activities in crop research. For this table, we used those research activities with a specific applied objective, excluding research activities aimed towards general technology development.
Source: IBS *BioServe* Database (1997).

The projects in crop biotechnology tend to be at the advanced end of the research spectrum. Around 30% of them involve crop transformation, 29% involve the use of molecular markers, while cell biology (micropropagation, regeneration) is applied in 31% of the total.

In summary, the crop biotechnology programmes aim at improved tropical food crop production with reduced levels of pesticides. This contrasts with 'mainstream' research in agricultural biotechnology which emphasizes temperate crops and mostly aims at developing herbicide-tolerant crops. Of the total number of field trials of transgenic crops conducted worldwide from 1986 through to 1995, the most frequent trait categories were herbicide tolerance (35%) and product quality (20%) (James and Krattiger, 1996).

In comparison to crop research, livestock research has a more modest position. It concentrates on developing new vaccines and diagnostics for tropical livestock diseases, such as trypanosomiasis, tick-borne diseases (e.g. theileriosis and cowdriosis), rinderpest, and foot-and-mouth disease (see Table 2.4). The major share of the livestock effort relates to cattle, although one programme exclusively concentrates on small ruminants (goats, sheep). The main player in livestock biotechnology is the International Livestock Research Institute (ILRI), with research programmes for trypanosomiasis and tick-borne diseases. Among the CGIAR (Consultative Group on International Agricultural Research) centres, ILRI was one of the first to develop biotechnology-based research and invests some US $5 million annually in animal biotechnology.

Human Resource Development

In parallel with their research projects, all crop and animal research initiatives have developed a strong component for human resource development. These training opportunities are primarily found at the IARCs and in advanced research institutions and universities in Europe and the USA. In a few cases, training positions are arranged in the laboratories of international private companies, such as Monsanto and ICI Seeds.

Table 2.5 provides an overview of training positions available to date in the 28 research programmes. Training activities are concentrated at the post-doctoral and doctoral levels, with fewer opportunities available for training at master level or for technicians.

In addition to the training opportunities provided through the crop and animal research programmes, there is one donor-agency programme which exclusively promotes human resource development. The UNESCO Biotechnology Action Council has designated a number of existing institutes as regional biotechnology education and training centres:

- The Agricultural Biotechnology Centre (Godollo, Hungary) for Eastern Europe and the Mediterranean.
- Bethlehem University (Bethlehem, Israel) for the Arab States.
- CINVESTAV (Irapuato, Mexico) for Latin America and the Caribbean.
- Ocean University (Qingdao, China) for Asia.
- The Vegetable and Ornamental Plant Institute (Pretoria, South Africa) for Africa.

From September 1991 to June 1995, the Council awarded 223 fellowships to scientists from developing countries (Vasil, 1995).

Advice on Policy and Management Aspects

In addition to research and technical training, most international research programmes provide advice and training on policy and management aspects of agricultural biotechnology. In this area, biosafety and intellectual property rights are priority topics, which will be discussed below. Some international initiatives have been established which concentrate entirely on policy and management issues, and play a useful complementary role to the research-oriented programmes. For example, the Biotechnology Advisory Commission (BAC), based at the Stockholm Environment Institute, provides on-request advice on biosafety aspects (also discussed below). The Intermediary Biotechnology Service (IBS) is an advisory service to developing countries on matters of biotechnology research programme management and policy, including biosafety and intellectual property rights. The IBS initiated a series of regional policy seminars that reviewed the issues

Table 2.4. Livestock programmes: research focus.[a]

Programme	Research focus
CIRAD - Animal Production and Veterinary Medicine Division	Vaccine for *peste des petits ruminants* Vaccine for *contagious caprine pleuropneumonia* (CCPP) Vaccine for *contagious bovine pleuropneumonia* (CBPP) Molecular probes and PCR to identify mycoplasmosis Monoclonal antibodies to identify mycoplasmosis
ICIPE - Biotechnology Research Unit	Development of vaccines for tick-borne diseases
International Laboratory of Molecular Biology for Tropical Disease Agents	Recombinant rinderpest vaccine Diagnostic kit for rinderpest Diagnostic kit for *vesicular stomatitis virus* Recombinant foot-and-mouth disease (FMD) vaccine Diagnostic kit for foot-and-mouth disease
ILRI - Tick-borne Diseases Programme	Recombinant vaccine for East Coast Fever Diagnostic tests for tick-borne diseases Genetic maps of protozoa
ILRI - Trypanosomiasis Programme	Vaccine development for trypanosomiasis Molecular analysis of genetic basis of trypanotolerance Quantification of drug levels using HPLC and ELISA technology Parasite characterization Diagnostic tests for trypanosomiasis Bovine genome mapping
ISCB	Diagnostic tests for CCPP and CBPP Foot-and-mouth disease (FMD) diagnostics (development of diagnostic kits for FMD virus and antibody detection)
International Programme on Vectors and Vector-borne Diseases	Recombinant vaccine and improved diagnostics for heartwater Development of attractant decoys for control of ticks
Small Ruminant Collaborative Research Support Programme - Animal Health Component	Vaccine for CCPP Identify vaccine genes for *Haemonchus contortus* (a parasitic nematode) Vaccine for Nairobi sheep disease

[a] The list above is illustrative, not exhaustive.
Source: IBS *BioServe* Database (1997).

involved in planning and policy formulation for agricultural biotechnology. These were organized for countries in Southeast Asia, East and Southern Africa, Latin America, and West Asia / North Africa. Based on the findings of the policy seminars, it is now providing specific in-country advisory services. In addition, IBS has initiated a management training course, with support from the government of Japan, aimed at agricultural research managers who are responsible for implementing biotechnology programmes.

Table 2.5. Human resource development (HRD) opportunities provided by international biotechnology programmes.[a]

HRD type	Number of opportunities (all programmes)
Post-doctoral	204
Doctoral	210
Master	82
Technician	101
Management	3
Internship	113
Other (mostly short-term courses)	146

[a] Figures are based on information gathered from 28 international research programmes that focus on research in crops and/or livestock.
Source: IBS *BioServe* Database (1997).

Expected Products

In the survey conducted by IBS, respondents were asked to indicate research progress for their research activities. Progress was charted through seven stages of development. The first three described progress achieved in the laboratory and greenhouse. The next three stages include testing, beginning with contained trials moving to large-scale, multi-year testing. Finally, each respondent was asked if a partner, public or private, had been identified with regard to technology transfer and product development.

The analysis of research progress shows that the major share of the research activities are in the experimental, laboratory phase. Still, some have already resulted in products which are ready for wider diffusion:

1. Disease-free planting material: Various tissue-culture techniques are applied for the micropropagation of disease-free planting material. The largest micropropagation components are found under the ABSP and CIRAD programmes. They involve mainly export crops such as coffee, cocoa, banana, oilpalm and sugarcane.
2. Biocontrol agents: The production of biopesticides is undertaken at pilot-scale in India as part of the ISCB programme. Products based on *Bacillus thuringiensis* and *B. sphaericus* have been provisionally registered. Bt-based pesticides are also developed by the International Centre of Insect Physiology and Ecology (ICIPE). A pheromone-based attractant decoy for tick-borne vector control was developed at the University of Florida under the International Programme on Vectors and Vector-borne Diseases.
3. Transgenic plant varieties: Virus-resistant potatoes have been planted in Mexico, as part of the CINVESTAV-Monsanto collaboration. Field trials of virus-free potatoes are being conducted in Egypt under the ABSP programme, and future trials are planned for transgenic tomatoes and cucurbits. As regards rice, transgenic rice materials with enhanced insect resistance are being evaluated and this should lead to new varieties before the year 2000 (Toenniessen, 1995).
4. New diagnostics and vaccines for livestock diseases: This appears to be the most significant area for product development to date, with diagnostic tests and rDNA vaccines for rinderpest, cowdriosis (heartwater), theileriosis (East Coast Fever) and foot-and-mouth disease. These are reported to be ready for large-scale testing or product development.

Client-country Perspectives: Maximizing the Benefits of Collaboration

The preceding sections show that an impressive range of activities has been generated in research and human resource development, while the total investment in these programmes is relatively small compared to the investments in agricultural biotechnology R&D made by international private companies. In this respect, the international programmes have clearly benefited from building on existing knowledge in advanced research institutions. It has also become clear that, apart from activities in collaborative research and training, a number of 'tangible' products can already be anticipated for testing and diffusion in developing countries.

Strong local partners are required in countries collaborating in international programmes, to expedite the adaptation of technologies and materials that are developed through collaborative research. Research and training are obviously essential elements for programmes aimed at transferring agricultural biotechnology to developing countries. Most international initiatives concentrate on these two elements, which is quite justifiable given their strengths in these two areas, and their limited resources. A few initiatives are directly concerned with product development. This section discusses some of the important factors involved in the process of technology transfer and product development which relate to: (1) national programme capacities; (2) biosafety; (3) private sector involvement; and (4) intellectual property rights.

Building National Programme Capacity

Most initiatives discussed in the preceding sections conduct research at the advanced end of the 'biotechnology spectrum': crop transformation, molecular markers, rDNA vaccines. Similarly, training opportunities are found at higher levels of achievement, at the doctoral and post-doctoral levels. This implies that countries with well-established scientific and technological capabilities are the best qualified partners in international programmes, and this is reflected in their geographic coverage. The number of developing countries taking part in international biotechnology initiatives is quite high, and efforts are more or less evenly spread among the different geographic regions. Within each region, however, international collaborative activities are concentrated in a small number of countries: Kenya, Egypt, Zimbabwe, and Côte d'Ivoire in Africa; Indonesia, Thailand and India in Asia; and Costa Rica, Mexico and Brazil in Latin America.

Even in the countries mentioned above, the physical research infrastructure can be a limiting factor in applying the results of international collaboration. Infrastructure development in partner countries is a secondary element of most international initiatives. On average, less than 10% of funding is directed towards, for example, the construction of facilities or the acquisition of equipment. An exception in this regard is the ISCB programme, in which around 45% of funds are directed towards infrastructure development at Indian partner institutes. This primarily includes the procurement and servicing of laboratory equipment which is not available in India, but also fine chemicals, documentation and communication appliances. The IRRI's Asian Rice Biotechnology Network (ARBN) also includes a strong component for infrastructure development, including the acquisition of supplementary equipment to support joint projects, and the construction of greenhouses. In addition, ARBN has an active backstopping programme for participating institutions, and this enables them to send researchers to IRRI to conduct 'shuttle research', for steps in a project that are not possible at their home institute.

Biosafety Review

A specific aspect of national capacity building relates to the capacity to review and manage the environmental and human safety aspects of recombinant DNA material of plants, animals, and microorganisms. Establishing a system for biosafety review has many facets in itself, including the formulation and adoption of safety guidelines, establishing national and institutional biosafety committees, and constructing additional infrastructure for small-scale, contained trials and large-scale field testing. As indicated above, a range of products can be anticipated as a result of international collaborative programmes, some of which require biosafety review such as transgenic plants and recombinant livestock vaccines. For example, the International Laboratory of Molecular Biology for Tropical Disease Agents at the University of California, developed a rDNA vaccine for rinderpest. Field trials of the vaccine have been conducted in Kenya. Guidelines for the field trials were defined by a joint committee, consisting of the UN Food and Agriculture Organization (FAO), the UN World Health Organization (WHO), and the Office International des Épizooties.

Donor agencies supporting international initiatives in biotechnology often require that the countries of collaborating institutes have a regulatory structure in place, before the transfer of transgenic material can take place. Consequently, the presence or absence of a national biosafety system has become an significant factor in international collaboration in biotechnology. As a result, most international initiatives discussed in this chapter are addressing biosafety issues. They provide advice on specific cases and practical exposure to the design and management of field trials, assist in the formulation of guidelines, and provide training in this area through individual internships and workshops.

A Biotechnology Advisory Commission (BAC) was established in 1993, based at the Stockholm Environment Institute, with financial support from the Rockefeller Foundation. The BAC provides advice on an on-request basis to governmental and intergovernmental authorities on the field testing and planned introduction of genetically modified organisms, and has organized biosafety and risk-assessment workshops in different countries. The ABSP programme initiated a biosafety internship programme in 1993, primarily, but not exclusively, for scientists participating in the programme. ABSP organized biosafety implementation workshops in Indonesia and Egypt. Under a collaborative project managed by ABSP, between the University of Arizona and AGERI in Egypt, a contained greenhouse facility has been constructed at AGERI. Similar facilities will be constructed in Indonesia. Regional collaboration and information exchange on biosafety in sub-Saharan Africa is supported by DGIS, through the Regional Biosafety Focal Point based at the Biotechnology Research Institute in Zimbabwe.

Private Sector Participation

The international biotechnology initiatives are primarily public-sector initiatives. The involvement of the private commercial sector is limited, both in terms of financial contributions and as suppliers of proprietary technology. This is essentially related to the focus of the international initiatives, emphasizing areas of crop and livestock improvement in which the private sector does not play a prominent role.

Some initiatives have started collaboration with the private sector for the transfer of proprietary technology. The International Service for the Acquisition of Agri-biotech Applications (ISAAA) aims at facilitating the transfer of proprietary biotechnology to developing countries. ISAAA functions as a 'broker' in identifying sources (public or private) of proprietary technology, and in negotiating arrangements for their transfer to developing countries. The ISAAA has brokered several 'technology donations', including the donation of coat-protein genes by Monsanto to the Centre for Advanced Studies (CINVESTAV) in Mexico for the control of viruses in potato varieties Alpha, Rosita and Norteña. The first generation of transgenic potatoes developed by CINVESTAV was planted in a field test in March 1993. Additional trials were undertaken in August 1995. A parallel project assisted Mexico in developing the infrastructure and regulatory procedures for introducing and testing recombinant products. Potato seed production and distribution is being done by private growers in Mexico. The Rockefeller Foundation provided financial support for technical training, biosafety capacity building, field experimentation, and laboratory supplies.

In the ABSP programme, private firms have been involved from the outset in the research and training activities. ICI Seeds participates in a research project aimed at developing stemborer-resistant maize varieties, which also involves the Central Research Institute for Food Crops (CRIFC) in Indonesia. Proprietary bioreactor technology for micropropagation has been transferred from the US company DNA Plant Technology to companies in Indonesia (Fitotek Unggul) and Costa Rica (Agribiotecnologia de Costa Rica).

Besides technology donations, the private sector may become involved in the development and diffusion of products resulting from international collaborative research. This is likely to be a more difficult issue, as the markets for those products are not very attractive to international private industry and are traditionally serviced by public extension programmes. A few examples exist in which the local private sector is expected to play a key role in product development. The attractant decoy technology developed in the International Programme on Vectors and Vector-borne Diseases has been patented in the United States, South Africa and Zimbabwe, and has been licensed to a US company for commercialization. The product has undergone registration trials in Zimbabwe, and a local company will also become involved in its commercial distribution.

Intellectual Property Rights

Issues related to intellectual property rights (IPR) have to be addressed in international biotechnology initiatives involving proprietary technologies or commercial product development. Private companies routinely seek patent protection for their inventions, and this practice is becoming prevalent at universities and advanced research institutes in industrialized countries. As a result, a single product developed through biotechnology may involve several proprietary technologies from different technology owners. While many developing countries have patent laws in place, many of those laws do not address plant breeders' rights or the products of agricultural biotechnology, particularly those developed through genetic engineering. For each international initiative, it will be essential to specify the rights and obligations related to intellectual property for the collaborating institutions.

The Rockefeller Foundation is one of the few donor agencies who established an IPR policy, specifically for its Rice Biotechnology Programme. Under this policy, grantees are allowed to pursue intellectual property rights on their discoveries and improved materials in industrialized countries. However, it also requires that they share materials and technologies with others in the programme at zero royalty for use in developing countries. Under the ABSP programme, a number of approaches are followed with respect to IPR. From the outset of the programme, external legal consultants have been involved to advise on IPR-related matters. In addition, the ABSP programme initiated an internship programme on IPR and technology transfer for collaborating countries, and has organized IPR workshops in those countries. Finally, as the programme enters a phase of product development and technology transfer, standard material transfer agreements and licensing agreements have been developed in conjunction with the Intellectual Property Office at Michigan State University.

In a few cases, the international initiatives have themselves resulted in proprietary technology. This has occurred in livestock research programmes, where patents are considered necessary to encourage private-sector collaboration in the large-scale production of vaccines and diagnostics. Under the International Programme on Vectors and Vector-borne Diseases, three patents have been taken out including one covering the attractant tick decoy. The patents are owned by the University of Florida, the programme's executing organization. The ILRI owns a patent for a novel vaccine for East Coast Fever.

Conclusion

The opportunities available through international collaboration in agri-food biotechnology are many, and can contribute to developing-country objectives for agricultural and social development. The hands-on experience gained through the involvement in collaborative research, training programmes, and international technology transfer projects are a great benefit as such. However, their ultimate success will often be judged on the basis of tangible products in farmers' fields. As discussed above, this is not a short-term process and will require strong national partners, from the public as well as the private sector. In addition, collaborating countries will have to review, and possibly adapt, their systems for biosafety review and IPR regulation. Such capacity building implies certain costs which must be weighed against the benefits expected from international collaboration.

References

Cohen, J.I. (1989) *Strengthening Collaboration in Biotechnology: International Agricultural Research and the Private Sector.* Agency for International Development, Washington, DC, pp. 480.

IBS *BioServe* Database (1997) IBS Database on International Agricultural Biotechnology. ISNAR, The Hague, The Netherlands.

IRRI (1985) *Biotechnology in International Agricultural Research: Proceedings of the Inter-Centre Seminar on International Agricultural Research Centres (IARCs) and Biotechnology, Manila, Philippines, 23-27 April 1984.* International Rice Research Institute, Manila, The Philippines, pp. 435.

James, C. and Krattiger, A.F. (1996) *Global Review of the Field Testing and Commercialization of Transgenic Plants, 1986 to 1995: The First Decade of Crop Biotechnology.* ISAAA Briefs No. 1. International Service for the Acquisition of Agri-biotech Applications, Ithaca, USA.

NRC (1990) *Plant Biotechnology Research for Developing Countries.* National Research Council, Board on Science and Technology for International Development. National Academy Press, Washington, DC, pp. 44.

Persley, G.J. (1990) *Agricultural Biotechnology: Opportunities for International Development.* CAB International, Wallingford, UK.

Sasson, A. and Costarini, V. (1989) *Plant Biotechnologies for Developing Countries. Proceedings of an International Symposium, Luxembourg, 26-30 June 1989.* Ede: Technical Centre for Agricultural and Rural Cooperation.

Toenniessen, G.H. (1995) The Rockefeller Foundation's International Programme on Rice Biotechnology. In: Altman D.W. and Watanabe, K.N. (eds) *Plant Biotechnology Transfer to Developing Countries*. R.G. Landes Company, Austin, Texas, pp. 193-212.

Vasil, I. (1995) UNESCO Centres on Biotechnology. *Bio/Technology* 13, 1165.

Chapter Three
International Initiatives in Medical Biotechnology

Tom Crompton

Signatories of Agenda 21 are required to contribute, through the environmentally sound application of biotechnology, to an overall health programme to help combat major communicable diseases, and develop and improve initiatives to assist in specific treatment of, and protection from, major non-communicable diseases (UNCED, 1992; 16.1, 16.12, and 16.22). Consistent with the working definition of international initiatives offered by John Komen in the first section of Chapter 2, here organizations or programmes are considered which conduct or support collaborative biotechnology research which addresses health improvement projects in developing countries. A report by the Commission on Health Research for Development (COHRED) found that an estimated US $1.6 billion spent on the health problems of developing countries represented only 5% of the total global investment in health research. Almost half of the funding for health research in developing countries originated in these countries themselves, and of the funding provided by industrialized countries, just one sixth was actually transferred to developing countries (COHRED, 1990).

The apparent paucity of projects which directly address the role of biotechnology in health in the developing world is, of course, in part a reflection of the proper concentration of medical research efforts on a 'vertical', problem-based approach, rather than a preoccupation with the development of specific technologies. International donor agencies and health organizations have therefore rarely developed policies which focus explicitly on the use of biotechnology in addressing health problems.

By comparison, exploitation of biotechnology for increased agricultural production has spawned a plethora of international research initiatives. This disparity is partly attributable to a local preoccupation with the strengthening of an established research base, which in the case of developing countries tends to imply agricultural rather than medical research (Mugabe, 1993). By contrast, the large majority of biotechnology investment in the developed world, both private and public, is earmarked for biomedical research.

Specific Research Areas

Vaccines

Recombinant DNA technologies have led to a variety of new approaches to vaccine research, beginning with the use of engineered microorganisms for vaccine production. Incorporation of a gene for expression of an antigen from the hepatitis B virus into *Saccharomyces* was one of the earliest examples of such engineering. Research is focused on:

- Deletion mutants, deficient in some virulence factor, though with an unaffected immunogenicity (the advantage is that these can be administered as a live vaccine, reducing the dosage requirement).
- 'Subunit' vaccines: a single antigen of pathogen is produced transgenically in large quantities, and purified.
- Use of a live recombinant vector, which can combine the benefits of both these approaches, using a virus or bacterium as a carrier, and which also offers the possibility of constructing multivalent vectors conferring immunity to several pathogens.

Several international initiatives are pursuing research on recombinant antifertility vaccines, focusing particularly on antihuman chorionic gonadotrophin (anti-hCG) vaccine. However, investment in antifertility research is controversial, and has been criticized on the grounds that such resources should be allocated for addressing the structural problems underlying population growth, rather than technological fixes that divest individuals, usually women, of control over their own fertility. More than four hundred groups advocating women's health, representing around forty countries, recently called for the termination of research on antifertility vaccines.

The case of vaccine development clearly illustrates the problems of encouraging private investment where returns are low, if products are to be affordable to countries with small health budgets. There is an important role for outside donors to provide grants to industry for the development of vaccines, on the understanding that these are then supplied to the World Health Organization (WHO) at production cost. This is an area where international cooperation is particularly important, serving to increase the potential market size of a successfully produced vaccine and thus improve the projected economic viability of a product. In fact, the availability of cheaper and more effective vaccines in developing countries may have ramifications extending beyond their immediate importance in preventing specific diseases. The experience of the Pan American Health Organization in the wake of the polio eradication scheme in South America was that the polio vaccine became a vehicle for raising the community standing, confidence and morale of the health structure right down to a local level (Walgate, 1992).

Diagnostics

Monoclonal antibody diagnostics are of importance to developing biotechnology industries, not just in themselves, but also, potentially, for capacity building for subsequent development of DNA probes or recombinant vaccines. Historically, there has been controversy over the ease with which monoclonal antibody diagnostics might be developed and exploited commercially within developing countries (Avramovic, 1996). In the 1980s, the United Nations Industrial Development Organization argued that the technology was accessible, both technically and economically, and pointed to the fact that the global diagnostic industry was in comparative infancy, facilitating entry of new countries (UNIDO, 1984). The enthusiasm of the World Bank, however, has been somewhat more muted (Avramovic, 1996).

Recombinant Proteins as Pharmacological Products

Applications include large-scale production of human insulin, clotting factor VIII, and growth and immunoregulatory molecules.

Donor Agencies Supporting Environmental and Health Biotechnology: Some Examples[a]

Private Foundations

The Wellcome Trust

The Wellcome Trust is a charitable foundation and the largest non-governmental source of funds for biomedical research in Europe, with a total research expenditure of something approaching half a billion US dollars. The Trust supports over 3000 researchers in 30 countries. An International Interest Group, which oversees exchange fellowships and collaborative schemes, advises the Trust on provision of support for biomedical research in developing countries. Most recently, the Trust has established fellowships and awards to support collaborative biomedical research between Australia and developing countries in South East Asia. A Tropical Medicine Interest Group has also been established with a particular responsibility for the promotion of training for researchers in tropical medicine, both in the UK, and in tropical countries.

The Wellcome Trust orchestrated the sequencing of the *Mycobacterium tuberculosis* genome. Completion of this sequencing project is thought likely to enable the production of a range of DNA and genetically engineered vaccines.

Transnational Government Organizations

The UN System

WHO

The World Heath Organization clearly recognizes the potential of biotechnology to help alleviate world health problems (in particular, the role of new vaccines) and convened a conference, Biotechnology and World Health, in November 1994. Delegates recognized the current restriction of most biotechnology research to industrial nations, whilst identifying vaccine development as the area in which the global impact of recombinant DNA technologies is likely to be greatest (WHO, 1994). Many millions of people die annually from diseases which are vaccine-preventable, and the hope is that vaccines developed by recombinant DNA technologies will be heat-stable, or will offer the possibility of incorporating several vaccines in a single product. Vaccine trials supported by the WHO include those of a mutated *Shigella flexneri* vaccine (following deletion of genes required for cell-to-cell spread of the bacterium, and growth of the bacterium within infected tissues), and, imminently, nucleic acid and genetically modified vaccines for tuberculosis. Trials of a recombinant enteric *Escherichia coli* vaccine are being conducted in Egypt and Bangladesh. Recently, steps have been taken by the WHO to coordinate research on the development of transgenic plants as new sources of vaccines appropriate for developing countries, and to offer support for contractual services for the preclinical development of such transgenic plants. The total projected budget for the Vaccine Research and Development unit of the WHO for 1998 is US $9 million, with major contributions from the US Agency for International Development (USAID) (US $2 million), the Rockefeller Foundation (US $600,000), and the United Kingdom (US $500,000).

---

[a]This section summarizes responses, where forthcoming, to a questionnaire circulated to major overseas development agencies and private foundations supporting the development of drugs or vaccines for use in the developing world.

Children's Vaccine Initiative

The Children's Vaccine Initiative (CVI) was founded in 1991, jointly between UNIDO, UNICEF, the Rockefeller Foundation, WHO and the World Bank. Subsequently, the International Vaccine Institute (IVI) was launched by UNDP, under the auspices of the CVI. The IVI will conduct laboratory, clinical and field studies of new and improved vaccines of particular concern to developing countries.

TDR (UNDP/World Bank/WHO Special Programme for Research and Training in Tropical Diseases)

In addition to engaging in vaccine and drug development and vector control techniques, the TDR is increasingly entering collaborations with both public and private sector research groups in developing countries with a view to strengthening research capabilities at a local level. For example, the TDR has orchestrated an 'Initiative for Biotechnology Implementation'. This aims to identify public and private sector researchers and laboratories with a view to establishing self-sufficient production units in developing countries. Frequently, these units are established through collaboration with research institutions in a developed country.

Amongst those projects funded by TDR which have involved use of recombinant DNA technologies are:

- Malaria: Development of transgenic mosquitoes resistant to malarial infection, and improvement of larvicides, by genetic engineering techniques. In collaboration with the MacArthur Foundation and the Wellcome Trust, research is conducted into the development of an engineered transposon for prevention of *Plasmodium* transmission.
- Schistosomiasis (or bilharziasis): Development of a recombinant vaccine.
- Leprosy: Recombinant antigens for diagnostic use.
- Leishmaniasis: Recombinant vaccines for both New and Old World strains, and genetically modified parasites for research purposes.
- Chagas' disease (trypanosomiasis): Genetic manipulation for new approaches to therapy.

Human Reproduction Programme

The Human Reproduction Programme operates under the auspices of the UNDP, UNFPA, WHO and World Bank, with support from the governments of the UK, Canada, Sweden, Norway, Denmark, Germany. The programme focuses on the development of contraceptive methods in developing countries.

ICGEB

Although now an autonomous international organization, the ICGEB remained a special programme of UNIDO from when operations began in 1987 until 1994. The ICGEB has over 200 full time staff in its laboratories in Trieste, Italy and New Delhi, India, and some 23 Affiliated Centres in developing countries. Major programmes in Delhi focus on recombinant diagnostics, vaccines and drugs for malaria, hepatitis B and E, and HIV. The ICGEB has also been involved in the development of a recombinant antifertility vaccine.

Bilateral Donors

The European Union (EU)

The European Development Fund will supplement the national development projects of EU member states to the tune of US $17 billion over the period 1995-1999. In addition

Table 3.1. Recent SAREC support for biotechnology-based projects in human health.

Project	Collaboration	Period of funding	Funding provided (US $ millions)
Chagas' disease, including recombinant vaccine development	Argentina Uruguay	1986-1994	3.524
Shigella vaccine production and simultaneous strengthening of biotechnology capacity at the National Institute of Hygiene and Epidemiology, Vietnam	Vietnam	1985-1995	1.077
Development of DNA probe for filariasis. Simultaneous capacity building in gene technology at University of Columbo	Sri Lanka	1988-1995	0.624
Development of recombinant mucosal vaccines	Mexico	1991-1995	0.500

specific programmes support collaborations with developing countries:

The International Scientific Cooperation programme promotes academic interchange between scientists from developing countries and the EU. The programme has funded, for example, the China-EC Biotechnology Centre in Beijing.

The Cooperations with Third Countries and International Organizations (INCO) programme is part of the Fourth Framework Programme, soon to be superseded by the Fifth Framework. Part of this programme, INCO-DC (Scientific and Technological Cooperations with Developing Countries), is funded with US $247 million from 1995-1998, and is open to all persons and organizations in developing countries to work in cooperation with EU member states. The European Community is urging greater collaboration with developing countries under new Framework 5 proposals which recommend research cooperation with developing countries in areas of both biomedicine and biotechnology (biotechnology was not an explicit objective of the Fourth Framework programme).

Japan

The International Centre for Biotechnology, Osaka University (ICBiotech) addresses the application of biotechnology to both environmental protection and human health in its policy statement, focusing on projects in ASEAN (Association of South East Asian Nations) countries. In the year 1996, a total of US $1 million was allocated to biotechnology projects, with half being earmarked for agricultural projects, and the remainder split equally between health and environmental initiatives. ICBiotech supported projects for the production of antibiotics, with the cooperation of universities in Malaysia, Indonesia, Thailand and the Philippines.

The Netherlands

The Netherlands Directorate-General for Industrial Cooperation (DGIS) explicitly addresses the application of biotechnology to both environmental protection and human health programmes, with an estimated expenditure on biotechnology-based health projects of US $250,000 during 1996 - contrasting with some US $2 million given for support of agribiotechnology projects (T. Van de Sande, Biotechnology Programme, Netherlands Directorate-General, Industrial Cooperation (DGIS), The Netherlands, 1997, personal communication). DGIS initiated a five-year Special Programme on Biotechnology in 1997 (Commandeur, 1997). However, the programme focuses on agricultural applications of biotechnology, adopting an 'Interactive Bottom-Up' approach which provides scope for the active involvement of small-scale farmers, as well as policy-makers, as stakeholders in the innovation process.

Sweden

The Swedish Agency for Research Collaboration with Developing Countries (SAREC) addresses biotechnology for both environmental protection and human health in their policy statement. Programmes are focused on eastern and southern Africa regionally, and India, Ethiopia and Zimbabwe specifically. Support is provided principally for human resource development and research, rather than infrastructure or product development. In 1995, around US $3 million was provided for biotechnology-based projects. The bulk of this was earmarked for health projects (44%) (principally medical diagnostics and vaccine development - see Table 3.1), with 'environmental and industrial' applications of biotechnology attracting 39%, and agricultural biotechnology 17%.

The United Kingdom

The Department for International Development (DFID) does not have a specific policy on the use of biotechnology in relation to their health programme. However, a small number of the projects funded by DFID involve biotechnologies. These include the development of new bacterial strains for the control of malarial mosquitoes and other disease vectors, the investigation of HIV-specific immunoglobulins in serum for the early detection of HIV in children, and the development of an ELISA test for quantifying pyrethroid deposits in mosquito netting (M. Smith, Knowledge Branch, Health and Population Division, Department for International Development, UK, 1997, personal communication).

The United States of America

The US Agency for International Development (USAID) does not have an official policy on the application of biotechnology to health or environment problems, with the majority of funding for biotechnology-related projects supporting agricultural biotechnology programmes. It is estimated that US $2 million were provided for the support of biotechnology-related agricultural programmes during 1996 and 1995. Biotechnology-based health programmes were supported with a further US $8.7 million over the period 1994-97. These human health programmes focus on the development of vaccines for malaria and rotavirus. The malaria vaccine development programme includes research on the culture and processing of recombinant organisms, prior to scale-up for commercial production. Field tests have already been conducted in Kenya, the Gambia, Tanzania, Colombia and Thailand. USAID has provided support for the evaluation of a malaria vaccine produced by the Walter Reed Army Institute of Research in collaboration with Smith-Kline Beecham. Rotavirus vaccine trials are also underway in Venezuela (J. Lewis, US Agency for International Development, 1997, personal communication).

References

Avramovic, M. (1996) *An Affordable Development? Biotechnology, Economics and the Implications for the Third World.* Zed Books, London.

Commandeur, P. (1997) The DGIS Special Programme on Biotechnology. *Biotechnology and Development Monitor* 31, 6-11.

COHRED (1990) *Health Research: Essential Link to Equity in Development.* Commission on Health Research for Development.

Mugabe, J. (1993) Africa must raise its budgets for human health research. *Biotechnology and Development Monitor* 17, 8-9.

UNCED (1992) United Nations Conference on Environment and Development, Agenda 21, http://www.igc.apc.org/habitat/agenda21

UNIDO (1984) *Biotechnology and the Developing Countries: Applications to the Pharmaceutical Industry and Agriculture.* UNIDO, Vienna, Austria.

Walgate, R. (1992) Making biotechnology appropriate - and environmentally sound. In: Da Silva, E.J., Ratledge, V. and Sasson, A. (eds) *Biotechnology: Economic and Social Aspects.* Cambridge University Press, U.K., pp. 284-308.

WHO (1994) Press Release WHO/87, 8.11.1994.

Chapter Four
Country Profiles

Belarus

N.A. Kartel, A.G. Lobanok, V.M. Dorogoush and O.A. Kazuchets

National Policy and Regulation

General Policy

A state programme (Biotechnology, 1996-2000) for fundamental research on new biotechnological processes is being carried out. The Biotechnology 1996-2000 programme gives support to the priority areas of biotechnology in the Republic. The Institute of Microbiology, National Academy of Sciences of Belarus is the coordinating body which supervises the realization of this basic research programme. The programme aims to investigate: (1) mechanisms controlling microbial synthesis of bioactive agents; (2) design of wasteless, ecologically safe, resource-saving bioprocesses; (3) biotechnological methods for producing stock material as the basis for plant cultivars; (4) animal and plant cell culture; (5) biotechnologies for the development of effective plant biocontrol agents; (6) production of veterinary preparations for prevention and treatment of pathologies in farm animals.

Funding

Shortages of funds for purchasing up-to-date equipment and training of scientific personnel cause major constraints. Subject to a general improvement in the economic situation of the country, research work on biotechnology in Belarus is likely to intensify.

Regulatory Structure

A government initiative for biosafety is currently under development and legislation for the control of genetic engineering activities is being prepared. Intellectual property legislation has already been adopted, including patents for plant varieties.

Research, Development and Marketing

Major Projects and Institutes

Biotechnological research is being carried out mainly in research institutes of the National Academy of Sciences of Belarus (Institute of Microbiology, Institute of Bioorganic

Chemistry, Institute of Genetics and Cytology, Institute of Experimental Botany, Institute of Photobiology) as well as the Belarussian Research Institute of Potato Breeding, the Belarussian Research Institute of Agriculture and Fodders at the Agrarian Academy of Sciences of Belarus, and the Belarussian Agricultural Academy.

The Institute of Microbiology at the National Academy of Sciences of Belarus is the leading research institution in the Republic responsible for the coordination of research and personnel training in the area of microbiology and biotechnology. The Institute houses the national collection of microbial and cell cultures, and supplies education and research institutions and industry with a broad range of cultures. The collection is comprised of over 2000 catalogued microbial strains.

Established research teams specialize in enzymology, lipid synthesis in yeasts and higher fungi, microbial transformation of nucleic acids, and the biodegradation of xenobiotics. Significant results have been obtained at the Institute in recent years:

- Founding the national collection of microorganisms, cell and animal tissue cultures, and viruses.
- Development of methods for the production of adenine and thymine, and the potential anti-AIDS drug, 2-chloro-2-deoxyadenosine.
- Development of an effective antileukaemia agent (virasol) and benzylaminopurinearabinoside (an antiviral compound applied in technologies of virus-free potato seeds) and bromovinyldeoxyuridine (a nucleoside possessing a pronounced anti-herpes effect).
- Systems for the immunochemical microassay of bioactive compounds.
- Production of enzyme preparations (pectomacerin, dextranase, amylases) and therapeutic, prophylactic, immunocorrecting and immunomodulating drugs based on bifido and lactic acid bacteria.
- Development of technologies for processing milk whey into alcohol and proteinaceous fodder concentrates.
- Development of industrial culture methods for edible fungi.
- Production of new agricultural preparations: 'Sillactim' (an ensiling agent), 'Sapronit' (a legume fertilizer), microbial 'Sapropel' preparations for vegetables, 'Rhizobacterin' for grain crops to enhance respective yields, agents for biological control of bacterioses ('Pentaphag' and 'Bactophil'), and 'Colepterin' and 'Dendrolin' insecticidal preparations.
- Design of growth-promoting and phytoprotective preparations based on active microbial strains - degraders of native and chemically modified lignocellulose allowing use of cheap and readily available feedstock (residual lignin wastes of agricultural and industrial processes).

In addition, the environmental behaviour of microorganisms decomposing xenobiotics has been examined, and the concept of microflora activity control was formulated to facilitate the breakdown of soil pollutants. Screening yielded unique microbial variants able to degrade a broad range of xenobiotics, including chloroaromatic mixtures at elevated concentrations. These findings broaden our notions of microbial degradative potential and their possible applications in xenobiotics disposal systems. A preparation ('Rodobel T') is also intended for the recovery of areas contaminated with oil and derivative products.

The effect of anthropogenic load on microbial diversity of major soil types in Belarus has been analysed, and a strategy has been developed for the biological remediation of soils. The necessity of introducing agents controlling soil fertility - nitrogen-fixing and phosphate-mobilizing bacteria - was demonstrated to increase harvest yields and quality in artificial phytocenoses. A collection of efficient nodulating bacterial strains symbiotic to main agricultural crops has been set up.

Research on genetic engineering (development of transgenic plants, and the construction of molecular and genetic maps of plants for breeding purposes) is at present conducted only at the Institute of Genetics and Cytology of the National Academy of Sciences of Belarus and partially at the Belarussian Research Institute of Potato Breeding, in collaboration with the Centre of Bioengineering, Russian Academy of Sciences in Moscow.

Closer collaboration with foreign scientific centres is desirable, and some joint projects have been pursued:

- The Institute of Genetics and Cytology at the National Academy of Sciences of Belarus took part in a project on the molecular and cytogenetic characterization of the rye genome, and the evaluation of new genes for plant breeding, in collaboration with the Institut für Pflanzengenetik und Kulturpflanzenforschung (Germany) and the John Innes Centre (United Kingdom).
- Since 1997, the Institute of Genetics and Cytology at the National Academy of Sciences of Belarus has participated in a project on the development of disease-resistant potato in collaboration with CPRO-DLO (Centrum voor Plantenveredelingsen Reproductie Onderzoek - Dienst Landbouwkunding Onderzoek [The Netherlands]) and the Austrian Research Centre (Austria) under a European Union INCO/COPERNICUS grant.

Private Sector Engagement

The microbiological industry in Belarus expanded in the 1970s following the achievements of researchers based in Moscow, Leningrad, Kiev, Minsk, Odessa and the Baltic republics.

Industrial products include fodder protein, amino acids, RNA and derivatives, antibiotics, vitamins and other biologically active substances, bacterial medicinal and veterinary preparations, enzymes and multi-enzyme compositions.

In 1994-1995, as a result of the high power consumption of the manufacturing sector, production prices rose and competitiveness fell. Since then, there has been considerable restructuring of the manufacturing sector. The association of microbiological and pharmaceutical branches within the framework of Concern Belbiopharm advises on the manufacture of, amongst other substances, medicinal and veterinary preparations. The enterprises of Concern Belbiopharm also promote cooperation and business relations with corporations and scientific establishments in Russia, Ukraine and other countries.

Brazil

Leila Macedo Oda, Bernardo Elias Correa Soares and Maria Cleria Valadares-Inglis [a]

National Policy and Regulation

General Policy

The Brazilian government is committed to placing Brazil amongst the developed countries of the world. In doing so, the country would have to face major challenges in order to

[a]This contribution represents an editorial synthesis of contributions by L.M.O. and B.E.C.S. (health biotechnology) on the one hand, and M.C.V. (agricultural biotechnology) on the other.

overcome great social inequities and modernize industry and production systems. This would mean that, inevitably, biotechnology will have to play an increasing role in most sectors of the economy. In Brazil, an expanding body of highly qualified science graduates is actively involved in broadening a range of disciplines including recombinant DNA, the cornerstone of modern biotechnology.

Funding

Throughout the 1980s, federal spending in science and technology was about 0.7 % of the GDP. In absolute terms, this represented about US $1.6 to 2.0 billion between 1981 and 1986, and about US $2.2 to 2.8 billion between 1990 and 1996. In 1995, government funds accounted for 78% of the investments in science and technology against only 22% from private companies. Since 1984, the Brazilian government has recognized the importance of biotechnological products and processes, leading the Ministry of Science and Technology to implement a national programme supporting scientific and technological development (PADCT). This programme concluded its first phase in 1991 and was aimed at training scientists at the postgraduate level, developing laboratory infrastructure and establishing mechanisms of interaction between universities, institutes and industry. During this first phase, PADCT invested US $172 million (US $72 million from international loans and US $100 million from the Brazilian government), and supported more than 2700 projects. The second phase of PADCT started in 1991 and was concluded at the end of 1996. The budget for this phase was US $290 million with about US $40 million allocated to the biotechnology subprogramme. The third phase of PADCT, for 1997, commenced with a budget of US $20 million, with US $3.4 million allocated for biotechnology.

PADCT has established a simple method of fund allocation; half is allocated to basic research projects and the other half towards the encouragement of collaborative projects between private companies and research institutes. This policy is being successfully used to direct development into new products or applied research and services. In the long-term, this policy should allow more integrative associations and greater efficiency in self-financing research programmes.

The creation of foundations directed towards sponsoring research and researchers is proving to be an excellent alternative for funding biotechnology projects. These foundations are generally located in the large state capitals and are supported by government, foreign and private capital. Examples of such foundations are: Fundção Oswaldo Cruz (FIOCRUZ), Fundção FEALQ, Fundção BIO-RIO, Fundção Dalmo Catauli Giacometti, Fundção de Amparo Pesquisa do Estado de São Paulo, Fundção de Amparo Pesquisa do Distrito Federal and many other state foundations located in Rio de Janeiro, Rio Grande do Sul and Minas Gerais.

Regulatory Structure

Biosafety

On 5 January 1995, the government approved and published the national biosafety law (No. 8974), which established the rules for development and use of genetically modified organisms (GMOs) and their release to the environment. A National Technical Committee for Biosafety was created in April 1996, and is responsible for the implementation of the law. This Committee, attached to the Ministry of Science and Technology, is composed of eight specialists in biotechnology, with representatives from the Ministries of Health,

Science and Technology, Environment, Agriculture and Education. Representatives of consumers' rights, occupational health and biotechnological industry associations also participate. The Committee is regulated by Decree No. 1752 of December 1995, which defines its affiliations and functions. Public and private, national and international and also financial organizations are required to apply to the committee for a licence for GMO development, use and commercialization. Developers and users also become responsible or co-responsible for possible problems due to incorrect or inappropriate use, as well as the effects on the environment and human health. Applicants for GMO licences are requested by law to produce risk assessments for GMOs and report these to the National Technical Committee.

The Brazilian Agricultural Research Enterprise (EMBRAPA) has formed its own internal biosafety committee, and its affiliated research centres working with biotechnology have scientists trained in risk assessment. CENARGEN (the National Research Centre for Genetic Resources and Biotechnology) maintains a national focal point for BINAS (Biosafety Information Network and Advisory Service), and a database on biosafety that can be accessed through CENARGENs internet home page.

In order to widen the biosafety capability, in 1996 the Oswaldo Cruz Foundation (FIOCRUZ) organized an 'International Course on Biosafety of Transgenic Organisms in Human Health' and a 'Workshop of Specialists on Biosafety', sponsored by several local agencies and with the technical cooperation of the United Nations Industrial Development Organization. This event gave rise to a better understanding of biosafety principles and their importance to ensure a proper management of the risks posed by biotechnology in Brazil.

Intellectual Property Rights

The use of natural resources and technology transfer is ruled by specific legislation governing patenting, aimed at protecting research property rights. Law No. 9279 of May 1996 deals with rights and obligations concerning industrial property and Article 18 of this law refers to new biological inventions and models. According to Article 18, whole or part of living beings (except transgenic microorganisms) are considered non-patentable. For the purpose of this law, microorganisms are taken as whole organisms except for plants and animals or their parts expressing, by human intervention, a character usually not achieved by the species under natural conditions.

The newly approved patent law (No. 824-1, 1996) is expected to have a major impact on technology transfer to Brazil. In 1997, a law (No. 9.456) relating to Cultivar Protection, was also approved, recognizing the rights of the producers of new plant varieties.

Biodiversity

Brazil is a signatory of the Convention on Biological Diversity, and established the Brazilian Biodiversity Fund at the end of 1996, with a view to building up the expertise required for accommodating biodiversity concerns whilst pursuing biotechnological advances. This is a joint venture between the Global Environmental Fund and the Getulio Vargas Foundation, a Brazilian private centre for financial and economic studies. The Brazilian Biodiversity Fund has a significant budget until the end of the century to support manpower training and research and development in strategic areas such as agriculture and biodiversity management.

Regulation of biotechnology and biodiversity in Brazil now comes under the proposed Law No. 306/95 of the Federal Senate, which deals with access to biological resources for use as raw material for biotechnological development. At present this proposed law is being discussed at the national level and the President of Brazil has assigned an executive group to review it. The law deals with, *inter alia*, the protection of knowledge, including the rights of traditional communities and the control of technology transfer. Its promulgation will have a major impact on access rights to medicinal plants, which are estimated to include about 25,000 plant species (mainly from the Amazon region). Most are being used in therapeutic preparations and constitute a potential source of new biotechnological products.

Research, Development and Marketing

Major Programmes

Health Biotechnology

Historically, Brazilian biotechnology advances are attributable to government institutes, and firms representing the pharmaceutical sector (such as the Oswaldo Cruz Foundation in Rio de Janeiro and the Butantãn Institute in São Paulo). Both of these institutes were founded at the beginning of this century, and their activities are linked to the production of sera and vaccines. They were modelled on research centres abroad, mainly the French Pasteur Institute, from where the most important methodological advances in scientific research at the time were brought to Brazil by the researcher Oswaldo Cruz. Since then, industrial technology and academic research have always developed hand-in-hand in Brazil, particularly since the Second World War, when many international pharmaceutical companies began setting up facilities in the country.

Biotechnological projects within the Brazilian health structure include a National Programme for Self-Sufficiency in Immunobiologics, a National Programme for Blood and Haemoderivatives and a National Scientific and Technological Capacity Building Programme on Biosafety, supervised by the Ministry of Health. These Programmes join several institutions working in different research fields using recombinant DNA methodologies to test and develop vaccines and to screen for blood-borne diseases, as well as to teach laboratory personnel the principles of biosafety. This last project was set up to meet the growing demand for biosafety practices due to the implementation of the National Programme for Emerging and Re-emerging Diseases.

Other important projects underway include forefront research in tropical medicine, with the identification of parasites and microbial agents through enzyme immunoassays, genome mapping by PCR or restriction fragment length polymorphism analysis, and the development of laboratory tests for early diagnosis of genetic diseases. So far, these areas have been mainly developed by the Oswaldo Cruz Foundation (Molecular Biology Improvement Project) and the University of São Paulo, together with its associated research centres. Another biotechnology landmark in Brazil is the National Biotechnology Programme, initiated in 1981 in order to integrate institutions and budgets related to technological activities in the agricultural, energy and health fields. This was the first initiative to set up directions for national capacity building in genetic engineering and is important for its complexity and revolutionary impact.

An important element in many projects is the simultaneous involvement of the public and private sectors. FIOCRUZ plays a prominent role in this regard, followed by the Butantãn Institute, the Adolpho Lutz Institute and the Institutes of Biochemistry and Microbiology from the Universities of Rio de Janeiro and São Paulo. Another institution deserving mention is the

Centre for Biotechnology of the University of Rio Grande do Sul, which follows several important research lines in health biotechnology. These institutions have already achieved considerable expertise in the area, incorporating procedures through interaction between public and private sectors in research and development.

Agricultural Biotechnology

EMBRAPA is the largest state agricultural research and development organization in Brazil and holds a major responsibility for directing investments and development of agriculture. Biotechnology, as a research interest, started at EMBRAPA in 1981, with the foundation of the Plant Genetic Engineering Laboratory at CENARGEN. The first five years were spent training people in techniques such as isolation, characterization and sequencing of proteins and genes, DNA synthesis and gene cloning and plant transformation and gene expression. During the period 1993 to 1996, the Brazilian Council for Scientific Research (CNPq), in association with CENARGEN/EMBRAPA, has funded a total of 154 scholarships via the Manpower Development Programme for Priority Research Areas (Programa de Recursos Humanas em Áreas Estratégicas - RHAE) programme. The programme generated or made possible 73 scientific papers published in national and international journals, 38 conference papers, nine technical visits, 21 seminars, 41 national and eight international scientific visits, 16 master degree theses and five international PhD theses, and six training/specialization courses. The RHAE programme has also allowed the interaction between EMBRAPA and 125 national and international research institutes. From a total of 154 RHAE programme trainees, EMBRAPA employed about 20%; 15% have gained appointments as academic lecturers in their states/towns of origin, 25% have started masters and PhD courses at Brazilian universities, 5% are studying for PhDs abroad, and around 20% of the trainees are preparing for postgraduate courses.

In 1988, EMBRAPA's biotechnology research programme had a total of 155 projects carried out by the National System of Agricultural Research (SNPA), involving national and state institutes and universities. The projects were concerned with animal health, molecular and cellular biology, food technology and fermentation, soil biology and phytopathology. Among the SNPA institutes conducting biotechnology research in 1988, 21 were segments of EMBRAPA, 13 were state companies, five were universities and one a foundation. EMBRAPA invested US $150,000 in these projects in 1985, increasing to US $500,000 in 1987. Agricultural products research is conducted by EMBRAPA in direct association with universities and private companies. EMBRAPA, through CENARGEN, has established cooperation with national and international companies such as Agroceres, Ciba Geigy-Brazil, Dow Chemical S.A. Brazil, Union Carbide, Pfeizer Chemistry, Biobras, Biomatrix, Embrabio, Bioferm, Microbiologica, Cyanamid, Monsanto-Brazil, Valle, Plant Genetic Systems and others. Also important are associations with institutes and foundations such as the Rockefeller Foundation, Pasteur Institute, Gerdat-France, Institut National de la Recherche Agronomique (INRA), Centre de Coopération International en Recherche Agronomique pour le Développement (CIRAD), the United States Department of Agriculture (USDA) and Centro Internacional de Agricultura Tropical (CIAT). Recently, research projects have been negotiated with the participation of private groups, which pay for research or establish cooperation for applied research, in return for a share in royalties. These mechanisms encourage the development of strong research collaborations, focused on specific subjects, such as the development of molecular marker technology, pre-breeding lines, plant disease resistance, pest and weed control as well as animal genetics, breeding and reproduction.

In the last five years, EMBRAPA has implemented a new planning and administration system with the aim of improving productivity. Associated with this, a Programme for Development of Basic Research in Biotechnology has been reorganized and established as an important instrument in redefining market-led policies for the development of biotechnology at EMBRAPA and SNPA. This represents a dynamic element in Brazilian agricultural research with the intention of improving cooperation between various scientific institutions and the private sector. Association and integration of groups has been stimulated facilitating cooperation, information exchange and technology transfer as well as the development of specialized products and new technology. The focus of the programme is basic research, and as a result of this reorganization, the programme has reduced the total number of projects, but promoted greater cooperation among institutes and companies. The demands for the programme are defined as production of technology for use in agricultural and forest production systems, development of technology in the area of molecular biology, development of biotechnology in human nutrition and the development of biotechnological procedures for native and exotic microorganisms. These objectives are important in order to make modern agriculture and forest production competitive, sustainable and of high quality and to develop and promote cooperation and transfer of information and technology between national and international institutions.

The new approaches for development of new biotechnology products, processes and services has stimulated a requirement for more investment in infrastructure. EMBRAPA, with financial support from the World Bank and the International Development Bank (IDB), has built new biotechnology laboratories at CENARGEN at a cost of around US $3.8 million, excluding equipment. Total equipment costs, in the past ten years are estimated to be approximately US $40 million. In 1997 EMBRAPA, with financial support from the International Bank for Reconstruction and Development, the Interamerican Development Bank (Banco Interamericano de Desarrollo) and other international financers, has invested US $19 million, including salaries, in activities related to biotechnology, US $4 million in equipment and US $4.6 million in infrastructure and refurbishing of laboratories. The new laboratory complex is designed to accommodate a group of researchers with the mission to develop techniques and methodologies to be used in development of biotechnological products. Other EMBRAPA centres, such as the Soybean, the Corn and Sorghum, the Cassava and Fruit Culture, the Soil Microbiology and the Viticulture Centres, are also building biotechnology laboratories which will allow the manipulation of bioengineered organisms and products. In 1997, EMBRAPA started PRODETAB (Project to Support Agricultural Development and Technology Transfer) with the financial support of the World Bank. This project aims to support research until the year 2002, and has a budget for advanced technology, including biotechnology, of US $18 million. The first projects approved are focused on development of diagnostic kits, and integrated pest management, including the rational use of chemicals in agriculture.

Major Institutes

Major institutes include:

- CENARGEN - Genetic Research and Biotechnology, which focuses on genetic resources, germplasm conservation, and biological control (*For more information about this institute, see Appendix I*).
- Corn and Sorghum Research Centre, concentrating on breeding, pest control and sustainable production of corn and sorghum. Also has a corn and sorghum germplasm bank.

- Soybean Research Centre, with principal research interests in genetic breeding of soybean, pest control in soybean crops, sustainable production systems for soybean, and soybean germplasm banks.
- Rice and Bean Research Centre, with principal research interests in the genetic breeding of, pest control in, and sustainable production systems for, rice and beans.
- ESALQ - Escola Superior de Agricultura Luiz de Queiroz, conducts undergraduate and postgraduate courses in phytopathology, entomology, genetics and zoology.

Private Sector Engagement

Investments in advanced biotechnology have increased in Brazil in the last five years in the areas of biological reagents, vaccines, plants, food products and pharmaceutical substances. Human insulin, the first Brazilian product developed through recombinant DNA techniques, is now fairly established on the market. The Brazilian biotechnological industry is organized mainly through the work of the Brazilian Association of Biotechnological Enterprises (ABRABI) and the Brazilian Association of Industrial Technology Research Institutes, which have crucial roles in conducting public policy review, as well as coordinating management and research guidelines for the main Brazilian technology parks. These associations comprise the vast majority of small enterprises dealing with genetic engineering or molecular biology, 60% of which have appeared over the last ten years or so. A few of these companies have formed joint ventures with overseas firms in order to fund research in infectious diseases, especially in the development of diagnostic kits using DNA probes and molecular reagents for the detection of animal and human tropical diseases.

Since 1980, a total of 79 biotechnology-related companies and research institutes have been established in Brazil. The majority started to operate in the period between 1981 and 1992, and among them, 36 enterprises are related to basic biotechnological processes and products. According to Craveiro and Guedes (1996), the creation of Biobras in the 1970s was a landmark in Brazil's biotechnological development. Conglomerated companies soon followed, such as Plo BIO-RIO, Centro de Biotecnologia do Rio Grande do Sul, Fundao Biominas, Simbiose, Genoma, Millirio, WL Imunoquimica, Backtron, Vectron, Biologica and many others. The activities developed by biotechnology enterprises in Brazil are broad, with some companies involved in more than one type of research or product. Most of these companies have their activities centred in information, technical support and contracted research. About 16% of the enterprises are concerned with animal production, health, genetics and reproduction. Human health products account for about 14% of the activities. The chemical industry which includes enzyme production, industrial chemistry and synthesis accounts for 13%, with plant production at a similar level of about 12.6%. Agro-industry accounts for only 9% of the total interest, followed by forestry, fishing and others. Strategically, the government has encouraged the development of a network among companies and larger purchasers of biotechnological products, chemical and environmental products and services. The Science and Technology Parks are responsible for maintaining connections between scientific groups and industries. Interactions have also been established with companies based in industrialized countries via contracting of specialized services, and technologies. Formation of the Science Technology Parks underlines the government's preoccupation with integration of medium and large biotechnological companies with a network of small dedicated biotech companies. These interactions may encompass research and development contracts, purchase of fine products and specialized services and/or commercial partnerships. Integration with international organizations is expected to improve the prospects of success of biotechnology in Brazil.

Environmental biotechnology has attracted attention and companies have been established to develop products, such as microbial mixtures and/or enzymes (Biobac, Microbac, Bactrat). Enzymes are also produced in Brazil by Knoll and Novo-Industry. Biotechnology research products and services are offered by companies such as Pharmacia and Sigma. Other companies have their activities focused on novel plant development, such as Multiplica and Proclone. Human and animal health-related products are represented by Genoma (humans) and Pitman-Moore (animal). Other companies produce inoculates and biological control organisms (Geratec and Solofix).

In addition, a more recent financial mechanism for small companies has been set up in the form of SEBRAE (Brazilian Support Service for Micro and Small Enterprise) which involves cooperation between private, commercial and industry associations. SEBRAE administrates the tax paid by such enterprises in Brazil, transferring these funds to Science Parks and technology incubators in universities and research institutes. SEBRAE focuses on small enterprises, allowing small farmers and fisheries, food businesses and pharmaceutical producers access to the market. One example of the success of this scheme is a small company, In-Vitro, working with banana micropropagation, which was started by two graduates in collaboration with EMBRAPA, using the facilities of the University of Brasilia. The technology was transferred to a small business operation, producing about 20,000 disease-free plantlets per month. Each plantlet is sold at US $1.15 with 2% in royalties paid to the University of Brasilia.

Constraints

Adequate funding of science and technology is vital for improvements in national competitiveness. According to ABRABI (Brazilian Association of Biotechnological Enterprises), the global average investment in scientific and technological research is about US $100,000 yearly per scientist in academia and twice as much in industry. The Brazilian national budget for biotechnology research and development is projected to be US $2 billion per year by the year 2000, with federal and state agencies contributing US $1.3 billion and the rest coming from industry. However, the situation is not yet properly resolved, with government releasing only US $150 million by 1996, which is insufficient to support the 4000 scientists currently conducting research (and represents around a third of the mean annual investment per researcher at a global level). Private companies have begun to be involved with cooperative research, probably as a result of the government offering tax incentives. Nevertheless, the system is not yet able to facilitate all the necessary interactions, and the incentives offered so far have failed to attract those industries which continue to buy technology from abroad.

With regard to human resources, researchers working on biotechnology are mainly based in the southeastern part of the country. About half of the researchers in biotechnology (54%) are involved in postgraduate teaching, or are attached to government institutions. However, there is a shortage of qualified manpower: only 26% of scientists working in molecular biology are qualified at doctoral level, a very low proportion compared to other countries, and not enough to meet the challenges faced by the research and production programmes.

Some Brazilian industries lack capital and depend on outside financial support to develop new products. Until recently, private companies competed with public research bodies for funding, but with the establishment of cooperative programmes, direct association has emerged. Such associations are expected to increase with the recent approval of the patent law. An example of this type of collaborative research is the development of genome studies in citrus, using biochemical and molecular markers, involving a university and a private juice producing company. Another example is the development of molecular marker technology and genetic mapping in eucalyptus, an association between EMBRAPA/CENARGEN and a cellulose producing company. These associations are vital in order to increase the competitiveness of Brazilian products and to exploit new markets. The outcome is expected to be a reduced dependency on expensive imports and the creation of new jobs.

According to the PADCT 1995 report, the successful conclusion of the programme has suffered due to insufficient government funding. By the end of 1996 the programme is expected to have spent US $300 million to support about 1700 projects. Much effort has gone into negotiating the next phase of this programme which is projected to have a budget of US $600 million which includes support for partnerships among the academic and industrial sectors. Projects approved by the biotechnology segment of the PADCT programme have concentrated on agricultural research which accounts for 14 of the 16 projects in this area. These projects involve private companies in association with universities and research institutes.

EMBRAPA's budget is insufficient to cover the full costs of projects. In recent years, due to reduced financial support from the government, EMBRAPA has had to compete with other institutes and universities for money, applying for support from PADCT or any other available financial programme. EMBRAPA has therefore encouraged its researchers to associate with and obtain financial support from private companies focusing on new markets and products, aiming to increase competitiveness and efficiency. Despite the restricted budget for biotechnology (about US $1.7 and 2.2 million for 1997 and 1998, respectively), EMBRAPA has, in recent years, attempted to increase the money allocated to research. From 1994 to date, the budget for the biotechnology programme has increased by 29% (this figure does not include salaries and infrastructure). Projects included in the programme generally have outside financial support which covers the costs of consumables and, in some cases, training. From 1994 to March 1996, CENARGEN/EMBRAPA applied for grants from such national and international financial institutions as the Ministry of Agriculture and Food Supply, Foundation to Support Research at Distrito Federal, PADCT, Brazil's Bank Foundation, National Environment Fund, Brazil and Argentina Biotechnology Centre, International Centre for Genetic Engineering and Biotechnology (ICGEB), Japanese International Cooperation Agency (JICA), World Bank, European Commission and the Common Fund for Commodities/International Bank for Reconstruction and Development. The total money applied for was US $12 million of which US $5 million has been approved and US $2 million already been released.

Biotechnology is changing the face of agriculture in Brazil. The potential to exploit new markets, to increase agricultural productivity and the social benefits of improving the nutritional content of food is stimulating research and investment in biotechnology. The recognition of demands and priorities in this area are extremely important for the next decade, which will allow more efficient use of current and future resources. EMBRAPA, in association with CNPq and Ministry of Science and Technology and other research bodies, is in a leading position in this effort to modernize and invigorate Brazilian agriculture, industry and society. The collaborations with international organizations are also important in the process, allowing the development of new products and minimizing wasted investment in the national development of technologies already available elsewhere.

Future Prospects

The Brazilian market for biotechnology-based products and services was estimated at US $17 billion in 1990, with a projected increase to US $32 billion by the year 2000. Agricultural and related agro-industries account for 24% of this figure in 1990, and 29% by the year 2000. The remainder is accounted for by research and development centres, pharmaceuticals, public health, biomass production, minerals, oils and biosensors.

In agricultural development, the use of biotechnology and biotechnology products shows promise in addressing Brazilian crop problems. Solutions include the development of plants resistant to pests, disease and adverse environmental conditions, animal disease control, and the development of useful microorganisms. Recent advances in research and the development of novel applications of biotechnological products and processes have led to increasing demand in the market, which in turn has contributed to a growing incentive for investment in this area.

Techniques such as molecular markers have been used recently in the study of cultivated plants and trees of commercial interest with the aim of making improvements in breeding programmes. In animal research, molecular techniques may be applied in the isolation of genes related to disease resistance, and to develop novel vaccination and diagnostic systems. The genetic vaccination of animals, using DNA instead of proteins, seems to have a great potential and is currently receiving wide attention. *In vitro* fertilization and embryo transfer are used even by farmers with basic infrastructure in animal breeding programmes.

The rich biodiversity of the country confers Brazilian agriculture with a great potential for the development of new products with increased nutritional quality. Many genes of exotic and wild species of plants, microorganisms and animals, encoding for important proteins and toxins, are not yet isolated and cloned or their products exploited. Exploration of this biodiversity by use of molecular techniques would make possible many new innovative products of great economic importance. These natural reservoirs of genes are still largely unexplored in Brazil, mainly due to a lack of trained researchers in this field and the sheer size of the undertaking.

Many biological pest control techniques have found successful field application. In Brazil, a number of new species and new pathotypes are being isolated and studied. A bacterial-based product developed to control mosquitoes is now under registration, two different species of fungi are being tested against important species of weeds, and other species of fungi are being field tested to control grasshoppers and white flies. In addition, CENARGEN holds an important collection of entomopathogenic fungi, which may be applied in the future against emerging pest problems. Such a collection must also be regarded as an important genetic resource for the future development of novel agents and products.

Bulgaria

Pavlin Petrinsky and Atanas Atanassov

Introduction

Bulgaria has a long tradition in exploiting biological processes. For instance, wine fermentation and bacteria-based yoghurt production have been practised for centuries and, according to many claims, may actually have originated here.

Currently biotechnology in Bulgaria is being applied in the field of plant breeding for the development of varieties with new characteristics. Bulgaria has produced its first transgenic plants which will be marketed locally in the coming years, placing the country in a leading position in development of biotechnology in the region.

National Policy and Regulation

General Policy

Biotechnology in Bulgaria is regarded in the context of the general economic strategy for the development and application of new technologies with a view to improving the quality and efficiency of production and competitiveness of exports.

National expenditures and investments in biotechnology are being made through the budgets of the R&D centres. These expenditures have been decreasing in recent years due to the financial difficulties the country has been facing during the transition period to a free market economy.

Funding

At present, central funding for biotechnology amounts to an average of about US $100,000 allocated annually to the national coordinating centre, the Institute of Genetic Engineering (IGE). This figure represents about 30% of the total amount spent on plant biotechnology - the remainder coming from international collaborative research under various international programmes and grant agencies, and contract research for major international companies and local corporations.

Regulatory Structure

GMO Regulations
The Ministry of Agriculture, Forestry and Agrarian Reforms of Bulgaria has established a Council for Safe Use of Genetically Modified Higher Plants acting as a competent authority and charged with the full responsibility of allowing releases (both for research and commercial purposes) on the behalf of the Ministry. On 16 August 1996 the Council was formally established. Thus Bulgaria became the first country in Central and Eastern Europe to have specialized formal regulations for GMOs. However, the regulation covers only the release of higher plants and products derived thereof. Field trials with transgenic plants began in 1991. This regulation is regarded as the first step towards the adoption of GMO regulations covering other biological entities, which will have to be ratified by the National Assembly.

The Council for the Safe Use of Modified Higher Plants
The Council is chaired by the Deputy Minister of Agriculture and has nine members, including representatives of the Ministry of the Environment and the Ministry of Health. It is obligatory that the Secretary of the Council be a scientist in the field of genetic engineering. Council members are appointed by the Minister of Agriculture. Foreign experts may be invited for consultations if needed. The Council has full authority to allow, reject or demand specific measures for the release of transgenic plants in Bulgaria. It is also charged with recording and controlling all permitted releases.

Procedure for Notification of Work with GMOs
The procedure for notification of work involving genetically modified higher plants follows the model of Directive 220/90 of the European Commission. The procedure requires the person submitting the notification to present the information required in the Directive's Appendix for higher plants. The Council is obliged to answer each notification within 30 days, indicating its approval or setting out additional conditions which should be met for approval of release.

Environmental release notification must contain a risk assessment, which the Council reviews in the light of potential benefits. There is no special protocol in place for conducting risk assessment: the principles that have been developed by the OECD and the EC are to be applied. In specific cases, the Council would rely on expertise resulting from international cooperation in biosafety and risk assessment. If there are no specific or relevant data on safety issues, the person submitting the notification can be asked to conduct field trials. Until the results from field trials are available, risk assessment is based on comparison of the possible risks with the expected benefits of a given release.

Other Regulations
Also relevant to biotechnology are sanitary and quarantine regulations enforced by the Ministry of Health Care and the Ministry of Agriculture. The Ministry of Health Care is responsible for:

- The Central Institute for Drugs, which approves new drugs and medicines, as well as imports.
- The Central Hygiene and Epidemiology Inspectorate, which controls the safe production and distribution of foods.

The Ministry of Agriculture is responsible for:

- The Central Veterinary Service, which supervises animal quarantine.
- The Governmental Commission for Plant Protection, which deals with pests and plant diseases.
- The State Variety Commission, which approves new plant varieties jointly with the Patent Agency. The Patent Agency protects plant breeders' rights (PBRs).
- The National Service for Plant Protection, which controls sanitary conditions, pests, agrochemicals and their use, etc.
- The General Inspectorate for Approbation and Seeds Control, which controls the quality of seeds.

Intellectual Property Rights
Bulgarian legislation has three laws defending intellectual property rights (IPRs) (see Table 4.1). All these laws are based upon similar legislation dealing with IPRs, which has been enacted in western European countries. Bulgaria is also party to the Biodiversity Convention and the General Agreement on Tariffs and Trade (GATT).

Table 4.1. Laws defending intellectual property rights in Bulgaria.

Law	Scope
The Authors Rights Act	Audio and video records, movies, books, software, etc.
The Patent Act	Licensing of innovations, technologies, know-how, trademarks.
The Animal and Plant Varieties Protection Act	New animal and plant varieties.

Research, Development and Marketing

Major Projects

The major biotechnology project to date is the development of disease resistant tobacco cultivars. Twelve local cultivars have been transformed with the nucleocapsid protein gene of the tomato spotted wilt virus, which is a major disease in the country, causing up to 80% loss in yield. The resistant cultivars were produced by *Agrobacterium*-mediated transformation. Field trials with transgenic plants started in 1991 and currently their sixth progeny has proved to have 100% resistance.

In the same project, development of tobacco genetically resistant to wild fire disease (a disease caused by the bacterium *Pseudomonas syringae* pv. *tabaci*) has also been achieved. The inserted *ttr* gene in the transgenic plants confers resistance to the so-called Tab toxin, produced by the bacteria. Field trials with seven genetically transformed cultivars started in 1993 and to date the presence of the trait in progeny has been confirmed.

The above project has been carried out by the Institute of Genetic Engineering in Kostinbrod under contract research for Bulgartabac Holding Group, Sofia. The aim of the contract research is the creation of new, genetically resistant varieties, the cultivation of which would minimize the use of pesticides in tobacco fields in Bulgaria. In 1997, seeds from the resistant transgenic tobacco plants were tested in numerous experimental stations in all of the main tobacco growing regions of the country. In parallel to this, procedures for registration and IPR protection will also be launched. Bulgartabac Holding Group is still a state-owned enterprise, but recently procedures for its privatization have been initiated and it is expected to operate privately as of 1999.

To date there are no biotechnology products on the market in Bulgaria. Nevertheless the above-mentioned varieties are expected to be widely sold in a few years' time.

Major Institutes

The national coordinator of plant biotechnology in Bulgaria is the Institute of Genetic Engineering (IGE). Another centre of excellence in plant biotechnology is the Institute of Wheat and Sunflower 'Dobroudja', near General Toshevo (*see Appendix I for further information*).

Constraints

Specialized biotechnology centres for each of the fields of application are the main resources of expertise and infrastructure in the country. For agriculture the IGE provides for all the crop-specialized breeding centres which are expected to be fully funded by the national R&D budget. This restricts their activities and does not allow extensive R&D. Although efforts are being made to set up links with industry, these are rather irregular and income from commerce is scarce. The lack of IPR and marketing expertise are further drawbacks discouraging collaboration with industry.

Bulgaria has at its disposal a great number of qualified staff within the academic sector. Contacts have been established with leading biotechnology centres worldwide, where Bulgarian scientists acquire knowledge and expertise in modern biotechnology methods. Nevertheless, owing to lack of modern biotechnology equipment and low salaries in Bulgaria, repatriation is a major problem.

The national R&D budget has been decreasing gradually since 1990 reaching a critically low point in 1996. Thus, involvement in international collaborative research programmes and joint projects is essential for the survival of the centres.

Although laws and regulations relevant to biotechnology are currently in place, most of them still do not function properly. This is due to the fact that most of them have been promulgated recently and the relevant capacity is still not in place. In terms of technology transfer, the problems of proprietary technologies are being addressed through discussions with the patent holders (mainly multinational companies) in order to reach agreements for joint commercial exploitation.

Future Prospects

Plant breeding is one of the main strengths of the Bulgarian biosciences. The application of biotechnology in this field creates the potential for the establishment of the country as a major world player in the creation of new plant varieties in the near future. The Institute of Genetic Engineering (IGE) has launched an ambitious programme aimed at the expansion of the application of biotechnology in plant breeding and the subsequent commercialization of the products obtained. For the latter purpose, IGE's Extension Service provides the entire spectrum of financial, legislative and marketing services required. It is planned that in about five years, Bulgarian crop varieties of the highest quality will be sold in various forms (licences, seeds, etc.) all over the world. The idea (depending on each particular case) follows the venture capital and/or strategic partnership paths to be used for attracting foreign investors, offering considerable stakes in the newly created joint ventures against financial and marketing commitment.

In view of the current economic difficulties being faced by Bulgaria, it is believed that this approach will have the widest possible positive impact and also prove to be the most cost efficient.

Conclusions

For many years, the goals of Bulgarian research lay in increasing academic prestige rather than being oriented at developing commercial products. This resulted in the current situation of an almost complete lack of relevant organizational, commercial and financial infrastructure needed to secure smooth technology transfer.

Several years ago, the IGE started to make attempts to commercialize some of its own products (for instance development of a technology for industrial production of rare medicinal plants) and those of other institutes (mainly new crop varieties with improved characteristics). Bottlenecks that have subsequently appeared as being severe drawbacks relate to lack of commercial experience and marketing information needed to assess the originality and technical viability of the projects.

It is believed that the above problems could be overcome if assistance during the technology transfer process were provided, including:

- General market information.
- Technical-economic feasibility assessment.
- IPR protection.
- Contacts with potentially interested venture capital and other commercial institutions.

Chile

Lionel Gil and Carlos Irarrázabal

Introduction

Chile, with a population of 14.5 million, an inflation rate of 5%, unemployment of 5.4% and illiteracy of 4.5%, is one of the leading economies in the region, exporting US $17.5 billion worth of goods in 1995. Although the Chilean scientific community is small, biotechnology is receiving increasing attention in the country. A survey undertaken by the Chilean Academy of Sciences has shown that the number of active researchers in the country (not including engineers) is 1300. Of these, approximately 200 work in the fields of chemistry and the marine sciences, whilst 800 are involved in research in the biological sciences.

National Policy and Regulation

General Policy

There is no official policy for biotechnology development, although the Ministry of Agriculture has recently formulated a National Plan for the Development of Agriculture, Forestry and Poultry Biotechnology. The programme, as yet without funds, has as its main objective the incorporation of advanced techniques for the development of new crop varieties for national and international markets. Some goals include: characterization of native species germplasm, isolation of genes of agronomic importance, improvement of plant and animal varieties, implementation of diagnostics for pathogenic diseases, and training at graduate level combined with investment in research facilities.

An important role in promoting biotechnology in Chile has been played by the National Committee of Biotechnology (NCB), which is an advisory committee to the National Commission of Science and Technology. The NCB fosters and coordinates academic activity in biotechnology and acts as an official link between Chile and international organizations such as the United Nations Development Programme (UNDP) and the International Centre for Genetic Engineering and Biotechnology (ICGEB). The NCB selects the Chilean research projects presented to the ICGEB, suggests candidates for international training courses, promotes visits of international experts to Chile as well as the training of Chilean scientists in advanced centres in developed countries. The Committee is also responsible for organizing the National Meeting of Biotechnology, which takes place every three years.

Chile has ratified the Convention on Biological Diversity. To date, regulations regarding the release of genetically modified organisms into the environment are not in place.

Funding

There is no special fund allocated by the government to support biotechnology research, and projects in this area must compete with projects from all other scientific areas for the funds available. The main funds available for research are:

Fondecyt. In recent years (1994-1997) this fund has approved 116 projects related to biotechnology, providing US $5.5 million. Most of these projects were assigned to universities. The most successful institutions were Universidad de Chile and Universidad Católica, which obtained financial support for 35 and 21 projects, respectively. Some of the successful projects featured participation by private companies, such as Tepual/Inual, Biosonda and Bios Chile although this was not a precondition for funding.

Fondef. This fund supported 153 projects between 1991 and 1997, with an investment of US $93 million. It is expected that these projects will generate benefits representing US $900 million based on the successful participation of the academic and productive sectors. A total of US $30 million has been allocated to 49 projects in biotechnology-related areas as follows: 27 in agriculture and poultry sciences, 13 in fishery, eight in forestry and one in mining.

Fontec. Up until September 1995, this fund supported 400 projects with a total investment of US $25 million. In the area of biotechnology, 17 projects, involving 12 companies, have been funded with a total outlay of US $900,000.

Fondo de Investigaciones Agropecuarias. This fund has provided US $10 million for 63 projects, but data are unavailable to assess the share of biotechnology projects alone.

Fondo de Investigación Pesquera. The objective of this fund is to support basic and applied research in the area of fishery and aquaculture. The fund is obtained through payment in advance of the fishery and farming patents. In 1996, this fund provided US $4.8 million for 56 projects, but no data are available indicating funding specifically for biotechnology projects.

Biotechnology is being successfully used in industry. Thus around 10% of Chilean copper production is obtained by a bioleaching process developed in Chile. Over US $600 million have been invested in bioleaching plants. The cellulose industry is using enzyme biotechnology to produce 1500 tons of bleached pulp per day.

Chile has one of the highest rates of foreign capital investment of all the Latin American countries (US $4.6 billion in 1996). Several factors account for this preference:

- Highly qualified human resources.
- Political stability.
- An open economy.
- The fastest-growing economy of the region (economic growth rate of 7% over the last ten years).

However, most of these investments have entered the country to exploit the natural resources in areas of mining, forestry, fishing, etc. Very little foreign capital has been invested in high technology or in biotechnology processes.

Research, Development and Marketing

Major Projects

Main areas of biotechnology research in Chile are: agriculture and forestry, marine biotechnology, mining, human and animal health, environmental biotechnology, and bioprocess engineering. Apart from the universities, biotechnology research is also carried out in some institutes, such as Institúto Nacional de Investigaciones Agronómicas which specializes in agricultural biotechnology and Instituto de Fomento Pesquero which specializes in aquaculture and marine biotechnology.

One of the fastest-growing areas is that of human and animal health; research in this area for humans is oriented towards development of diagnostic kits for various conditions (Chagas' disease, rotavirus, pregnancy, syphilis, blood typing and gene mutations) and vaccines (rabies, hepatitis B, meningitis). In the case of animal health, vaccines are being developed for cattle, swine and fish.

Research in marine biotechnology is directed towards aquaculture of molluscs, fish and algae, as well as towards diagnostics, nutritional supplements and hormones useful in sea-farming. Several products obtained from marine species are already in production such as agar, carrageenans, alginates, bioadhesives, chitin, chitosanes and high quality proteins. Chile is one of the largest world producers of fish meal and by-products.

In agriculture and forestry, scientists are working in micropropagation, pathogen diagnosis, biological control of plant diseases and genetic diagnosis of cultivated plants. In collaboration with institutions from Canada and the USA, field studies involving transgenic plants are in progress. Other projects are related to bioleaching of minerals, bioremediation, industrial waste treatments and processing of fine chemicals.

Major Research Institutions

The number of scientists and institutions related to research in biotechnology is quite small and is concentrated mainly in universities such as Universidad de Chile and Pontificia Universidad Católica in Santiago, and also in regional universities such as Universidad Austral de Valdivia, Universidad Católica de Valparaíso, Universidad de Talca and Institúto Nacional de Investigaciones Agronómicas.

Biotechnology Network

CamBioTec is an international network dedicated to the promotion of biotechnology-based applications and products in the agri-food and environmental management fields in Latin America and Canada. The network is supported by the International Development Research Centre of Canada, and works with focal points in Argentina, Canada, Colombia, Cuba, Chile and Mexico. CamBioTec Chile has organized several seminars, workshops and business meetings, and has created a directory of biotechnology firms. It has also established links and joint ventures between Canadian and Chilean firms, has participated and organized entrepreneurial visits to Chile and Canada, and has brought Canadian experts on biotechnology regulatory policy to Chile.

Private Sector Engagement

Around 30 biotechnology companies have been created in the last ten years. Most of these are small companies; many of them have links with universities; some of them have been commercially unsuccessful and have disappeared. Among the successful biotechnology companies are the following:

- Bios Chile, a company with over 40 employees, most of them professionals, which produces biological reagents, diagnostic tests, monoclonal antibodies, nucleic acids, peptides, probes, etc. The company markets products in the USA, Japan, Europe and Latin America and participates in joint ventures with foreign companies.
- Tepual/Inual, a company with over 100 employees that produces fish meal and krill enzymes, and is very active in the commercialization of biological process control. Tepual/Inual is one of the main Chilean exporting companies in biotechnology products.

Constraints

Factors limiting the development of industrial biotechnology might explain the low foreign investment in this area. These include:

- *Patents.* Chile has only recently subscribed to international agreements on patents and copyrights.
- *Technological Transfer.* Lack of experience in product development and in technological transfer.
- *Biosafety.* Lack of regulations regarding commercialization and management of recombinant organisms which can be released into the environment.
- *Incentives.* Lack of investment incentives in new technologies.

China

Wei Jiewei

National Policy and Regulation

General Policy

Historically, traditional biotechnology has played an important role in a variety of economic activities in China, especially in agriculture and medical care. As early as the 16th century, Chinese doctors understood that rabies could be spread by mad dogs. During the 10th century, a live vaccine against smallpox was developed, and by the Ming Dynasty, the variolation technique had been transmitted to European countries via the Silk Road.

Following the beginning of recombinant DNA technology in the early 1970s, serious biotechnology research started in China at the end of the same decade. Both the government and the people were aware of the importance of biotechnology, especially for the next century. As a developing country with a huge population of about 1.3 billion, one fifth of the world's population, food production and human health care pose important problems for China. The Chinese government has singled out biotechnology as the key technology to solve these problems. In 1987, the government, together with the State Science and Technology Commission of China (SSTCC) and the China National Centre for Biotechnology Development (CNCBD) of the SSTCC, jointly published a National Blueprint Book for Biotechnology Development in China, to clarify the government policy in promoting biotechnology and to define the priority research and development fields for this century (up to the year 2000).

China's first High Tech Research and Development Programme, the so-called '863 Programme', was formulated in March 1986, with the final approval of the State Council of China. The SSTCC is responsible for the organization and implementation of the eleven subjects comprising the five areas of the programme (biotechnology, information technology, automation technology, energy technology and advanced materials). Biotechnology is a very important part of the '863 Programme'. The main purpose of biotechnological research is to meet the challenges of improved nutrition and health care for the Chinese people. The biotechnology programme focuses on the following areas:

- New varieties of high-yield, high quality plants and animals which are resistant to biotic and abiotic stresses. Special emphasis is given to mapping the rice genome.
- New medicines, vaccines and gene therapy technology.
- Protein engineering technology.

Funding

Funding for modern biotechnology in China occurs mainly at the following four organizational levels:

1. The central government, such as the SSTCC, CNCBD and other ministries through national programmes.
2. Local government in the provinces and cities.
3. The national science foundations and academy associations, such as the National Natural Science Foundation, the Chinese Academy of Science, the Chinese Academy of Medical Science.
4. Research institutes and laboratories.

Funding for biotechnology from central government sources totals about US $25 million annually. This funding is not intended to cover research workers' salaries, instrument maintenance, large laboratory equipment or building costs; researchers are obliged to find other sources to cover these costs.

There is also considerable investment in biotechnology from the private sector. Industrial companies and international concerns are investing large sums in biotechnology R&D in China.

Regulatory Structure

China has only one regulation related to biotechnology: the 'China Safety Administration Implementation Regulation on Genetic Engineering' (or the 'China Biosafety Regulation') which was reviewed and passed at the Conference of the Standing Committee of the Ministry of Agriculture, on 25 April 1996. The major points of the regulation include:

- General principles.
- Safety classes and safety evaluation.
- Application and approval.
- Safety control measures.
- Legal responsibilities.
- Supplementary articles.

Research, Development and Marketing

Major Projects

Agriculture

The majority of research work in agricultural biotechnology concerns the development of new varieties of plants and animals with high yield or other desired traits. Major projects deal with:
- Hybrid rice and genome mapping.
- Transgenic plants.
- Nitrogen fixation.
- *In vitro* fertilization of calves.
- Transgenic fish and livestock.

Pharmaceutical Products, Vaccines and Gene Therapy

Research work in this area centres around the following:
- Recombinant peptides for biotherapy, such as interferon and interleukins.
- Recombinant vaccines against major infectious diseases, such as hepatitis A, B and C and haemorrhagic disease.
- Gene therapy.
- Antibody engineering.
- Protein purification technology.

Protein Engineering

Research efforts in this area are concentrated on:
- Insulin.
- Molecular modelling and design.

Major Institutes

Major institutes carrying out biotechnology research in China include:
- Life Science College, Beijing University.
- Basic Medical Research Institute, Chinese Academy of Medical Science.
- Beijing Institute of Virology, Chinese Academy of Medical Science.
- Shanghai Institute of Biochemistry, Chinese Academy of Science.
- Life Science College, Beijing Agricultural University.

Private Sector Engagement

Commercialization and industrialization of the results of biotechnology is considered the ultimate goal of research work. During the last ten years, Chinese scientists have developed new techniques in biotechnology to complement the rapid economic growth of China. For example:
- Several new combinations and varieties of two-line hybrid rice have been obtained with an average increase in yield of 5-10% as compared with that of the original variety. Approval has been granted by the Ministry of Agriculture of China for scale-up of the two-line hybrid rice technique to the commercial level. It is envisaged that these new varieties will be propagated to cover 1 million hectares within two years.
- About 40 genes for lymphokines and cytokines have been cloned or synthesized, characterized and expressed in prokaryotic or mammalian cells. Some have already been com-

mercialized and marketed, such as recombinant interferon α1B and recombinant interleukin IL-2.

- A recombinant hepatitis B virus (HBV) vaccine has been developed using the Chinese hamster ovarian cell system. The new vaccine is cheaper and more effective compared to the plasma-derived vaccine. Annual production of HBV vaccines in China now amounts to about 15 million doses.
- Monsanto has entered an agreement with a domestic seed company in Hebei province for development of pest-resistant cotton.

Future Prospects

Biotechnology promises to be a very important technology in the next century. The Chinese government has already recognized this and has put a lot of effort into planning and organizing high-technology R&D, in the hope of meeting the future nutrition and healthcare needs of the people of China. At present, efforts are focused on two areas: agricultural improvement and new medicine and therapeutic technology.

In the area of agricultural improvement, it is expected that transgenic plant technology will be applied to obtain new varieties of virus-resistant tobacco, virus-resistant tomato and pest-resistant cotton with a view to improving production yield and quality. In parallel, the biosafety issues related to release of such transgenic plants will be investigated in order to ensure the safety of human health and the environment. It is also hoped to increase the number of new varieties of two-line hybrid rice available to farmers, thus improving production yield and quality. The hybrid technology will be used to develop new varieties of other crop plants.

In the medical sciences, the following areas have been chosen for exploitation:

- Vaccine technology is considered very important for the improvement of the quality of life of the Chinese people. Production of existing vaccines will be increased, aimed at the eradication of viruses such as HBV, HAV and HCV in future generations.
- Development of new biomedicines and their scale-up to industrial level production. Examples are: interferons, colony stimulating factors, interleukins, tumour necrosis factors (TNF), growth factor, tissue plasminogen activator, etc.
- Gene therapy technology will focus on cancer.
- Large-scale cell culture technology will supply products in sufficient quantities for use in disease treatment in hospitals and also lower the cost of production of biomedicines.

Constraints

As far as basic research is concerned, work initiated in some Chinese departments has attained international level. However, in the area of applied research most Chinese projects are derived from investigations initiated by Western scientists. Insufficient funding is the chief limitation for further progress. Chemicals and disposable laboratory supplies are inadequate and not up to international standards. Much of the equipment and materials essential for biotechnology research must still be purchased from foreign suppliers.

The 'Brain Drain' is now a serious problem. It is estimated that more than 5000 Chinese students and scientists are planning to go abroad for further training in biotechnology or related subjects. Half of these will go to the United States of America. It has become common practice for Chinese students to pursue postgraduate studies abroad, leaving local research laboratories short of research workers.

Colombia

Rafael H. Aramendis and Elizabeth Hodson de Jaramillo

National Policy and Regulation

General Policy

COLCIENCIAS is the main agency entrusted with the implementation of the biotechnology programme in Colombia. The strategic plan for biotechnology in Colombia is connected with the following aspects:

- Integration of fundamental biotechnology techniques in research.
- Human resource development in areas extending beyond basic research (to engineering, marketing and commercialization, and technology management).
- Promoting venture capital investment in biotechnology by introducing appropriate fiscal policy instruments.
- Promoting and supporting the establishment of small biotechnology-based businesses.
- Designing and implementing appropriate intellectual property regimes and biosafety regulations and policies.

Priority action areas include:

- Development of biopesticides and biofertilizers.
- Development of diagnostic, prevention and treatment systems in connection with pre- and post-transition health problems of critical national importance.
- The study, conservation and utilization of biodiversity, clearly defining patterns of exploitation with high social and economic impact.

Funding

In terms of overall resource allocation under the COLCIENCIAS biotechnology programme, the main purpose of the initial phase of investment carried out (1992-1994) was to stimulate the creation of basic capabilities of research centres and teams in the management of core technologies, infrastructure and equipment. The second phase of investment (1995 to date) aims to strengthen the capacity of these research centres and teams with a view to defining, within a medium-term time-frame, their agendas and work priorities in terms of appropriate linkages with the industrial sector and the development of socially beneficial research strategies.

From 1992 to date, biotechnology resource allocation by COLCIENCIAS has constantly increased (in terms of current pesos) and now stands at the overall figure of US $6,853,000. The number of projects submitted and approved is following a similar trend.

Regulatory Structure

Biosafety

COLCIENCIAS is currently directing an interdisciplinary team from different national entities that are seeking to establish a national norm in biosafety. The Instituto Colombiano Agropecuario has produced, with the collaboration of other institutions involved, 'a national biosafety protocol with emphasis on transgenic plants'. Also, similar working groups have been planned in the fields of health, industry and the environment, and domestic legislation on biosafety is expected to be enacted in the coming months.

Intellectual Property Rights

The country's intellectual property regime is governed by Andean Decision No. 344, or the common regime applying to industrial property, a framework standard enforceable in the countries of the Andean community of nations (Bolivia, Colombia, Ecuador, Peru and Venezuela); the applicable norm in regard to plant breeders' rights is Andean Decision No. 345, or the common regime applying to breeders of varieties of plant species.

In the commercial field, the country has adopted the agreements of the World Trade Organization (WTO) and is currently working on adapting domestic law to the framework agreement on Trade Related Aspects of Intellectual Property Rights (TRIPS Agreement) established by WTO.

Biodiversity

Access to genetic resources (an issue of vital importance for Colombia) has been regulated though Decision No. 391 of the Andean Pact whose purpose is to establish criteria for access to and utilization of the country's resources under terms laid down by the agreements emanating from the Biodiversity Convention.

Research, Development and Marketing

Colombia's existing infrastructure for developing biotechnologies is located primarily at research centres and universities which are the main recipients of State funds for the pursuit of such activities. In the agricultural and livestock sector, the presence of public research centres and also of research centres linked directly to farmers' associations is significant. One example of the former is the Colombian Agricultural and Livestock Research Corporation (CORPOICA), a public institution operating under the private-law system. It is engaged in research and development work of socioeconomic benefit to the country. Prominent within the latter group are research centres which are financed and supported by specific farmers' associations and whose task it is to solve individual problems of their members. These organizations include the coffee, sugarcane, potato, rice and barley growers' associations.

The main strength of these centres lies in their ability to interact directly with users and producers, thereby shortening the cycle between problem identification and solution. It should be pointed out that biotechnology research networks are promoted and supported by the State as being of strategic importance for the country.

Notable among these is the work in biological control, with emphasis on *Bacillus thuringiensis* that is being carried out by the Colombian Biopesticides Network, which brings together public and private institutions (the National University, the Biological Research Corporation, CORPOICA, and the Colombian Industry of Veterinary Products (VECOL S.A).

The main advantage of this mechanism of inter-institutional alliances is the fostering of multi-sectoral and interdisciplinary work with a view to providing solutions to problems regarded as crucial by specific sectors of the community. Its drawbacks lie in the low level of institutional development, the researchers' limited experience in tackling problems using interdisciplinary methodologies, and the legal and technical difficulties that arise when research projects lead or may lead to results that can be marketed or be subject to legal protection through patents or other intellectual property protection mechanisms.

Private Sector Engagement

The traditional low level of industrial research is reflected by the following data. Industry contributes 10% for plant/agricultural and human health applications, 8% for the manufacturing sector, 7% for the livestock sector, and virtually nil for the environmental field of the funds that go into the national research effort.

A few specific cases of positive interaction between the academic sector and industry in the country are worthy of special mention. Within the agricultural sector, reference can be made to the cooperation between the National University and a number of firms in the flower-growing sector in developing diagnostic systems and between the National University and the potato growers' association in producing virus-free seeds, and also to the cooperation agreements between banana producers and the country's universities for the supply of virus-free material. In the environmental sector, university-enterprise cooperation is a recent phenomenon, which is directed primarily at waste-water management in the chemicals and food industries (e.g. the cooperation of both Alpina and Sucromiles with St. Xavier's University).

It should be pointed out in connection with biotechnology applications in the health sector that, while this field does not have the highest percentage resource allocation or the largest number of fully qualified personnel, it shows the most promising results with regard to the development of marketable products or processes. One of these cases relates to the work of developing diagnostic systems for blood group identification and is being undertaken by a national company in association with a number of national research teams.

This case calls for special attention, since it combines some of the factors conducive to the success of university-enterprise linkages. The researchers involved in the project hold master's degrees or doctorates and have extensive research experience at universities and research centres, both at home and abroad; it also includes entrepreneurs who are willing to invest venture capital and who possess experience in the marketing of technology-based products - State aid has been provided by means of soft funding and support through the technology-based enterprise incubator schemes (e.g. INNOVAR).

Human Resources

The main features of the existing national biotechnology research and development capacity in the country are as follows: According to data for 1997, there are currently 74 active research teams in Colombia, of which the majority (57%) still deal primarily with biotechnology applications in the plant and agricultural sector, followed by human health (17%), the environment (14%) and the livestock sector (7%), the smallest field of application being the manufacturing sector (5%).

These research teams are based mainly at universities and in the industrial sector (35% and 33%, respectively) but their presence is significant also at research centres (27%) and, to a lesser extent, in other entities, such as foundations or semi-public institutions.

As regards the current standard of overall academic qualification of Colombia's biotechnology community, there are at present 532 researchers, of whom some 17% hold doctorates and 26% possess master's degrees. These professionals are engaged primarily in work in the plant and agricultural sector (42%) and in the fields of human health applications (31%), followed by the environment (14%), livestock (7%) and manufacturing (6%).

A comprehensive analysis of postgraduate training programmes offered in Colombia to support biotechnology development suggests that, within the short term (some five years), there will be a considerable number of researchers trained in core biological technologies but also a noticeable lack of new researchers competent to handle biotechnology processes from the viewpoint of production process engineering in addition to a lack of managerial and specialist training in areas related to the commercialization and marketing of biotechnology-derived products.

An analysis of the profiles of doctorate candidates studying at home and abroad suggests that the high demand for researchers in the plant and agricultural spheres is likely to be met, but that a greater number of personnel and a higher standard of academic qualifications will be required in the fields of industrial, environmental, livestock and human health applications.

In addition, there will be an acute shortage of expertise in bioinformatics, technological management, technology transfer and intellectual property. It is necessary to define a strategy and mechanisms of international cooperation that will enable the country to begin developing its human resources in these areas.

International, Bilateral and Multilateral Activities

Historically, cooperation in biotechnology with high-quality teams and centres at the international level was vertical in nature. Thus top-level institutions, primarily in Europe and the United States of America, offered training for personnel, particularly university-educated staff, in advanced techniques and methodologies through their participation in internships and workshops. The main objective of these training courses was to provide tools that would enable students to apply their acquired skills upon their return to their laboratory in their home country. However, this task was not always carried out. On many occasions, the equipment, resources or research teams that were to be introduced in the home country were in the end not available and consequently the training acquired did not result in any benefit for that country.

This situation began to change in the mid-1980s, with the setting up of the first regional biotechnology programmes (UNDP/UNESCO/UNIDO Regional Biotechnology Programme for Latin America and the Caribbean, and Organization of American States Multinational Food Technology and Biotechnology Programme), which started to offer horizontal-type participation for countries and institutions, in particular university establishments. Involvement in this experiment has enabled the country to train teams with consultative capability and to locate laboratories with which similar work programmes were shared. Towards the beginning of the 1990s, regional biotechnology programmes started to enter their output evaluation phase, but a new tendency emerged whereby support is provided to selected teams involved in priority research. This approach appears to be particularly suitable for specific thematic areas that will meet the needs not only of universities and research centres but also of enterprises.

It can be further optimized by cooperation with international centres and agencies of recognized high standing and quality in areas of interest for the country (e.g. technology and enterprise management, support for regulatory systems, training in high technologies, participation in multilateral projects).

Constraints

Three principal constraints on the development of the biotechnology sector in Colombia can be identified:

- Absence of clear and appropriate regulations and lack of enterprise involvement.
- The public's lack of awareness and/or poor perception of the benefits of biotechnology.

- Monopoly of the transnational corporations in virtually all biotechnology product market segments, with increasing tendency by these enterprises not only to maintain their market share but also to capture remaining niches.

Future Prospects

COLCIENCIAS has identified national priorities and competitive niches. In health care, these involve the development of vaccines and diagnostic systems where multinational corporations have little or no interest. In addition, there is a clear demand for goods and services in environmental management, agricultural production and livestock breeding.

Although appropriate biotechnology in Colombia does not require major investment, there are powerful economic groups with sufficient venture capital to invest in biotechnology as a means of diversifying their capital assets.

The main strengths of biotechnology in Colombia can be summarized as follows:

- A reasonable infrastructure and availability of human and material resources, although it is clear that these need to be strengthened.
- Experience in joint interinstitutional and interdisciplinary activities and a good record of cooperation with internationally recognized centres of excellence.

The development challenges facing the country's biotechnology community are concerned with three key aspects: (1) training human resources (see above); (2) defining technological and market potential for the development of selected products and/or processes; and (3) adopting legislative and regulatory systems.

With regard to technological and market aspects, two key types of action are necessary. The first involves defining competitive strategic niches on the basis of patterns of needs, capacities and opportunities, and the second accessing public or private venture capital for biotechnology joint ventures.

Finally, there is clearly also an urgent need to adapt and develop regulatory and legislative systems - extending beyond the domestic scene and with a view to the strengthening of activities relating to the transfer and adaptation of technologies which major multinational corporations wish to develop in the country. It is thus necessary to modify and update the legislative systems currently in force in the area of intellectual property, adopt national norms in conformity with international standards on biosafety and exercise regulatory oversight with regard to access to genetic resources.

Cuba

Pedro A. López-Saura

National Policy and Regulation

General Policy

Education, science and technology development have been high priority items for the Cuban government since 1959. National education and science policy had to cope with almost 20% illiteracy and the absence of any scientific infrastructure apart from four small experimental agricultural stations. The current policy is based on four simultaneous approaches:

1. The development of indigenous scientific potential.
2. The orientation of scientific and technical activities to support the country's social and economic development.
3. The accelerated acquisition of international knowledge and technologies.
4. The generation of domestic products and technologies (Simeûn-Negrìn, 1996).

The first modern research centres were founded in the 1960s. They played a very important role in infrastructure formation and familiarization with modern scientific methodology, and produced the first results and scientific contributions. In 1981, a special policy on modern biology and biotechnology was launched. A huge investment programme started and several research and production centres were built and equipped. Besides those institutions listed in Table 4.2, the biotechnology programme includes a centre for immunoassay reagents and equipment for mass screening programmes, a site with large-scale good manufacturing practice production and dispensation facilities, and a centre specialized in the production and breeding of high quality laboratory animals. Education in biological sciences, chemistry and related technological specialities has been reinforced in all universities. Postgraduate training has been given high priority in international bi- and multilateral collaboration projects. This effort has led to a significant outcome, as shown in Table 4.3 (Academy of Sciences of Cuba, 1996; Ministry for Science, Technology and Environment, 1996; Centre for Genetic Engineering and Biotechnology [CGEB], 1997).

Some of these projects have had a notable social impact. Since 1992, all Cuban newborns, people at risk and selected age groups have been vaccinated against hepatitis B so that the whole population under 20 years old will be protected by 2000 (Ministry for Public Health, 1996a). The anti-meningitis B-C vaccine is administered under the Extended Immunization Programme. Since 1993, all patients with acute myocardial infarct capable of undergoing thrombolytic therapy have been receiving recombinant streptokinase. This initiative represents at least 200 lives saved per year (Cuban Group for the Study of Recombinant Streptokinase, 1996). The national AIDS, Hepatitis B and C, neural tube defects and other screening programmes (600,000 blood donors, 150,000 pregnant women, and other samples per year) use the recombinant proteins, synthetic peptides, monoclonal antibodies, and antisera-based ELISA as well as original ultramicroanalytical devices developed at these centres. Fetal and newborn screening procedures have contributed significantly to the low infant mortality index (below 8/1000 in 1996, an estimated reduction of at least 1/1000) of the country in the last years (Ministry for Public Health, 1996b). Safe blood and production of haemoderivatives is assured by the AIDS and viral hepatitis programmes. More detailed information on Cuban institutes and products can be obtained from the references (BIOMUNDI Counsellors, 1994). Research and development activities in Cuba are performed by 11,115 researchers (11% PhD) in 210 institutions (Díaz-Otero, 1996). Thirty-three of the latter are universities or institutes that are part of the higher education system. The rest are attached to various government agencies or directly to the State Council. There is no private R&D activity. Biotechnology-related disciplines (agricultural, technical and biomedical) represent 75% of the institutions, 87% of the scientific workers, and 82% of the researchers with PhDs.

The Ministry for Science, Technology and Environment (CITMA) controls these activities through national agencies and regional offices. This ministry was created in 1994. It assumed most of the functions of the former Academy of Sciences and other competent authorities (such as the Environment and Nuclear Energy Commissions)

Table 4.2. Major research and development institutes of biotechnology in Cuba. (All belong to the West Havana Scientific Pole. More details and information about other institutions are available in BIOMUNDI Counsellors (1994).)

Institute	Main research and/or production activity
Centro de Ingeniería Genética y Biotecnología (Centre for Genetic Engineering and Biotechnology)[a]	Recombinant DNA products for human and animal therapy, vaccines and diagnosis, industry and agriculture; transgenic animals; monoclonal antibodies for diagnosis, analysis and therapeutics; protein structure; GMM fermentations and recombinant proteins purification
Instituto Finlay (Finlay Institute)	Immunology; conventional and recombinant vaccines; therapeutic sera
Centro Nacional de Investigaciones Científicas (National Centre for Scientific Research)	Multidisciplinary centre. Pharmaceuticals; natural products derivatives; environmental research; microbiology; neurosciences
Instituto de Medicina Tropical 'Pedro Kourí' (Tropical Medicine Institute)	Microbiology and molecular biology for diagnosis and epidemiological research; WHO collaborative centre
Centro de Inmunología Molecular (Centre for Molecular Immunology)	Monoclonal antibodies for diagnosis and therapeutics; cancer immunology and immunotherapy; recombinant products in mammalian cell expression systems
Centro Nacional de Sanidad Agropecuaria (National Centre for Agricultural Health)	Animal and crop health; animal-derived products for diagnosis and therapeutics
Institúto Cubano de Investigaciones de los Derivados de la Caña de Azucar (Cuban Research Institute of Sugar Cane By-Products)[a]	Sugar cane derivatives for human and animal consumption, alcohol, industry, paper, wood, etc.

[a] More information about these institutes can be found in Appendix I.

involved in the control and management of R&D activities. The Academy of Sciences is now a non-governmental, elected institution with only academic and advisory functions.

According to the changes that the Cuban economy has adopted during the 1990s, a new management system for science and technological innovation was established in 1995 (Díaz-Otero, 1996). It is based on the principles of: (i) prospective analysis of science and technology development priorities; (ii) organization of R&D activities in programmes at the national, sector and regional levels; (iii) project presentation and peer review evaluation; (iv) contractual relationships among the different parties involved in a project; (v) state financing of projects within each programme; (vi) alternative funding sources such as enterprises, bank credit and foreign financing; (vii) decentralization of fund administration; (viii) promotion of interface activities among the scientific, technological, productive and financing systems; (ix) improvement of employee and wage conditions of human resources linked to R&D; (x) organization and development of an efficient information system; (xi) strategies for international collaboration; and (xii) establishment of an appropriate regulatory environment. The above are aimed at ensuring the commercial orientation of the system as a whole.

New coordinating centres, set up in the1990s, are the scientific 'poles', which exist in all 14 provinces of the country. They comprise all related research centres, higher education institutes, and the main potential users of the R&D results in a province (relevant health institutes and production enterprises). These focal points play a very important role in the coordination of multi-institutional approaches and the introduction of results into social and economic practice. The first and most important focal point is that of West Havana, with 42 of the major research, education, health and economic institutions related to the biotechnology field. This group of institutions produced most of the achievements listed in Table 4.3, 67% of the Cuban articles indexed by the Science Citation Index in 1990-1995 and 42% of all the publications generated in Cuba (BIOMUNDI Counsellors, 1997).

An important part of the national science policy is international collaboration. In 1996, the country participated in 204 bilateral and 66 multilateral projects (Ministry for Science, Technology and Environment, 1996). Among the latter, the country is a founding member of the International Centre for Genetic Engineering and Biotechnology (ICGEB). The Centre for Genetic Engineering and Biotechnology, Havana is an ICGEB affiliated centre. Cuba is an active member of the UNESCO/UNIDO Latin American Network on Biotechnology.

Funding

Despite the severe economic slowdown during 1990-1994, the government's relative expenditure in R&D has been kept to roughly 1% of the GDP. There was an increase from 0.8% in 1990 to 1.43% in 1994, corresponding to approximately US $25 per inhabitant per year (Simeûn-Negrìn, 1996). The central state budget covers infrastructure support, current operation expenses, and salaries. Direct financing of projects under the national programmes (discussed further below) comes from the national agencies. It reached 88.5% of the planned funds in 1996 in Cuban pesos, evenly distributed among the programmes. Regarding convertible currency, the programmes on vaccines, the pharmaceutical industry, and agricultural biotechnology received 70%, 17% and 7%, respectively, of all the funds allocated in 1996 for the national programmes (Ministry for Science, Technology and Environment, 1996). These figures are indicative of the national research priorities.

The nationally-allocated funds are approximately 32% of the total government expenditure in R&D. Sector programmes take an important proportion of funding, particularly in Public Health, Agriculture, Metallurgy, Energy and Chemical Industry. Institutional projects receive 20-25% of the government funds.

Product commercialization, scientific services, postgraduate teaching, and international collaboration are other sources of funds. The latter reached the level of US $4.2 million in 1996. In recent years, as part of the economic changes in the country, foreign investment has been introduced in various joint ventures for product development and production.

Regulatory Structure

The regulatory environment has developed in parallel to the R&D activities. The State Centre for Quality Control of Medicaments regulates good manufacturing, laboratory and clinical practices on diagnostics, pharmaceuticals and vaccines for human use. It approves

Table 4.3. Major recent achievements of modern biotechnology in Cuba.

Product/procedure[a]	Registered	Undergoing clinical trials (Phase)
Vaccines (for human use if not specified)	Anti-hepatitis B (rec)[b] Anti-meningitis B and C Anti-tick for cattle (rec) Anti-E.coli for swine (rec) Anti-avian coccidiosis	Anti-leptospirosis (III) Anti-HIV (rec) (I) Anti-cholera (rec) (I)
Pharmaceuticals	Interferon alpha-2b (rec) Streptokinase (rec) Epidermal growth factor (EGF) (rec) Natural leucocyte interferon alpha Sugarcane-derived hypocholesterolemic agent Anti-CD3 Mab[c]	Interferon gamma (rec) (II-III) Interleukin-2 (rec) (II) Anti-EGF receptor Mab (I-II) Anti-CD6 Mab (II-III) Anti-cancer vaccines (pilot): ganglioside, modified EGF, anti-idiotype Mab
Scintillographic imaging diagnostics	Anti-carcinoembryonic antigen Mab Anti-EGF receptor Mab	Anti-colon cancer Mab
In vitro diagnostics (registered only)	Large-scale screening systems for AIDS, hepatitis B, hepatitis C (synthetic peptides), neural tube defects, neonatal hypothyroidism, phenylketonuria Anti-CD3, CD6, CD4 and CD8 Mabs Blood classification Mabs ELISA and visual multi-step serological diagnostic kits for HIV (rec), hepatitis C (synthetic peptides), syphilis (rec), and lipoprotein A (Mabs) One-step visual devices for pregnancy-testing in urine and rotavirus in faeces (Mabs)	
Agriculture; transgenic plants (in field trials or initial extension studies)	Insect-pest resistant transgenic cabbage, sugarcane and sweet potato Virus resistant transgenic potato Growth-promoting synthetic brasinosteroid-like molecules Modern propagation techniques in sugarcane, coffee, pineapple and sweet potato Molecular immunological diagnosis of infectious agents in tomato, citrus fruits and potato	
Transgenic animals (in field trials)	Accelerated growth fish (*Tilapia*)	

Table 4.3. Continued.

Product/procedure[a]	Registered
Industrial biotechnology	Cloning, expression in yeasts and technology development for the production of recombinant invertase, chymosin, α-amylase, dextranase and lactase Yeast immobilization technology for the production of fructose liquors

[a] Products or procedures referred to are registered or undergoing clinical or field trials.
[b] (rec) refers to recombinant products or procedures that use them.
[c] Mab refers to monoclonal antibodies or procedures that use them.
Sources of information: CGEB (1997); Academy of Sciences of Cuba (1996); Ministry for Public Health (1996a).

sanitary registrations and certificates for export. A centre with similar functions as regards medical devices has also been established. Both operate independently from all the production enterprises and are attached to the Ministry for Public Health. A series of guidelines and audit procedures have been established, following international criteria (WHO, International Harmonization Conferences, European Community Guidelines, etc.) and have been sanctioned by resolutions of the Minister for Public Health. Products for veterinary use are licensed by the National Institute for Veterinary Medicine, from the Ministry of Agriculture, under similar procedures.

The National Assembly approved the Environment Law on 11 July 1997. It regulates all environment and biodiversity-protection activities. The Centre for Environmental Management and Inspection takes care of environmental audit and licensing. The National Centre for Biological Safety, affiliated to the CITMA, regulates all aspects concerning biosafety, genetic manipulation, containment of GMOs, release of transgenic organisms, and biodiversity following compulsory guidelines according to international criteria.

Intellectual Property Rights

The Cuban Office for Intellectual Property, under the auspices of the CITMA, issues patent rights and author certificates according to the Decree-Law No. 68 on Inventions, Scientific Discoveries, Industrial Models, Trademarks and Denominations of Origin of 14 May 1983. Some articles of this Decree-Law were modified by Decree-Law No. 160 of 10 June 1995. This accepts patent rights on 'vegetable varieties, animal races, microorganism strains, alimentary and medicinal substances, methods of prophylaxis, diagnosis and curing diseases of human, animals and plants, and substances obtained by chemical procedures'. The country is a full member of the Paris Convention for the Protection of Industrial Property, the World Intellectual Property Organization, the World Trade Organization, and, since July 1996, the Patent Cooperation Treaty.

Research, Development and Marketing

The Cuban government is committed to development of biotechnology. A clear national policy and political will towards this purpose have created a significant potential. The priority level of this activity has been kept high despite the economic crisis of the 1990s. This potential includes modern laboratory and production facilities and an active and relatively young scientific population working on various fields in contact with or within first line laboratories at home and around the world. Given the degree of coordination and integration among different institutions, the higher education system and a strong international collaboration policy, the high-level of pre- and postgraduate education is guaranteed. The regulatory environment, although still missing some definitive legislation and guidelines, has evolved to adopt internationally accepted criteria.

Government investment is still important in areas such as the pharmaceutical industry, agriculture and food production, but the necessity of scaling-up some of these processes and the development of new products, opens space for other sources of investment and joint ventures. Another interesting area is that of industrial enzymes. Several patent-owned technologies and products, based on recombinant proteins, require scaling-up investment to become commercially interesting prospects. The Foreign Investment Law, approved by the National Assembly in 1994 gives the legal support and guarantee to this kind of activity. It is controlled by the Ministry for Foreign Investment and Economic Collaboration.

Major Projects

Thirteen national programmes (Table 4.4) include 513 projects (approved out of 1006 presented) which are controlled and partially funded by the Agency for Science and Technology from the CITMA. Twenty-five per cent of these projects are basic, 62% applied and 13% correspond to technological innovation. Six of the programmes and 22% of the projects are directly related to biotechnology.

The programme on sugarcane agriculture and the sugar industry has four main goals: field yield increase, harvest technology, development of hybrid seeds and their application to production conditions, and improvement of industrial efficiency. The programme on food production is aimed at productivity increases in grains, vegetables and other products, and at attainment of domestic demand for rice and seeds.

The programme dealing with the pharmaceutical industry and 'green' medicine is aimed at the production of drugs from indigenous genetic resources and at reducing dependence upon imports. The vaccines programme addresses a range of health problems (AIDS, cholera, dengue, leptospirosis, meningitis, viral hepatitis). Work is underway on multiple vaccines and the so-called 'infancy' vaccine. The programme on animal feed is focused on obtaining useful derivatives from the sugar industry for this purpose.

The rest of the national programmes are not completely unrelated to biotechnology since they comprise R&D work on biogas and biomass technologies, mountain ecosystems and crops (for example, coffee), natural resources and biodiversity. Even programmes on social sciences include economic, commercial, management and ethical aspects of biotechnological research.

Sixteen ministries or other government agencies approve and control 78 sector programmes. Approximately 100 programmes are controlled directly in the provinces. Finally, institutions can run projects, not included in programmes, financed from their budgets or by external funding sources.

Private Sector Engagement [a]

Commercialization of biotechnology products is undertaken by the producing centres themselves, directly or through associated companies. Some of them are: HEBER Biotec S.A. which deals with products from the Centre for Genetic Engineering and Biotechnology,

[a]Although the heading 'Private Sector Engagement' has been retained here, for the sake of consistency, it should be interpreted with care. Companies associated with the research and production centres are legally private, and operate as such, but are state-owned.

Table 4.4. National scientific and technical programmes in Cuba.

Programme	Number of projects
Development of sugarcane agriculture and sugar industry (B)[a]	64
Food production by sustainable methods (B)	58
Agricultural biotechnology (B)	59
Development of biotechnological, pharmaceutical and 'green' medicine products (B)	67
Human and veterinary vaccines (B)	17
Sustainable development of energy sources	47
Sustainable development of mountain areas	47
Development of tourism	24
Animal feeding by biotechnological and sustainable procedures (B)	30
Cuban society: Challenges and perspectives for the 21st century	30
The actual Cuban economy: Challenges and perspectives	19
Current tendencies of world economy and the international relations system	17
Global changes and the evolution of the Cuban environment	34
Total	513

[a] (B) indicates programmes related to biotechnology.

CIMAB S.A. which deals with those from the Centre for Molecular Immunology, DALMER Laboratories, dealing with the National Centre for Scientific Research, ERON S.A. with pharmaceuticals from the National Pharmaceutical Industry and other centres, and TECNOSUMA for ultramicroanalytical diagnostic systems. These companies generally have local partners in the countries to which they export their products. Products being marketed are those listed in Table 4.3. More information about these and other companies can be obtained from the references (Díaz-Otero, 1996).

Scientific Output in Terms of Publications in Peer-reviewed Journals

One of the frequently quoted criticisms of Cuban science has been the scarcity of publications. Cuba's level of publications is low compared to other Latin American countries even taking into account their lower level of scientific potential and development in terms of results impacting onto social practice (Ayala, 1995). This situation is being changed. During the last five years of economic difficulties, the number of Cuban scientific journals diminished significantly. In contrast to this, the number of articles indexed in international databases has increased by 60% during the same period, indicating that Cuban scientists have tended to use international journals more frequently in recent years. Biomedical and biotechnological articles comprised 57% of all international scientific publications from Cuban authors. Physics and other exact sciences comprised 38%. As mentioned above, biotechnology centres from the West Havana Scientific Pole account for most of this work. The most frequently internationally indexed Cuban publication in the biotechnology field is *Biotecnologia Aplicada* (Applied Biotechnology), a bilingual journal published by Elfos Scientiae Editorial from the Centre for Genetic Engineering and Biotechnology, Havana (BIOMUNDI Counsellors, 1997).

Limitations and Constraints

Major limitations and constraints on biotechnology development arise from the country's economic situation, characterized by a shortage of freely convertible currency. Despite the high priority given to biotechnology by the government, investment and funding have not been enough to cover the increasing needs of development and renewal of equipment. This is illustrated by the fact that the national programmes related to biotechnology received only 40% of the funds planned for in 1996. This caused delay in some projects although the execution of 57% of them was considered normal or advanced at the end of 1996 (Ministry for Science, Technology and Environment, 1996).

Another consequence of fund shortage is that some academic and technical infrastructure and equipment are becoming obsolete. Even though this constraint is partially mitigated by interinstitutional cooperation its effects are felt in some areas (e.g. protein structure and modelling).

Additional difficulties arise from the reduction in the number of institute subscriptions to journals in the last few years and delays of several months in receiving some journals. The country was connected to the Internet as recently as 1996 and most centres require investments in hardware to exploit the network efficiently. Nevertheless, researchers and organizations use electronic communication and information systems as much as possible.

Lack of marketing expertise is a notable limitation. Cuban commercial enterprises dealing with biotechnology products are young and require considerable training effort. Promotion activities are poor, partly due to funding shortages. Products are thus disadvantaged in relation to similar ones from financially stronger, well-experienced competitors. This lack of expertise is partially ameliorated through partnerships with foreign companies.

An unquestionable constraint on Cuban scientific development is the US economic blockade to the country; its effect on healthcare and food supply have been analysed by external observers (Kirkpatrick, 1996). Cuban institutions, which cannot purchase reagents, materials, and equipment from the US market, must buy them in Europe, Japan or through intermediaries, increasing all costs significantly. Access to several international, even regional, funding organizations is limited. The scientific exchange between both countries is virtually nonexistent, despite the status of development attained by Cuba and the fact that it is the second closest Latin American country to the USA. Attendance of Cuban researchers at scientific meetings or training activities in US territory is hampered by slow and complex visa procedures. At the same time, visits of US scientists to Cuba are discouraged by severe regulations. In reverse, US companies are prevented from profiting from the active Cuban biotechnology market, commercial exchanges with Cuban companies and investment opportunities.

Socioeconomic Impact of Investments on Biotechnology for the Country

The social impact of Cuban biotechnology has already been commented upon. The economic impact has been significant too. Some products such as the anti-meningitis and anti-hepatitis B vaccines, pharmaceuticals and diagnostics are exported. Savings of imports in items like diagnostics for large-scale screening programmes, seeds for agriculture, or chemical anti-tick baths for cattle increase every year. Although difficult to calculate, some applications like massive hepatitis B vaccine, interferon treatment of chronic hepatitis and screening for hereditary-metabolic disorders have a positive cost-benefit balance. Expectations of ongoing and future work are higher. Many projects, covering a wide range of fields are ongoing. They range from global health problems like AIDS and cholera, and first line therapeutic procedures, to improvements in agriculture and industry efficiency by the introduction of new (transgenic or otherwise) varieties, hybrid seeds and recombinant enzymes.

Concluding Remarks

Cuban biotechnology has developed in the last 16 years under *sui generis* conditions characterized by (i) a well-defined favourable policy consisting of large investments and governmental support; (ii) interinstitutional approaches to scientific and technological problems including pre- and postgraduate infrastructure education; (iii) economic difficulties that have intensified in the last five years; and (iv) adaptation to the new world-economical order. Despite the difficulties, perspectives can be viewed optimistically. This is shown by social impact and cost-benefit analysis, potentialities created in infrastructure and human resources, international cooperation, and increasing commercial activity.

Egypt

Hassan Moawad and Magdy Madkour

National Policy and Regulation

General Policy

Egypt has recognized that biotechnology will play a major role in many areas, particularly in health, agriculture, industry and environment. The development of biotechnology in ways which provide for the sustainability of resources is of particular strategic importance.

According to the classification of countries in relation to their involvement in biotechnology, Egypt has moved from the category of the developing countries with an interest but no direct involvement in biotechnology, to the category of countries with a national strategy and research programme in modern biotechnology, complemented by strong overseas collaborations in both public and private sectors. This coincides with a governmental effort directed towards economic reform which has proved more successful than expected.

The Ministry of Higher Education and Scientific Research of Egypt has been assigned responsibility for the formulation of the National Strategy for Genetic Engineering and Biotechnology. This assignment was based on a strong belief in the role of biotechnology as one of the most promising areas of scientific and technological progress available in support of economic development, either directly or indirectly.

There has been complete agreement, since the initiation of the national strategy, between all science and technology organizations in Egypt on the importance of specific projects that address health problems endemic to Egypt, and that enhance competitiveness in the post-GATT global market whilst facilitating environmentally sustainable development.

Egypt laid the foundation for the development of biotechnology through the establishment of research centres that can serve its interests and possibly those of the region. These biotechnology centres were established by the Ministry of Agriculture, the Ministry of Higher Education and Scientific Research and the Ministry of Higher Education (the latter including a number of universities). The Ministry of Higher Education and Scientific Research has also identified the technological focus for the Genetic Engineering and Biotechnology Research Institute (GEBRI), one of the two National Research Institutes that are currently under construction in the Mubarak City for Science and Technological Applications near Alexandria.

Regulatory Structure

In Egypt, as in other developing countries, a national biosafety system ensures the safe development of biotechnology products and facilitates collaborative research activities with other countries. Until 1995, existing regulations in Egypt did not include guidelines for handling transgenic material under contained conditions, nor did they cover the release of GMOs into the environment.

The Egyptian National Biosafety Committee (NBC) was established according to Ministerial Decree No. 85 in January 1995. This committee is responsible for putting together policies and procedures to govern the use of modern biotechnology in the country. To strengthen the structure of expertise in the Committee, and to cover broadly all sectors interested in the safe application of genetic engineering, the decree has been modified by amendments No. 322 (March 1995), No. 174 (February 1997), No. 902 (July 1997) and No. 1055 (August 1997).

To formulate a biosafety system for Egypt, information was gathered from different countries regarding their regulations, guidelines and system design. A draft document entitled *The Establishment of a National Biosafety System in Egypt: Regulations and Guidelines*, was prepared by the Agricultural Genetic Engineering Research Institute (AGERI) including regulations and guidelines adapted to an Egyptian context. This document was revised by the NBC and approved by the governmental authorities as the binding code of conduct for biosafety in Egypt (Ministerial Decree No. 136 in February 1995).

In Egypt, the protection of intellectual property rights is carried out through the Egyptian Patent Office. This office was first established in 1949 by Governmental Decree No. 132. Later in 1972, the office became affiliated with the Academy of Scientific Research and Technology by governmental decree.

Research, Development and Marketing

Major Projects

Plant tissue culture work is being carried out in various research centres, universities and private companies. In the governmental sector, this activity takes place in the Ministry of Agriculture's Horticultural Research Institute, Desert Research Institute, and Agricultural Development Systems Project. The plant tissue culture work is also conducted in the National Research Centre (NRC), the Universities of Assiut, Ain Shams, Suez Canal, Zagazig and Al Azhar, and the American University in Cairo. In the private sector, three commercial companies (PICO, DANTON and EL-Shiaty) develop tissue culture products.

Animal and fisheries biotechnology research is carried out in Agricultural Research Council institutes, the faculties of agriculture and veterinary medicine at Egyptian universities, and at NRC laboratories. Biotechnology research in the area of human health is carried out at the NRC, the Theodore Bilharz Research Institute, the National Cancer Institute, and the medical faculties of most universities.

In the areas of environmental biotechnology, the Centre of Research and Graduate Studies at Alexandria University is involved in the use of microorganisms for the purification of sewage water with the aim of its reuse in irrigation. In the Faculty of Sciences at Alexandria University work is being carried out on the application of plants as bioindicators for some air pollutants. In the Faculty of Agriculture at Ain Shams University, certain microbial and insect strains carrying multienzyme loci which are sensitive for monitoring pollutants, especially agrochemicals, are being developed.

Major Institutes

Over the last 15 years, scientists working in different universities and research centres in the country have initiated research and technology transfer efforts in the area of genetic engineering and biotechnology, and current research activities cover a wide range of scientific applications. The major Egyptian institutions involved in genetic engineering and biotechnology research and development are:

- The Agricultural Genetic Engineering Research Institute (AGERI), Agricultural Research Centre.
- The Genetic Engineering and Biotechnology Research Institute (GEBRI), Mubarak City for Scientific Research and Technological Application.
- The Genetic Engineering and Biotechnology Division, NRC.
- The Genetic Engineering Institute, Menofia University.
- The Centre for Research in Genetic Engineering, Faculty of Agriculture, Cairo University.

The Agricultural Genetic Engineering Research Institute (AGERI)

AGERI represents a vehicle within the agricultural arena for the transfer and application of genetic engineering technology. The original establishment of AGERI in 1989 was the result of a commitment to the development of expertise in agricultural biotechnology. At the time of its establishment, AGERI was named the National Agricultural Genetic Engineering Laboratory. The rapid progress of its activities during the following three years encouraged the Ministry of Agriculture and Land Reclamation to authorize the foundation of AGERI, which is a centre of excellence in agricultural genetic engineering and biotechnology. The main goals of AGERI can be summarized as follows:

- To broaden the R&D capabilities and scope of the Agricultural Research Centre in the public and private sectors (i.e. initiation of new programme areas and application to a wider array of crop species).
- To expand and diversify the pool of highly qualified professionals in the area of biotechnology and genetic engineering.
- To promote opportunities for private sector development.
- To provide opportunities for university trained professionals (e.g. faculty researchers and teachers), private venture companies and the Ministry of Agriculture (researchers) to cooperate in agricultural genetic engineering research.
- To achieve the desired level of self-reliance and self-financing within AGERI through fund mobilization.
- To provide agricultural genetic engineering capabilities to meet Egypt's current problems.

AGERI has based its projects on the concept of maintaining a programme that is focused on the problems of Egypt. The main objective is to build a national capacity within Egypt for the sustainable production of crops crucial to the economy and a safer, cleaner environment by adopting the most advanced genetic engineering and gene transfer tools.

Hence, AGERI aims to improve resistance to biotic and abiotic stresses in a limited number of crops: maize, potato, tomato, cucurbits, fava bean, wheat and cotton. The projects carried at AGERI have the potential to substantially improve crop productivity and the environment by:

- Reducing dependency on pesticides through the production of transgenic cotton (resistant to insect pests) and Bt-modified potato (resistance to potato tuber moth).
- Saving substantially on foreign currency from the importation of virus-free potato seeds by producing transgenic potato plants resistant to potato virus X, potato virus Y and potato leaf roll virus.
- Cloning the genes encoding for important economic traits in tomatoes, especially those related to stress-tolerance and disease resistance.
- Developing an efficient regeneration and transformation system in maize for the production of transgenic plants resistant to the corn borer insects, a major limiting factor for maize production in Egypt.
- Mapping the rapeseed genome in order to develop cultivars adapted to the constraints of the Egyptian environment, thus securing a good source of edible oil.
- Production of transgenic fava bean plants with resistance to bean yellow mosaic virus and fava bean necrotic yellow virus.
- Developing tomato yellow leaf curl virus resistant tomato germplasm through an *Agrobacterium*-based transformation system.
- Engineering coat protein genes to develop virus resistant cucurbits.
- Developing efficient diagnostic tools for the identification and characterization of major viruses in Egypt.

These projects are important for Egyptian Agriculture, and will have a positive impact on agricultural productivity and foreign exchange. By way of illustration, Egyptian Bt transgenic cotton, resistant to major insect pests, would result in substantial savings on the US $50 million spent annually on the purchase of imported pesticides. Mapping of rapeseed oil has the potential to substantially reduce the 400,000 tons of edible oil which is imported into Egypt annually. Similarly, transgenic potato varieties resistant to selected viruses and insect pests will reduce the expenditure of approximately US $33 million per annum on the import of seed potatoes.

In addition, because almost all arable land is currently in production, researchers are eyeing the desert as the only new place to cultivate. This will require the development of crops resistant to salinity, heat and drought.

While all these efforts are directed toward bridging the food gap, they also focus on the production of healthier food with minimal use of agrochemicals and pesticides. This will guarantee the production of environmentally friendly agricultural products in a sustainable manner.

The Genetic Engineering and Biotechnology Research Institute (GEBRI)

GEBRI, previously named the National Institute for Genetic Engineering and Biotechnology, was reoriented to avoid duplication of the efforts of organizations working in advanced biotechnology research in other countries. It is envisaged that this new institute will lead the field in modern biotechnology applications. GEBRI will be of undisputed value to the advancement of genetic engineering and biotechnology by national, regional and international standards. The GEBRI mission is to establish a centre of excellence in advanced biotechnology able to introduce different valuable biotechnology products to the market through R&D and technology transfer, to produce commercial biotechnology products jointly with private companies. The targets of the institute include:

- Improvement and development of expertise and knowledge in biotechnology and genetic engineering.
- Financial sustainability through product development and the provision of services.
- Improvement of the image of scientific performance in Egypt and paving the way for cooperation between Egypt and industrial countries.

GEBRI will undertake the responsibility of establishing a national directory of Egyptian capability in biotechnology, including manpower, equipment, market needs, industrial companies and programmes of biotechnology in Egypt. This survey will also help in the identification of potential areas for R&D.

The GEBRI business development office will be responsible for marketing the research activities and services. In addition, the office will bring the market needs to the attention of research programmes at GEBRI. The role of this office will be primarily devoted to the creation of links between the Institute and industry.

The work at GEBRI is focused in five main programmes:

1. Protein research.
2. Nucleic acid research.
3. Biomedical research.
4. Bioremediation research.
5. Biotechnology product development.

Other Institutes

The Genetic Engineering and Biotechnology Division, NRC, comprises six departments: Human Genetics, Molecular Biology, Microbial Genetics, Cell Biology, Plant Cell and Tissue Culture, Microbial Biotechnology. The main research focus of the NRC Biotechnology divisions are related to agriculture, biology, microbiology, food and pharmaceutical research areas.

The Genetic Engineering Institute, Menofia University, was established recently. This institute is involved mainly in conducting research with emphasis on agriculture and development of basic molecular biology techniques.

The Centre for Research in Genetic Engineering at Cairo University focuses its activities mainly on agricultural biotechnology.

Private Sector Engagement

Several industrial biotechnology-based companies are successfully operating in Egypt for the production of alcohol, acetic acid, alcoholic beverages, fermented foods, bakers yeast and solvents (Cabinet publication, 1994). All major fermentation companies use traditional strain improvement techniques and mutant selection. Among the industrial biotechnology corporations in Egypt, the Organic Chemical Industries company produces all the major organic solvents and yeast, using molasses and rice bran as feedstocks. The South Egypt Drug Industries Company produces pharmaceuticals, nutrition products and cosmetics (about 200 products in total).

Hungary

Ferenc Rudan

National Policy and Regulation

General Policy

In the mid-1980s, the Hungarian government drew up a framework of special programmes supporting various areas of R&D, one of them being biotechnology. This initiative provided considerable potential for capacity building, particularly in the fields of plant and animal biotechnology, diagnostics, the fermentation industry (including the production of antibiotics) and the food industry.

At the beginning of the 1990s, a new 'bottom-up' grant application system was introduced. Under this system, different projects, covering all possible areas of R&D, are given equal weight and compete with each other for granting of funds. No special priority areas have been established. In 1996, a special biotechnology programme was initiated by the National Committee for Technological Development and is intended to become fully operational in 1998.

Funding

National R&D efforts receive government support from two sources: the National Scientific Research Fund for basic research, and the Central Technical Development Fund for product and technology development. Since both these funds are available for all kinds of R&D activities, only a small proportion has been allocated for specific biotechnology projects. Owing to fluctuations in exchange and inflation rates in recent years, it would be misleading to give actual figures for the amount of funding received from the government. A realistic estimate would be that about 5 to 10% of total government R&D support is given to biotechnology projects; approximately the same amount is provided by private companies. For both private companies and government sources, there are no data available for expenditure on biotechnology R&D specifically.

Regulatory Structure

In the area of regulation the picture is somewhat clearer. The Environmental Protection Act required that a law be formulated to regulate the release of genetically modified organisms (GMOs). This took effect as of January 1999. The law, which draws heavily upon the European Directive 90/220, covers the modification, contained use, release, marketing, transport, and import or export of genetically modified organisms and products derived from these. The law also requires compulsory labelling of products containing genetically modified organisms. However, it does not cover work related to the human genome or human health, which are to be covered by a separate law.

A 17-member commission advises the Minister of Agriculture whether or not to approve applications for activities with genetically modified organisms. The commission is comprised of five members nominated by the academy of sciences, six by different ministries, four by environmental protection non-governmental organizations (NGOs), and two by health-protection NGOs.

As concerns intellectual property rights, Hungary has adopted an approach in line with the EC regulations. Hungary is also a member of the international Union for the Protection of New Plant Varieties (UPOV).

Research, Development and Marketing

Major Projects

Major biotechnology projects are presently underway in the agriculture and pharmaceutical industries. In the former, the main emphasis is on plant breeding for special purposes (for example, high or specific protein content of wheat varieties, shorter growing season for corn, virus or disease resistance in potato and tobacco, and herbicide tolerance in corn). The other major area in agriculture is animal breeding. Information gained from genome analyses has provided the necessary basis for effective animal breeding programmes. In certain areas, embryo transfer has become a routine procedure. *In vitro* fertilization (IVF) and sex determination of embryos are techniques that will be applied in the near future.

In the area of pharmaceutical research, the privatization of companies over recent years has resulted in a lack of information regarding the progress of some promising earlier work. One such case is that of recombinant hirudin, which in the 1980s and early 1990s was the focus of much attention. Similarly in the field of antibiotic production, little is known of the outcome of some earlier activities.

Major Institutes

The following are the most important institutions in Hungary where R&D in biotechnology is carried out:

- Biological Research Centre of the Hungarian Academy of Sciences, Szeged. Major research areas include transgenic plants, immunology recombinant protein production and tissue plasminogen activators. (*For more information about this institute, see Appendix I.*)
- Agricultural Biotechnology Centre, Godollõ. Major research areas include transgenic plants, protein engineering, immunology and transgenic animals.
- Institute for Drug Research, Budapest. Major research areas include microbial biotechnology and antibiotic research.

- Central Food Research Institute, Budapest. Major research areas include food starter cultures. (*For more information about this institute, see Appendix I.*)
- Lajos Kossuth University, Debrecen. Major research areas include microbial biotechnology for antibiotic production and fermentation processes.

Private Sector Engagement

The private sector was virtually non-existent in Hungary until the beginning of the 1990s. Prior to that, there were many small companies, mostly inadequately financed and with little knowledge of marketing, which were unwilling to invest in products and markets where the returns were uncertain. Pharmaceutical companies were no exception. The investors, who were mainly foreign, seldom wanted to maintain the R&D capacity of a company, and often invested solely in market and production facilities instead. The private sector in Hungary remains comparatively weak, especially in areas such as biotechnology, which are characterized by high initial investment and uncertain long-term returns. Companies with traditional activities in plant breeding have been able to integrate biotechnological methods into their research activities and have shown good results. The problem is that there are only a few of these companies and they mostly lack market access and marketing expertise. Only one company has had success with production and marketing of diagnostics.

Constraints

The major limitation for the development of biotechnology in Hungary is the lack of a uniform and well-founded government policy. The absence of such a policy is becoming increasingly detrimental to the long-term progress of research. Coordination between the Hungarian Academy of Sciences, the National Committee for Technological Development, and the Ministry for Education is not always optimal and the responsibilities of each overlap in some areas. On the other hand, lack of legislation does not present an obstacle. Rather, it is hoped that the new law on GMOs will clarify the situation and allow research activity to progress unrestricted.

Another general problem is lack of adequate funding. As there are no designated R&D priority areas, all areas compete on an equal basis for the generally limited resources, which are frequent targets of budget cuts and restrictions. Although not an immediate problem, it is foreseeable that in some years time, the acquisition and maintenance of expensive instruments and equipment will be increasingly difficult, owing to the absence of an established system of financing for the basic expenditures of research institutions.

International collaboration has been a regular feature of the Hungarian research establishment over the last 20 years or so. There are presently eight projects run under the European Union's Fourth Framework Programme for Research and Technology Development with Hungarian participation. The Hungarian research community has established a quiet but intensive partnership with the international research community. However, a frequent problem is that some of the best scientists leave for extended periods, taking advantage of better opportunities abroad.

Future Prospects

At present, the social environment is favourable for biotechnological research. There is no public opposition to biotechnology R&D, or to field releases for research purposes.

The new law mentioned above may reinforce this attitude. There is an abundance of highly qualified personnel for plant breeding and antibiotic production projects.

The agricultural sector is seen as holding the greatest potential for biotechnology R&D. In the field of plant breeding, there are considerable resources and a wealth of knowledge which could and should be put to future use. The production of new plant varieties with disease and herbicide tolerance is of great commercial value and the Hungarian research community is able to carry out this type of work.

In the field of pharmaceuticals, the picture is not so clear. The Hungarian pharmaceutical companies have either been privatized, or are in the process of being so, and it will take time before the situation stabilizes and new research demands can be formulated.

India

Manju Sharma and Renu Swarup

National Policy and Regulation

General Policy

In 1982, following detailed deliberations with the scientific community, the then Scientific Advisory Committee to the Cabinet set up a National Biotechnology Board to foster programmes strengthening indigenous capabilities. The Government subsequently decided to set up a separate Department of Biotechnology (DBT) in February 1986. Since the establishment of the Department, a strong infrastructure for biotechnology research and services has been created in both national laboratories and academic institutions. An integrated human resource development programme and research and development programme which aims to support basic, applied and product-oriented research, both from the viewpoint of attaining excellence and for product and process development, has resulted in good research leads and about 20 technologies being developed and transferred to industry (diagnostic kits for various diseases representing a prime example). A unique feature of the DBT has been the provision of support to areas which could give quick results, and to encourage more and younger scientists to join this field through manpower development efforts.

Regulatory Structure

Recombinant DNA Safety Guidelines were issued in 1990, and revised in 1994. Further revision is presently underway. Contained use of GMOs began in 1993, with the first field trials following in 1995.

India is not a member of the Union for the Protection of New Plant Varieties (UPOV), and Plant Variety Protection has not been introduced. There is therefore no official position on breeders' rights. India has, however, signed GATT, though the parliament has yet to ratify this. IPR (intellectual property rights) registration in India does not allow the patenting of GMOs.

Research, Development and Marketing

Major Projects

Demonstration programmes in biological pest control, biofertilizers, aquaculture, embryo transfer technology and plant tissue culture have been very successful. These have resulted in:

- Fifty-five thousand hectares of land under biological pest control.
- Development of new biofertilizer technology packages.
- New techniques for semi-intensive prawn aquaculture.
- Standardization of embryo transfer technology.
- Setting up of two tissue culture pilot plant facilities (at the National Chemical Laboratory, Pune and the Tata Energy Research Institute, New Delhi).
- Research on birth control vaccines (both for men and women).
- Setting up of a unique bioinformatics system through ten Distributed Information Centres and 23 Sub-Centres all over the country.
- Evolving biosafety guidelines.
- Support for R&D projects in microbial biotechnology.

Biomedical research has concentrated on all major infectious diseases, and more recently initiatives have been taken to address genetic disorders. One area of particular concern has been the development of antigen/antibody and DNA-based diagnostics. These have been developed for almost all major infectious diseases (including AIDS, tuberculosis, malaria, leishmania, amoebiasis, cholera), as well as those for cattle and poultry diseases. Some of these diagnostics have been transferred to industry, and a few have reached the market.

Many laboratories are now competent in gene transformation. A salt-based *E. coli* expression system, Baculovirus and silkworm larva-based expression systems, and yeast are all being used. Work with streptokinase, epidermal growth factor, insulin, cholera antigen and hepatitis-S-antigen has been successful. Links have been made with industries, and the next few years may see at least a couple of recombinant DNA products on the market. There are also proposals to produce cell culture-based vaccines such as those for rabies, measles and polio. The concept of human fertility control vaccines both for men and women has been pioneered in India. However, careful scrutiny of the efficacy of these in a large-scale trial, and the development of successful recombinant DNA strategies to produce commercially useful amounts of the vaccine would be necessary if this approach is to be exploited for population control. Conventional strategies based on the use of related or unique strains of mycobacterium or vibrio to develop vaccines against leprosy and cholera have been successfully demonstrated. Public institutions are engaged in this type of R&D, since multinational companies are largely disinterested in vaccines for diseases confined to the Third World. The efforts of the DBT and the Indian Council of Medical Research have led to the establishment of a few units/centres to diagnose haemoglobinopathies, Duchenne muscular dystrophy and a few other genetic disorders. Such diagnosis has become an integral part of genetic counselling.

In the area of agricultural biotechnology, the DBT has established seven centres of plant molecular biology in universities and national laboratories. In addition, two network projects for the development of transgenic cotton and the improvement of quality in wheat were initiated during 1993-1995. The expertise generated in the area of new biotechnology has led to the application of RFLP/RAPD/simple sequence repeat analysis of markers linked to quantitative trait loci, in a variety of crops and plants. Transgenic plants with marker genes have been obtained and successes with transgenic rice and cotton containing insecticidal genes have been reported. Two Tissue Culture Pilot Plants, set up by DBT in 1989, have successfully demonstrated the technology for large-scale production of elite forest trees through tissue culture. Three million plantlets have been planted over an area of 2500 ha. R&D projects have been supported for development of a regeneration protocol. Twenty such protocols have been developed for a number of horticultural, plantation crops and forest trees. Micropropagation and plant tissue culture are used on a commercial scale, the notable examples being disease-free cardamom, cut flowers, exotic ornamental plants, vegetable seeds, sunflower, and tomatoes. Many export-oriented companies with technology and buy-back arrangements with foreign companies have expanded their activities. The DBT has also set up two micropropagation technology parks for effective transfer of proven technology from the Institutes to the industry. These parks will also serve as centres for skilled manpower training.

In short, the Government of India has created a large research and development base, with adequate expertise in modern biotechnology. Several meetings with the industry have led to a change in attitude, and have generated more enthusiasm in the industry.

Major Institutes

The DBT has three research laboratories and institutions under its administrative control:

- The National Institute of Immunology, New Delhi, which is a premier institute in the country, and enjoys a very high international rating for its work in immunology.
- The National Centre for Cell Science, Pune, which has a repository of cell lines, plasmid vectors and other types of tissues, is the first of its kind in the country.
- The Centre for DNA Fingerprinting and Diagnostics at the Centre for Cellular and Molecular Biology, Hyderabad acts as both a referral centre for DNA Diagnostics, and a DNA fingerprinting and analysis laboratory.

The DBT has also supported the development of the International Centre for Genetic Engineering and Biotechnology (ICGEB), a component of which is fully operational at New Delhi. This focuses on biotechnology research in agriculture and human health care.

Essentially, the last 15 years have seen the establishment and consolidation of active research groups in modern biology and biotechnology. In addition to the establishment of the three institutes listed above, the DBT has also been able to support and establish strong research groups in the universities and national laboratories. Major infrastructural facilities have also been established in different parts of the country. Along with DBT, other research agencies such as the Department of Science and Technology, Indian Council of Medical Research, Indian Council of Agricultural Research and the Council for Scientific and Industrial Research have also contributed to many joint ventures.

The DBT, with participation from other agencies, has set up several national infrastructural facilities. These include a Microbial Type Culture Collection, a Blue Green Algae Collection, a Marine Cyanobacteria Collection, Parasite Gene Banks, a major animal house incorporating primate facilities, genetic engineering units, a Biochemical Engineering and Process Development Centre, and national facilities for X-ray crystallography and NMR. A network of biotechnology information systems has been established in the country. MSc (Biotech) and postdoctoral fellowship programmes have been introduced in selected universities, together with provisions for training abroad. International collaboration through the Indo-US Vaccine Action Programme and the Indo-European Commission, interaction among SAARC (South Asian Association for Regional Cooperation) countries, and many bilateral programmes have been promoted. The DBT is the Coordinator for the G-15 Countries Programme on Gene Banks, and the Regional Coordinator for UNDP/FAO programmes in agricultural biotechnology. The Department has also established a Technology Transfer and Patent Unit, under which 20 technologies have so far been transferred to industry, and 17 patents filed. Two US patents have been granted for DBT supported research.

Future Prospects

Biotechnology has already demonstrated its impact in many important sectors, particularly from the viewpoint of environmentally sustainable development. The DBT will continue to support well defined, focused, result-oriented and time-bound research and development projects of relevance to the country, especially to develop products and processes. Taking up the leads obtained so far, the following new initiatives have been planned for the next five-year plan:

- Bioprospecting of biological resources.
- Coastal zone monitoring and bioindicator research.
- Facilities for testing of transgenic plants in collaboration with the Indian Council of Agricultural Research.
- Establishment of a Micropropagation Technology Park.
- Mapping of silkworm genome.
- Edible vaccines.
- Establishment of a Centre for Brain Research.
- Drug design and drug development.
- Gene therapy research.
- Facilities for chemical testing and validation of diagnostics and vaccines to be set up with the Indian Council for Medical Research.
- Transgenic animals and plants.
- Establishment of a Centre for Bioresource Development.
- Biotechnology programmes for women.

Indonesia

Sugiono Moeljopawiro

National Policy and Regulation

General Policy

The Indonesian government has given high priority to the application of biotechnology in agriculture. In the government's latest five year development plan it is stated that the ultimate goals of the agricultural biotechnology research activities are to sustain self-sufficiency in food production, to develop agro-industry, to increase efficiency in using biotic and abiotic resources, and to increase crop and animal production. Although agricultural production in some parts of Indonesia has benefited from the use of improved technologies, low yields still persist in certain growing areas due in part to wide environmental diversity. This situation negates the value of general recommendations for varieties and management practices, and hence requires a location-specific approach. Breeding new varieties is complex because of the need for a range of varieties, which are also resistant to pests and diseases and tolerant to environmental stresses, to fit widely varying agricultural production systems. The intensity of agricultural production, particularly of rice, brings about the build-up of pests and diseases, which precludes higher productivity and causes yield instability. It is anticipated that techniques in molecular biology can be used to develop molecular diagnostic kits to isolate genes proven to be effective against various pests and diseases, and to construct new genes such as coat-protein and antisense genes.

The Agency for Agricultural Research and Development of the Ministry of Agriculture supports agricultural research projects aimed at increasing production, productivity and the utilization of agricultural wastes. It is believed that in order to achieve more efficient and effective use of available resources, the agricultural biotechnology research programme should focus on areas in which biotechnology is most likely to provide new breakthroughs in agricultural development.

Funding

One major initiative was the launching of the 'Riset Unggulan Terpadu' (Integrated Superior Research) grant programme by the National Development Planning Agency (NDPA). This research programme will provide funding for research in biotechnology and other sciences to participating research organizations. Scientists can submit proposals for research focusing on priority issues, with grants being awarded by an evaluation committee. This committee, in cooperation with the State Ministry for Research and Technology and the National Research Council, has been set up to give an unbiased assessment of each proposal, whilst ensuring strict compliance to the needs of the national development programme in Indonesia.

The progress of each research project is monitored every year and funding is discontinued if progress is unsatisfactory. A new grant system called 'Riset Unggulan Kemitraan' (Cooperative Research) was initiated in 1995 in order to accelerate the commercialization of products from this programme. Under this system, any proposal submitted has to be jointly developed by public institutions and the private sector in order to enable efficient scaling-up from the laboratory to a commercial level.

Since the beginning of the integrated superior research, the NDPA has spent US $7,639,152 on a total of 241 research proposals (Table 4.5).

A US $3,347,619 loan from the World Bank for the fiscal years 1995/1996 and 1996/1997 has been allocated for human resource development to the Bogor Agricultural University, Gadjah Mada University and 'Institut Teknologi Bandung' as the biotechnology centres for agriculture, medical health and industry, respectively.

Regulatory Structure

Biosafety

Biosafety guidelines were released under Ministerial Decree as 'Provisions on Biosafety of Genetically Engineered Agricultural Biotechnology Products', September 1997. In addition to these provisions, a containment facility has been constructed at the Research

Table 4.5. Yearly budget granted by the National Development Planning Agency of Indonesia.

Year	Amount (US $)	Number of research proposals
1993/94	1,158,701	47
1994/95	2,271,304	73
1995/96	2,470,970	81
1996/97	1,738,177	40
Total	7,639,152	241

Institute for Food Crop Biotechnology (RIFCB) for the risk assessment of transgenic plants, and a Biosafety Committee (BC) and Biosafety Technical Team (BTT) have been established. The BTT will conduct risk assessment analysis to determine the risk management to be adopted. The results of the risk evaluation will be forwarded to the BC, which, in turn, make recommendations to the Ministry of Agriculture as to whether or not the product is safe to be released to the environment.

As a result of last year's transgenic test in the containment facility, there are ongoing confined field tests for the evaluation of Bt corn, Roundup ready corn, Bt cotton, Roundup ready cotton and Roundup ready soybean.

Intellectual Property Rights

Intellectual Property Rights (IPR) legislation has been implemented in Indonesia, but does not cover animals for meat or milk production, or food crops such as rice and corn. However, in order to broaden the coverage of protection and promote more collaborative research, this legislation was amended in early 1997.

Indonesia is not a member of the UPOV (Union for the Protection of New Plant Varieties) Convention. However, in order to protect the work of plant breeders and support the seed industry, a plant variety protection law has been drafted, and is under review by the Department of Agriculture.

Due to recent developments in biotechnology and in order to provide an environment that is conducive to innovative research to support national development goals, the Agency for Agricultural Research and Development of the Ministry of Agriculture, is preparing for the establishment of a technology transfer office. This office will advise the researcher in dealing with the private company with respect to issues of commercialization.

Biodiversity

Indonesia ratified the Convention on Biological Diversity in 1994 (Law No. 5). Additional biodiversity legislation falls under Law No. 23 on Environmental Management (1997).

Research, Development and Marketing

Major Institutes

At the end of 1988 three institutes were designated as focal centres for a biotechnology research network, each specializing in a different area. These are:

- The University of Indonesia in Jakarta for medical biotechnology (Eijkman Institute for Molecular Biology, Jakarta).
- The Central Research Institute for Food Crops (CRIFC), of the Ministry of Agriculture, for agricultural biotechnology. [Subsequently, in 1993, the Central Research and De-

velopment Institute for Biotechnology (CRDB) of the Indonesian Institute of Sciences was designated as the second centre for agricultural biotechnology.]
- The Agency for Assessment and Application of Technology, of the State Ministry for Research and Technology, for industrial biotechnology.

The Research Institute for Food Crops Biotechnology (RIFCB) of the CRIFC and the CRDB has the following mandates:
- To coordinate and to develop agricultural biotechnology of crops, fisheries and livestock.
- To develop a national agricultural biotechnology network which includes research institutes, universities and private sectors.
- To develop an international collaborative network with advanced biotechnology laboratories of developed countries and International Agricultural Research Centres.

Collaborative Research and Partnerships

Collaboration with the International Agriculture Research Centres (IARCs) and other donor agencies such as the International Rice Research Institute (IRRI), The US Agency for International Development (USAID), the Rockefeller Foundation, Japan International Cooperation Agency (JICA), and the Australian Centre for International Agriculture Research (ACIAR) are important in enhancing agricultural biotechnology development in the country. In addition, more emphasis will be given in future to public-public, public-private and private-private collaborations.

Kenya

Ahmed Hassanali

National Policy and Regulation

General policy

Over recent years, the government of Kenya has formulated various acts, policies and regulations aimed at supporting scientific research and development in the country and, to a certain extent, these have enabled Kenya to embark on research and development in biotechnology. Examples of this legislation include:
- The Science and Technology Act of 1979.
- Sessional Paper No. 4 of 1981 on Food Policy.
- Livestock Policy of 1986.
- Sessional Paper No. 1 of 1986 on Economic Management for Renewed Growth.
- Sessional Paper of 1992 on Recovery and Sustainable Development to the Year 2010.
- National Environment Action Plan of 1994.
- National Development Plans (five-year plans).

Although traditional biotechnology has been practised for a long time in Kenya, policies on biotechnology *per se* were not developed until the late 1980s. At this time, a committee was formed under the then Ministry of Research, Science and Technology, in order to plan a strategy on how to exploit biotechnology for national development. This committee, the National Advisory Committee on Biotechnology and its Applications (NACBAA), produced a report in 1990 which indicated the priority areas of biotechnology and made recommendations for their development. Despite the production of the NACBAA report, there is as yet no comprehensive policy on biotechnology. Policy in biotechnology is implicit in a number of scattered statements.

Funding

The national expenditure and government investment in various sectors is contained in the annual government estimates. These estimates are general and do not list expenditures in specific sectors. As a result, it is difficult to give a precise figure for expenditure and investment as far as biotechnology R&D is concerned. However, if the total research funding in the country is considered, a rough picture of the situation for biotechnology research may be obtained. Table 4.6 shows a comparison of the total research budget against the national budget and GDP for the years 1974, 1984 and 1989-1992.

The pattern that emerges from Table 4.6 is that there was an increase in the total research budget from K£3 million in 1974 to K£60 million in 1991. However, the level of allocation of funds to R&D in relation to the national budget remained low, ranging from 0.45 to 1.6%. Similarly, R&D funding as a percentage of the GDP remained low at under 0.5% (except for 1991 when it reached 0.53%).

The government's allocation to the six research institutes which were established under the Science and Technology Act, and where biotechnology activities are expected to take place, shows that the total funding rose slightly over the period 1986 to 1991 (Table 4.7). However, this was due to a substantial increase in the research budget of only one of these, the Kenya Agricultural Research Institute (KARI).

A closer look at the current situation of KARI shows that K£5.6 million are required between 1994/95 and 1999/2000, if all the biotechnology R&D projects proposed under the National Agricultural Research Programme II are to be implemented. These funds for biotechnology constitute 1.05% of KARI's consolidated budget and are expected to be provided by external donors. Based on current financial allocation trends, KARI requires about K£131.51 million from the Government of Kenya, while the balance of K£402.39 million is expected to be provided by donors. However, the actual funds expected or committed by the government and donors in 1995 were K£262.20 million (KARI-NARP, 1995).

Several organizations have shown a keen interest in developing biotechnology in Kenya. One such is the Special Programme for Biotechnology and Development Cooperation of The Netherlands Government, which has pledged to give Dfl 10 million for a period of five years, aimed at helping Kenya develop agricultural biotechnology. These funds are earmarked for projects on maize, biofertilizers, horticultural tissue culture, root and tuber tissue culture, agroforestry tissue culture, as well as providing necessary financial support for technology transfer, training activities and the development of a National Biosafety Mechanism. The World Bank also supports projects at KARI, and WHO supports projects at the Kenya Medical Research Institute (KEMRI). An example is a project on molecular mechanisms of drug resistance in malaria, which has received US $100,000 from WHO. The Special Programme on Biotechnology and Development Cooperation has also provided funding for projects in medical biotechnology.

Table 4.6. Analysis of total R&D budget against national budget and GDP (K£ millions).

	1974	1984	1989	1990	1991	1992
Total R&D budget	3	17	37	41	60	56
Total national budget	745	1,263	2,810	3,256	3,829	4,686
Estimated GDP	908	4,056	8,452	10,032	11,362	12,000
R&D budget as a percentage of national budget	0.4	1.3	1.3	1.3	1.6	1.2
R&D budget as a percentage of GDP	0.33	0.42	0.42	0.40	0.53	0.47

Regulatory Structure

Biosafety

The National Council for Science and Technology (NCST) has embarked on a programme for the development of regulations and guidelines for biosafety in biotechnology for Kenya. These regulations and guidelines have taken into account the safety aspects related to research, development and use of genetically engineered organisms. The NCST is the national focal point for safety in biotechnology and is a member of the African Regional Focal Point for Biosafety based in Zimbabwe.

Other general aspects of safety are dealt with in several acts and regulations, amongst which are the Seed and Plant Varieties Act and the Plant Protection Act; the Crop Production and Livestock Act; the Agriculture Act; the Food, Drugs and Chemical Substances Act; the Dangerous Drugs Act; the Public Health Act; the Radiation Protection Act; the Use of Poisonous Substances Act and the Narcotic Drugs Act.

Table 4.7. Funding of research institutes as a proportion of GDP 1986/87 - 1989/90.

Year	KARI[a]	KEFRI[b]	KEMFRI[c]	KEMRI[d]	KETRI[e]	KIRDI[f]	Total
1986/87	0.18	0.03	0.03	0.07	0.04	0.02	0.37
1987/88	0.25	0.10	0.03	0.06	0.03	0.03	0.50
1988/89	0.35	0.05	0.03	0.06	0.02	0.02	0.54
1989/99	0.37	0.05	0.03	0.07	0.03	0.02	0.58

[a] Kenya Agricultural Research Institute.
[b] Kenya Forestry Research Institute.
[c] Kenya Marine and Fisheries Research Institute.
[d] Kenya Medical Research Institute.
[e] Kenya Trypanosomiasis Research Institute.
[f] Kenya Research and Development Institute.

Intellectual Property Rights
Kenya signed the Uruguay Round of Agreements (URA) in 1994 and ratified them in the same year. In effect, Kenya became a founder member of the World Trade Organization (WTO). Prior to the URA, the General Agreement on Tariffs and Trade (GATT), established in 1947, had been applicable.

The Kenya Industrial Property Office was established under the Industrial Property Act CAP 509. The Industrial Property Act deals with the rights of invention and recognizes the part played by the inventor. It also provides for the patenting of microorganisms and products of genetic engineering such as the replicating material, parts of a plant and products of biotechnological processes such as vaccines, diagnostic materials, pharmaceuticals, pesticides and herbicides. However, the Act excludes the patenting of living organisms such as animals and plants, and the patenting of methodologies for the treatment of animals by surgery or therapy or for disease diagnosis.

Biological Diversity
Kenya has about 35,000 known species of plants, animals and microorganisms. Conservation measures are supported by the various policies which the Government has formulated and implemented. Additionally, the Government has signed, and ratified the Convention on Biological Diversity.

Research, Development and Marketing

Major Projects

Tissue Culture
The importance of tissue culture techniques for Kenya cannot be underestimated. As a result, KARI is conducting projects employing tissue culture for virus elimination in Irish potato, banana, cassava, sweet potato, ornamentals and strawberry. Tissue culture and micropropagation work is being undertaken on Irish potato and pyrethrum. Tissue culture for *in vitro* selection is also being carried out on cassava. At the Jomo Kenyatta University of Agriculture and Technology, tissue culture work on bananas aimed at producing clean planting material is in progress. Similarly, the University of Nairobi and the Oserian company have projects on tissue culture for horticultural crops and ornamentals, respectively.

Plant Disease Diagnosis
Detection of viruses using polyclonal antiserum is being carried out at the Plant Quarantine Station of the Plant Pathology Laboratories in Muguga and at the National Agricultural Laboratory.

Animal Biotechnology
The Kenya Agricultural Institute, the Kenya Trypanosomiasis Research Institute, the International Livestock Research Institute (ILRI) and the University of Nairobi all have projects underway in animal biotechnology. Among the main projects are development of monoclonal antibody-based ELISA assays for rinderpest, rabies, lumpy skin and Nairobi sheep disease. Use of DNA probes for diagnosis of theileriosis, trypanosomiasis, babesiosis and heartwater is being investigated at KARI. Development of virus-vectored multivalent vaccines and evaluation of recombinant vaccines produced outside Kenya is also in progress at KARI.

At the International Centre of Insect Physiology and Ecology (ICIPE), recombinant DNA techniques are being used in the characterization of tick and protozoan parasites. Work on the production of biopesticides is also in progress. The International Livestock Research Institute (ILRI) is involved in bovine genome mapping, development of subunit theileria vaccine and development of diagnostic tests for protozoan infections. The Institute of Primate Research uses non-human primate models to study candidate vaccines against leishmaniasis.

The Kenya Medical Research Institute started its biotechnology programme in 1993. Completed projects include identification of schistosomiasis antigens which could have possible use in vaccine development, evaluation of vaccine development, evaluation of a DNA probe for diagnosis of malaria, detection of point mutations associated with drug resistance in malarial parasites, development of DNA probes and monoclonal antibodies for diagnosis of filariasis, and the study of the incidence of tuberculosis in HIV-infected patients. At the University of Nairobi, as part of the MIRCEN (Microbiological Resources Centre) Programme, work on the development of biofertilizers has been underway for over 15 years.

The government has given the responsibility for the manufacture of livestock vaccines to the Kenya Veterinary Vaccines Production Institute (KEVEVAPI). In 1993, KEVEVAPI was producing a total of 15 vaccines, comprising two animal-produced vaccines, four egg-derived vaccines, three broth culture derived vaccines and six tissue culture derived vaccines. However, the demand for vaccines is high and some of the more specialized vaccines such as those for pet animals, horses and poultry have to be imported. Bacterial vaccines against anthrax are also imported.

Major Institutes

Major institutes carrying out biotechnology research in Kenya include:
- KARI.
- ILRI.
- KEMRI.
- ICIPE.

Kenya has tertiary education institutions comprising teacher training colleges, polytechnics, technical institutes, institutes of technology and several public and private universities. At the University of Nairobi several postgraduate programmes related to biotechnology are offered in the fields of agriculture, veterinary science and medicine. Similar programmes exist in other universities. In addition, the Jomo Kenyatta University of Agriculture and Technology offers a course in biotechnology.

In 1995, the population of Kenya was estimated at 27.5 million and it is projected to grow to 31.9 million by the year 2001. Enrolment in tertiary level education is currently around 4% of the relevant population, of which half is at the university level. Specific examples of human resources in biotechnology R&D are given by the Kenya Medical Research Institute biotechnology programme which employs 39 scientists and five technicians and KARI which in 1993/94 employed 30 scientists and four technicians.

Private Sector Engagement

The participation of the private sector in biotechnology development in Kenya has been very limited and confined mainly to the manufacture of livestock products and pharmaceutical products for human use. Companies involved in these activities include Wellcome Kenya Ltd., which manufactures livestock products, Dawa Pharmaceuticals, which manufactures pharmaceutical products, East African Industry, which manufactures household products, and Kenya Breweries, which produces beer. The Monsanto Company, in a collaborative project with KARI, has successfully developed a variety of sweet potato resistant to the feathery mottle virus using plant transformation techniques.

Constraints

The main constraints to R&D in Kenya are summarized in the National Development Plan for 1997-2001 as follows:

> 'Because Kenya's R&D efforts are dominated by public bodies, the constraints faced by these bodies largely encompass the constraints facing the sector as a whole. Kenya's R&D establishment is faced by a range of operational constraints, ranging from inadequate quantity and quality of research personnel, mainly due to the relatively poor remuneration and inadequate incentives and rewards, inadequate management and personnel practices that have led to inappropriate ratios of support to scientific staff, to capital intensive research practices and poor maintenance of research equipment, inadequate incentives for local production of consumables for R&D equipment and overemphasis on donor-funding. This approach has led to a research sector that is largely divorced from industrial production as characterized by the limited number of industrial companies using patented information from the local R&D institutions.'

The Government, in the same National Plan, has outlined measures to address the above constraints.

Future Prospects

In the agricultural sector, R&D investment opportunities lie in tissue culture for micropropagation of crop plants, such as bananas and pyrethrum. In addition, the use of tissue culture for disease elimination in root crops such as cassava, sweet potato and Irish potatoes will prove increasingly important in future. Of equal importance is the development of biopesticides and biofertilizers.

In the area of animal health, the development of bioengineered vaccines for control of livestock diseases holds considerable potential. Monoclonal antibody production using cell fusion could also be commercially exploited.

Other products of potential commercial value include:

- Diagnostic kits for hepatitis.
- Hepatitis vaccines.
- Plant-derived products.
- Diagnostic kits for leishmaniasis.
- Rotavirus vaccines.

Korea

Cherl-Ho Lee

National Policy and Regulation

General Policy

Efforts to develop biotechnology within Korea were based on the food and pharmaceutical industries. The former drew upon a long tradition of fermentation technology, and the latter, which has been expanding rapidly since the 1960s, is also fairly advanced. Nevertheless, it was not until the 1980s that a systematic effort to develop biotechnology was launched at a national level. At the beginning of the 1980s, an informal circle of scientists emerged to advocate biotechnology. Having returned home to Korea following advanced training abroad in biotechnology-related fields (such as molecular biology, genetic engineering and microbiology) they were working in universities and government research institutes. At this time the Korean economy was undergoing a transition from traditional manufacturing sectors into high technology industries and the group were thus able to convince policy-makers and industry leaders of the urgent need to prioritize the development of biotechnology.

The first achievement of such concerted efforts was the establishment in March 1982 of the Korean Genetic Engineering Research Association, a consortium of 13 companies with active interests in biotechnology. In the same year the Ministry of Science and Technology included biotechnology in the list of seven strategic areas of national technology development, and in April 1982, the Korean Research Council for Applied Genetics was created to promote and coordinate research activities in universities.

An important milestone in government policy for biotechnology was the passage of the Genetic Engineering Promotion Law in 1984. Article 1 of the law declares the government's responsibility for the development and industrialization of genetic engineering, and stipulates the establishment of a national centre for genetic engineering research. The law also prescribes the duties of various agencies for the promotion of genetic engineering.

In 1993, the government established the Biotech 2000 programme, led by the Ministry of Science and Technology. Under this programme, a comprehensive governmental biotechnology promotion plan was formulated, with the aim of promoting the standard of national biotechnology to that of advanced countries, and to occupy 5% of the world market share by the year 2000.

As a result, R&D manpower has grown dramatically. There are now over 1200 PhD students and 1800 Masters and BSc students working in biotechnology R&D. The number of industrial patent applications from bioindustry increased 28% annually, amounting to 559 in 1994 alone.

Funding

Spurred on by governmental, academic and industrial activities, biotechnology R&D has made rapid progress in Korea over the last decade. National R&D expenditure reached US $9399 million in 1994, corresponding to 2.45% of the GNP (Table 4.8). Biotechnology R&D expenditure has increased more than 40% annually. Of this 1994 total of US $9399 million, the government contributed US $53 million, with the industrial sector investment contributing the remainder (see Table 4.9).

Table 4.8. R&D expenditure in Korea (in US $ millions).

	1993	1994	Increase 1993 to 1994 (%)
(A) Gross national product (GNP)	329,826	383,258	16.19
(B) Total R&D expenditure	7,691	9,399	22.2
(C) Public sector R&D expenditure	1,299	1,515	16.63
(D) Industry biotech R&D expenditure	109	180	65.1
B/A	2.33%	2.45%	
C/B	16.89%	16.12%	
D/B	1.42%	1.91%	

Regulations

A set of Guidelines for Recombinant DNA Experiments were prepared by the Ministry of Health and Welfare in 1995, and regulations for the environmental release of genetically modified organisms are in preparation by the Ministry of Agriculture, Forestry and Fisheries (MAFF). The Ministry of Environment is in charge of the provisions for the Convention on Biological Diversity.

Research, Development and Marketing

Major Institutes

Government Institutes and Associations

Ministry of Science and Technology

- Korean Research Institute of Biosciences and Biotechnology.
- Microorganism depository organization, comprising the Korean Collection for Type Cultures, the Korean Culture Centre of Microorganisms and the Korean Cell Line Research Foundation.
- Korean Biotechnology Research Association, with 21 member companies.

Ministry of Trade, Industry and Energy

- Biotechnology Laboratory, National Industrial Research Institute.
- Life Science Division, Korea Institute of Industry and Technology Information.
- Bioindustry Association of Korea (BAK), with 52 member companies (19 pharmaceutical, 13 food, eight chemical, five textile, four supplier, three other).

Ministry of Health and Welfare

- Division of Biotechnology, National Institute of Health.
- National Institute of Safety Research.

Ministry of Agriculture, Forestry and Fisheries

- Department of Bioresources, Rural Development Agency.
- Korea Food Research Institute.

Ministry of Environment

- National Institute of Environmental Research.

Academic Institutions

Academic institutions include 39 biotechnology-related departments, 21 biotechnology research institutes, nine Science Research Centres and Engineering Research Centres,

Table 4.9. Government support for biotechnology R&D in Korea in 1994.

Ministry	Research area	Number of projects	Expenditure (US $ millions)
Ministry of Science and Technology	Screening technology for biomaterials, transgenic animals and plants, genome research, protein engineering, biopharmaceuticals, biopolymers	45	42.34
MAFF	Agricultural genetic engineering, biodiversity and environmental protection, ocean biotechnology, forest breeding, food biotechnology	43	2.72
Ministry of Trade, Industry and Energy	Bioenergy and mineral resources, biotech industrialization	15	2.46
Ministry of Health and Welfare	Biopharmaceuticals, clinical trials, effectiveness and hazard analysis	14	1.95
Ministry of Environment	Biodiversity management, organisms for waste treatment	14	1.01
Ministry of Education	Basic research in biotechnology	166	2.88
Total		297	53.36

and one nationally-supported graduate school, the Graduate School of Biotechnology, Korea University.

Private Sector Institutes and Associations

Industrial research institutes

- Mogam Biotechnology Research Institute.
- Asan Institute for Life Science.
- Samsung Biomedical Research Institute.
- Lucky Biotechnology Research Institute.
- Sunkyung Research Institute.

Industrial consortia

- BAK.
- KBRA.

Table 4.10. Biotechnology products on the market in Korea in 1994 (US $ thousands).

Area	Product[a]	Market Domestic	Imported	Per cent
Pharmaceuticals	Antibiotics	52,922	1,200	
	Vaccines	4,029	21,068	
	Anti-cancer	9,951	3,929	
	Insulin	-	3,973	
	Anti-inflammation	4,256	-	
	Anaemia	-	823	
	Hormone	14,720	4,893	
	Blood products	2,673	436	
	Other drugs	2,558	3,050	
	Diagnostics	10,298	35,060	
	(Subtotal)	(101,407)	(74,432)	73.30
Foods	Functional foods	625	-	
	Sweeteners	9,699	-	4.30
	(Subtotal)	(10,324)	(-)	
Environment	Microorganisms for waste treatment	11,799	1,189	
	Waste water treatment	7,188	-	
	(Subtotal)	(18,987)	(1,189)	8.41
Chemicals	Bio-cosmetics	1,250	-	
	Enzymes and reagents	1,523	2,028	
	Biopolymers	1,450	12	
	(Subtotal)	(4,223)	(2,040)	2.61
Agriculture	Animal drugs	518	351	
	Feeds and pesticides	2,211	3,225	
	(Subtotal)	(2,729)	(3,576)	2.63
Bioprocessing	Instruments	8,075	12,894	
		(8,075)	(12,894)	8.74
	TOTAL	145,745	94,131	
	GRAND TOTAL	239,876		99.99

[a] Examples of domestic products: human growth hormone, hepatitis B vaccine, α-interferon, γ-interferon, diagnostic kits, oligosaccharides, phenylalanine.

Private Sector Engagement

The KBRA comprises 21 member companies and its work is coordinated by the Ministry of Science and Technology. The Ministry of Trade, Industry and Energy supports the BAK which has 52 member companies comprising 19 pharmaceutical, 13 food, eight

Table 4.11. Budget planning for strategic R&D area of biotechnology in Korea (in US $ millions).

		Phase I			Phase II	Phase III
Area		1994	1995-1997	Total	1998-2002	2003-2007
Biomaterials	Government	16	122	138	324	973
	Industry	82	374	456	1,240	2,410
	Total	98	496	594	1,564	3,383
Health care	Government	13	88	101	250	825
	Industry	61	283	344	430	1,900
	Total	74	371	445	680	2,725
Agriculture and foods	Government	9	98	107	374	825
	Industry	35	206	241	870	1,800
	Total	44	304	348	1,244	2,625
Environment, safety and conservation of bioresources	Government	7	59	66	212	687
	Industry	25	137	162	590	240
	Total	32	196	228	802	927
Alternative energy sources	Government	4	36	40	110	325
	Industry	17	67	84	490	650
	Total	21	103	124	600	975
Basic life sciences	Government	12	112	124	350	970
	Industry	-	9	9	170	400
	Total	12	121	133	520	1,370
Supportive functions and infrastructure	Government	4	45	49	110	325
	Industry	-	4	4	60	300
	Total	4	49	53	170	625
	Government	65	560	625	1,730	4,930
Total	Industry	220	1,080	1,300	3,850	7,700
	Total	285	1,640	1,925	5,580	12,630

chemical and five textile companies as well as four suppliers and three other companies. Table 4.10 shows the biotechnology products market in Korea in 1994. Pharmaceutical products constituted the major market covering 73.3% of the total US $240 million market and foreign imports accounted for 40% of the total sales value.

Constraints

The present level of R&D is insufficient, and the country continues to rely on foreign biotechnology products. Of all biotechnology-based products (especially in new-biotechnology products), foreign imports accounted for 40% of the total sales (valued at US $239 million) in 1994 (see Table 4.10). This heavy dependence upon foreign basic research and advanced technology is a major constraint upon Korean bioindustries. The cost of technology importation is rapidly increasing, and technology transfer itself is becoming limited.

Future Prospects

The biotechnology related market size has expanded rapidly during the last decade. The market size was US $25 million in 1989 and grew to US $200 million in 1993. It is expected to reach US $4 billion by the year 2000 and US $17.5 billion by 2005. Domestic production was 56% of the total market in 1993, and is expected to cover 62% of the total market by the year 2000. The investment plan for the Biotech 2000 programme allocates US $1925 million during Phase I (1994-1997), from which 32% is government input and 68% comes from the private sector. This will increase to US $5480 million during Phase II (1998-2002) and US $12,630 million during Phase III (2003-2007). Table 4.11 details this budget planning.

Kuwait

Sabah AlMomin

Major Projects and Research Institutions

In Kuwait, scientific research, particularly in the field of biotechnology and molecular biology, lies mainly within two institutions: Kuwait University and Kuwait Institute for Scientific Research (KISR). The major source of funding for both is governmental, though in addition to other financial sources such as the Kuwait Foundation for the Advancement of Science (KFAS), the Environmental Protection Council, the Public Authority for Agriculture and Fisheries and private industry.

The Kuwait Institute for Scientific Research (KISR)

KISR was established in 1967 to support national industries by conducting scientific research in problem areas. A special Department for Biotechnology was established in 1980. The first major biotechnology project was on the production of single cell protein as animal feed. This was followed by a related project on the conversion of cellulosic products. The total investment for these projects amounted to US $4.3 million, including the establishment of a pilot plant facility of 1500 litres fermentor capacity.

KISR is the only scientific institution in Kuwait which has a defined biotechnology programme and has set up a national strategic plan for 1995-2000 (KISR, 1995). The programme identifies areas which solve national problems and hold the promise of economic benefit. Priority areas include genetic engineering, cell and plant tissue culture, food technology, bioremediation and fermentation, with emphasis on agriculture, animal health and environmental safety. The current investment in the projects underway in the Department for Biotechnology is US $6 million with client funds totalling US $1.6 million. Current research objectives of the biotechnology programme are:

- Isolation of stress tolerant genes from salt-tolerant and pest-resistant plants.

- Bioremediation and rehabilitation of oil-contaminated soils.
- Establishment of techniques to ensure trueness-to-type of the date palm.
- Development of production technology for the date palm through tissue culture.
- Advancement of plant tissue culture developments for food and vegetation plants.
- Evaluation of horticultural crop yields in greenhouse systems including saltwater greenhouses.
- Application of embryo transfer techniques for sheep production improvement.
- Development of DNA fingerprinting techniques for marine species.
- Carrying out field tests and development of an action plan for remediation of oil-contaminated soils.
- Development of microbial-based products, e.g. biosurfactants and bioinsecticides.
- Preparation of long-term solid/semisolid/liquid waste master plan.

The Biotechnology Department has achieved significant accomplishments in several areas. It is now feasible to plan a pilot commercial delivery of plantlets of date palm and determine trueness-to-type through tissue culture, in addition to the enhancement of the production of native plants. Similar advanced stages of project development have been reached with the initiation of mangrove growth in a pilot test region and the bioremediation of oil-contaminated soils. In addition to technology transfer, the Department has made recommendations for improving the quality and safety of food products.

The Kuwait Foundation for the Advancement of Science (KFAS)

An Amiri Decree was issued in 1976, by the State Government, for the establishment of the KFAS. This is viewed as a scientific base for promoting quality research and development through the support of Kuwait corporations, which contribute 5% of their annual profit to KFAS. The Foundation's main objective is to extend timely and broad scientific support, by implementing a variety of scientific programmes through grant-awards, in order to provide Kuwaiti scientists with the opportunity to conduct hands-on technological research, whilst simultaneously training aspiring young Kuwaitis.

In 1995, the Research Projects Directorate reviewed 124 research proposals, out of which 35 projects were approved. These were valued at US $6.5 million, with KFAS contributing almost 50% of the total budget. Through the period 1990 to 1995, the foundation invested over US $1 million in research projects in genetic and molecular studies.

The sponsored research projects are multidisciplinary and classified into general categories. Biotechnology or genetic engineering are not recognized as separate fields in themselves, but fall under the field headings of agriculture, environmental or medical sciences. Research projects in the area of biotechnology and genetic engineering funded by KFAS in 1992 include (KFAS, 1995):

- Effect of salinity on various plant characteristics of selected tolerant and sensitive genotypes of barley.
- Analysis of cell proliferation and death in the stomach of normal and transgenic mice developing gastric diseases.
- Molecular genetic studies of insulin-dependent *diabetes mellitus* in the State of Kuwait.
- Molecular genetics of muscular dystrophy.
- Identification of new tumour markers and their role in the development and progress of colorectal carcinoma.
- Pharmacologically active agents from Arabian Gulf catfish.
- Radiopharmaceutical studies of cyclosporin-A toxicity in rats and correlation with biochemical measurements.

National Committee on Genetic Engineering and Biotechnology

The Government of the State of Kuwait recognized the potential long-term benefits of biotechnology, and has formed a National Committee on Genetic Engineering and Biotechnology in consultation with UNIDO in 1989 (Kuwait, 1989). The Committee is to act as a National Safety Committee and will set up a National Plan for the development of Biotechnology and Genetic Engineering in Kuwait. Kuwait also has been closely involved in the establishment of the International Centre for Genetic Engineering and Biotechnology (ICGEB). In this regard, the Committee has identified six research areas as priority fields of research:
1. Use of DNA-based genetic diagnostics related to clinical genetics and health.
2. Production of biomass and fine chemicals from methanol and petroleum by-products.
3. Use of microorganisms for biosurfactant production and pollution control.
4. Marine genetic engineering and biotechnology.
5. Agricultural plant tissue culture and crop improvements.
6. Production of industrial enzymes.

Work in some of the above areas has not been initiated due to the discontinuation of the Committee since the Iraqi aggression on Kuwait in 1990. This event changed the priorities in the research areas as new problems have evolved due to the postwar reconstruction, such as food security and environmental safety, which became the major concerns of the country. Increased food self-sufficiency became the main objective of the biotechnology programmes and the problem is to be tackled through the use of biotechnology and genetic engineering to modify and increase production in plant and animal agriculture, fisheries and aquaculture. In addition, the Iraqi invasion, in 1990, resulted in the destruction of oil production facilities where vast amounts of crude oil were released into the environment. This has caused marine, terrestrial and atmospheric pollution. These newly emerged areas have reoriented the research direction and priorities.

Kuwait University

Kuwait University, the only higher academic institution in the country, has 189 multidisciplinary research projects, according to the latest sponsored research report of 1992 (Kuwait University, 1994). Recently, research activities were diverted towards priority areas of local needs. The University has neither a defined policy nor strategic plans regarding research priorities. Projects are initiated depending upon the speciality and background of the faculty members, rather than on a well-defined plan of the institution. According to the 1995 Kuwait University Research Abstracts Publication (Kuwait University, 1995), 18 new scientific areas have been added to the already existing fields of research and studies, one of which is molecular biology. This indicates the significance placed upon newly emerging fields of research. The total investment for the research projects for 1996 was in excess of US $1.65 million.

The current main research areas in molecular biology and biotechnology in Kuwait University are categorized into three main areas, as follows:

1. Environmental Biotechnology:
- High-temperature hydrocarbon degradation by *Bacillus stearothermophilus* from the oil-polluted Kuwaiti desert.
- Oil-degrading rhizospheric microorganisms as potential contributors to the self-cleaning and bioremediation of the Kuwaiti desert.

- Pharmacologically active agents from the Arabian Gulf catfish (*Arius thalassinus*, Ruppell).
- Assessment of environmental pollution in Kuwait (Marine Environment).
- Microfungal response to acute stress conditions in the environment, including increased salinity, temperature and oil pollution.
- The effects of crude oil pollution on *Vicia faba* leaf structure and ultrastructure.
- Induction, purification and characterization of keratin-degrading enzymes from fungi.
- Isolation and characterization of halophilic and keratin-degrading actinomycetes from saline areas in Kuwait.

2. Plant Biotechnology:
- Effects of salinity on the characteristics of tolerant and sensitive barley genotypes, and the application of anther culture in breeding for salt tolerance.
- Induction and cytological analysis of different chromosome-doubling approaches of microspore derived wheat haploids.

3. Medical Biotechnology:
- Human T-cell recognition of mycobacterial heat-shock proteins in the context of multiple HLA-DR molecules.
- Long lasting T-cell reactivity to *Mycobacterium leprae* antigens in human volunteers vaccinated with killed *M. leprae*.
- Study of adaptation response of glioma cells to high temperature and metabolic stress in culture.
- Somatic cell variability of unstable trinucleotide repeats.
- Cloning and expression of a growth factor gene in muscle.

Constraints

Kuwait experiences constraints common to developing countries, and scientific contributions and achievements remain moderate. The following causes are identified specifically:
- Lack of trained local laboratory technicians.
- Lack of well-defined research policies and planning in biotechnology.
- Budget and financial constraints.
- Limitations of access to international scientific data centres.
- Scarcity of industries which promote scientific research.
- Lack of awareness of the uses and benefits of biotechnology and its applications.
- Contracting of majority of projects to clients.
- Partial destruction of the agri-industrial sector due to the Iraqi aggression on the country.
- Shortage of highly qualified and trained Kuwaiti manpower in certain scientific disciplines, especially the areas of biotechnology and molecular biology.
- Limitation and delays in importing necessary chemical reagents.

Future Prospects

Following the Iraqi invasion, and the environmental catastrophe this entailed, the attention of Kuwaitis has turned to the biological resources of the country and the necessity for their conservation and development. The signing of the Convention on Biological Diversity in Rio de Janeiro in 1992 emphasized the value of biotechnology as a tool to execute such a commitment and the government at present is setting out legislation and plans of action to this end, a policy hitherto lacking but critically needed.

The State of Kuwait has recognized that biotechnology has an important role to play in economic and industrial development. The identification of national problems, such as improvement of natural resources, environment, food and health, has put a great demand on the utilization of biotechnology to solve such issues. Regionally, Kuwait is considered to be a leading country in the biotechnology sector. However, for further improvements and the application of biotechnology, there is a need for a well-defined long-term scientific policy, and legislation. A plan of action is also needed to put more emphasis on manpower training through academic education, public awareness and international collaboration. There is no doubt that Kuwait, although having an experience of just 15 years in the sector, is nonetheless well placed in the international biotechnology arena.

Latvia

Martin Beker, Alexander Rapoport and Uldis Viesturs

Introduction

Following the Second World War, Latvia developed an industry for the production of citric acid, lysine feed concentrate, enzymes, antibiotics, plant growth stimulators, fodder yeast, bacterial fertilizers and other biopreparations. Most of these new technologies were developed at the Institute of Microbiology of the Latvian Academy of Sciences.

The transition to a market economy has given rise to major changes in the Latvian biotechnology industry. Practically all of the bioindustry organized after World War II was stopped, including lysine production at the Biochemical Factory in Livani, and the production of biochemical reagents and enzymes at 'Biolar' in Olaine.

National Policy and Regulation

Funding

Research in molecular biology, microbiology, virology and biotechnology, are mainly financed by the Latvian Council of Sciences, through about 30 grants (amounting to around US $400,000 per annum). Some additional funds are also received annually from EC programmes (INCO-Copernicus and TEMPUS) and from some other funds (Soros Foundation and Volkswagen).

Some financial support is also received from special funds in Scandinavian countries. Shortages of equipment and chemicals, arising from inadequate funding, are acute in Latvia. Furthermore, because of the low salaries of researchers, science has become unpopular amongst the younger generation. As a result, talented university graduates prefer careers other than science.

Main publications prior to 1991 appeared in Russian, in Latvian and Former Soviet Union journals, as well as in locally published scientific books. At present, there is a significant increase in the quantity of papers published by Latvian scientists in English, in European and American journals. The quantity of annual scientific papers in the peer-reviewed, international journals is one of the main criteria used by the Latvian Council of Science in allocating funds.

Regulatory Structure

According to the Environmental Health Action Plan for Latvia (drawn up by the Ministry of Welfare and the Ministry of Environmental Protection and Regional Development) a number of regulations should be developed during the period 1998-2000. Among them will be new national legislation concerning human and environmental safety with regard to genetically modified organisms (GMOs). Although there are no GMO-plants or animals developed in Latvia, such organisms or their products are expected to be imported. Laws and bye-laws will be drafted following review of the existing 'gene' laws in other countries, and in accordance with the relevant EC and United Nations Environment Programme (UNEP) directives and recommendations.

Research, Development and Marketing

Major Projects

Research in virology is carried out in three major directions: protein engineering, plant virology and oncovirology. Research in oncovirology in its turn was carried out in two areas - investigations of retroviruses found in Latvia and their characteristics, and a search for new ways to induce antiviral immunity. Results suggested the possibility of obtaining a protective vaccine against bovine leukaemia virus, based on anti-idiotypic antibodies.

During 1994-97, in the field of applications of heterologous gene expression, some human papilloma virus proteins were synthesized in yeast, and the *E. coli* expression system was optimized for the expression of a wide range of foreign genes.

Microbiology in Latvia focuses mainly on physiological and biochemical properties of primary bacterial metabolites - ethanol, lactate, lysine and storage polymers, fructose polymer levan and polyhydroxybutirate. The structural and functional role of intracellular stress protectors has been investigated using the model of yeast cell dehydration-rehydration.

Many bacterial strains - destructors of oil components in polluted soil and water - were isolated, identified and investigated. Research on the liquid-cell interaction was performed with the anaerobic culture *Zymomonas mobilis*. A method of testing growth control action, based on measurements of the guanosine-5'diphosphate-3'diphosphate content changes in cells, has been developed. Methods of biosynthesis of some natural insecticides of biological origin have been pursued, and two new strains of *Aspergillus niger* for production of citric acid were developed.

To promote the further development of biotechnology, a collection of microorganisms was founded at the Faculty of Biology, University of Latvia. At the moment this is the sole internationally approved collection in the Baltic countries of the Former Soviet Union.

Major Institutes

Latvia has significant scientific potential, with specialists working in a number of research and educational institutions:

- Biomedical Research and Educational Centre, University of Latvia, Riga.
- Institute of Microbiology and Biotechnology, University of Latvia, Riga.
- Institute of Microbiology and Virology, Riga.
- Institute of Wood Chemistry, Riga.
- Faculty of Biology, University of Latvia, Riga.

There are about 100 researchers (including about 70 scientists with doctoral degrees) working in biotechnology-related disciplines in these institutions. Their areas of interest include molecular biology, genetic engineering, microbial physiology, biochemistry and cytology, as well as fermentation and downstream processing technology.

Private Sector Engagement

There has been a recent privatization of previously public industries, particularly in brewing and distilling. In addition, a pharmaceutical company, 'Gri-dex', which was developed on the basis of a small biotechnological production facility (l-asparaginase enzyme preparation) has been privatized. The Latvian government promotes this process of privatization through a specially appointed Privatization Agency, and although there are no special national policy priorities regarding biotechnology, there are instances where state support is offered. One such example is the national programme on the production and use of biofuel. This addresses the possibility of the future production of ethanol for blending with benzene and biodiesel from rape seed, and biogas from industrial and municipal wastes. Special scientific and technical programmes of a similar orientation are supported financially by the Latvian Council of Sciences. A number of small to medium-sized enterprises are active in the field.

Lithuania

Arvydas Janulaitis

National Policy and Regulation

General Policy and Funding

Within the Lithuanian government the overall political responsibility for research policy and financing rests with the Ministry of Education and Science. Financing of biotechnological research from the state budget comes through general financing of the research institutes and universities without any particular accent on biotechnology. Further allocation of resources depends on decisions within the appropriate institution itself. A major part of state finances (over 80%) are supplied by this route. The Ministry consults with the Lithuanian Science Council which has no funds of its own. Individual grants or science projects are distributed by the Lithuanian Science and Studies Foundation on a competitive basis. The major institution - the Institute of Biotechnology - is supported on the basis of contracts and grants from two local biotechnological companies, AB Fermentas and AB Biofa. Lithuanian biotechnologists from the Institutes of Biotechnology, Biochemistry and Vilnius University have also received grants from international sources such as the European Commission (INCO-Copernicus programme), the International Science Foundation, Volkswagen Stiftung, Howard Hughes Medical Institute, etc. Estimated total financing of Lithuanian Biotechnology research is presented in Table 4.12.

Recently the policy of the Government has changed and special attention has been given to biotechnology. The five-year state programme 'Molecular basis of biotechnology: gene structure, function and regulation', was approved and funded from the state budget in 1997. The following key points of biotechnological research are emphasized: gene structure and molecular mechanisms of gene regulation; enzymes for genetic engineering; new expression systems for heterologous proteins.

Regulatory Structure

The Convention on Biological Diversity was ratified by Parliament in 1995. The Law of Environmental Protection was passed by Parliament in 1992. The latter includes regulations for manipulation, multiplication and utilization of microorganisms. Additional rules, 'Procedures for registration, use, storage, introduction and transportation of microorganisms and other sources of biological pollution' were ratified in 1993, and include rules concerning recombinant microorganisms. There are no special rules concerning the use and introduction of recombinant plants and animals into the environment. The Law of Biomedical Research Ethics is under confirmation by the Parliament. The Ministry of Environmental Protection is responsible for preparation of the rules concerning GMOs.

Each institution working on biotechnology in Lithuania has internal regulations complying with national laws and EU recommendations. An Advisory Committee on Genetic Manipulation was established in 1996 under the auspices of the Lithuanian Society of Genetics and Breeders. The preparation and adoption of the final regulations modelled according to international standards is being facilitated by the participation of Lithuanian scientists in a number of international and regional cooperation activities in biosafety.

Research, Development and Marketing

Major Projects

The most important projects at present under way in Lithuania are:

- Restriction-modification enzymes: screening, purification, molecular mechanisms of site-specific interaction with DNA.
- Cloning and sequencing of genes of restriction-modification enzymes, analysis of gene structural organization and regulation of gene expression, construction of overproducing strains of restriction endonucleases.
- Bacteriophage T4 gene structure and regulation of expression.
- Differentiation dependent gene expression in *Bacillus subtilis*.
- Development of new gene expression systems in *Kluyveromyces* and *Saccharomyces*.

Table 4.12. Financing of biotechnology research in Lithuania.

Year	Amount US $ millions
1994	0.58
1995	0.88
1996	1.0

- Recombinant protein purification, immunoenzymatic test systems, chemical modification and investigation of the relation between structure and biological activity, stability and stabilization studies.
- Biosensors for detection and assay of various metabolites.

Major Institutes

The major institutes involved in biotechnology R&D in Lithuania are listed in Table 4.13.

Private Sector Engagement

Main achievements in commercialization of research results are attributable to the Institute of Biotechnology. Founded in 1975, the initial activities of the Institute were aimed at the development of classical fermentation and purification technologies for the production of enzymes for scientific research, clinical analysis, industry and agriculture. The most successful investigations have been made in the field of restriction endonucleases - many new restriction endonucleases have been discovered and commercialized. Since 1983, the enzymes have started to sell in Japan and other countries. In 1991, the first sales affiliation for the distribution of restriction endonucleases and other products for molecular biology was founded in Germany. Research and development of recombinant proteins started in 1984. In 1990, Reaferon, the first human recombinant interferon-α2b to be produced in injection dosage form, was registered in the former USSR. In 1992, a joint venture between the Institute and Tianjin Amino Acid Corporation (P.R.China) was established for the production of interferon-α2b. In 1994, two spin-off companies, AB Biofa and AB Fermentas, separated from the Institute of Biotechnology and were privatized. The latter remains as a state institute with a staff of 128, of which 76 are involved in scientific research.

The Institute of Biotechnology continues to maintain close links with AB Biofa and AB Fermentas, and this is reflected in collaborative projects carried out at the Institute, such as:

- Investigation of highly purified recombinant proteins aimed at the development of technologies for the manufacturing of products used in medicine, agriculture and research.
- Research in the field of genetic engineering and the development of producers of biologically active substances.
- Molecular genetic investigation of restriction modification enzymes.

AB Biofa is an R&D and manufacturing company operating in the field of pharmaceuticals. The total staff of the company numbers 116, of which 27 are R&D personnel. AB Biofa is already a recognized manufacturer of recombinant protein therapeutics - human interferon-α2b and human growth hormone. Some other recombinant proteins are also expected to be available in the near future. In addition, the company supplies a variety of recombinant proteins and immunochemicals for research purposes and offers custom development of recombinant DNA technology. Dosage forms of recombinant protein drugs were registered in the People's Republic of China, South Korea, the Russian Federation, Ukraine, Kazakhstan, Bulgaria and others, and are under registration in, amongst other countries, India and Indonesia.

AB Fermentas is an R&D and manufacturing company acting in the field of products for genetic engineering and molecular biology. Total staff number 109, of which 20 are research personnel. The company produces restriction endonucleases (over 120 items),

Table 4.13. Institutions carrying out basic research in biotechnology in Lithuania.

Institution	Research activities
Department of Natural Sciences, Vilnius University	Covalent links between nucleic acids and proteins; pre-RNA processing; molecular basis of *Escherichia coli* response to pH stress
Institute of Biotechnology	Restriction-modification enzymes; screening, purification, structural organization of genes and regulation of expression; molecular mechanisms of site-specific interaction with DNA; recombinant proteins (human, virus); relation between structure and function; gene expression in *Kluyveromyces marxianus*
Institute of Biochemistry	Bacteriophage T4 gene structure and regulation of expression; differentiation dependent gene expression in *Bacillus subtilis*; identification of proto-oncogenes involved in cell proliferation and differentiation; structure of genes and regulation of expression; biosensors for determination of various metabolites
Institute of Botany	Structure and function of yeast killer systems

DNA methyltransferases, DNA polymerases, DNA and RNA ligases, reverse transcriptase, DNA topoisomerase, phosphatases and other enzymes, sets and kits for molecular biology, nucleic acids and markers, oligonucleotides, nucleotides. Products of AB Fermentas are exported to 35 countries including the USA, Germany, the United Kingdom, France and Japan. AB Fermentas, as a part of its policy to sponsor Lithuanian science, provides its products free to local scientists.

The Institute of Biotechnology and share-holding companies AB Biofa and AB Fermentas play a supportive role for the research activities of other Lithuanian institutions providing training, facilities for research and products.

Constraints

Lithuania is passing through a period of major transition, characterized by severe financial constraints. These constraints also apply within research and development in biotechnology. Low national funding and a paucity of investors who are willing and able to support research projects mean that it is barely possible to maintain an adequate standard at research institutions. Three factors in particular are likely to have a negative influence on the development of biotechnology research:

- Research personnel have been reduced considerably as a result of qualified people leaving for other jobs, emigrating, or being made redundant because of pressure of short-term economic problems.
- Equipment is often outdated, worn out or insufficient. This applies to both laboratory equipment and computer software.
- Libraries are poorly supplied with up-to-date international literature and journals.

Future Prospects

As indicated above, a national programme for the development of biotechnology has been implemented for 1997-2001. Under this programme most investigators will be involved with modern biotechnology.

The size of the country does not warrant commercial activities targeted at the domestic market. The most important areas of biotechnology for Lithuania appear to be those which have good prospects for manufacturing products for export.

In some areas, ability to compete successfully in the international market has already been proven, in particular, in research and commercialization of: a) recombinant proteins, particularly cytokines and protein hormones for medicinal use; b) techniques for gene manipulation and development of recombinant DNA technologies. These fields are competitive both commercially and scientifically. The successful activities of AB Fermentas and AB Biofa have proved their viability, and prospects seem encouraging.

Malaysia

A.H. Zakri

Introduction

The development of biotechnology in Malaysia is still in its infancy. Research and development is confined to universities and public-supported research institutions. Apart from some investment in R&D put up by the plantation-based companies, there has been until now little interest or investment shown by the private sector. Malaysia is also one of the 12 megadiversity countries in the world and it harbours a vast range of flora and fauna, some of which are endemic. This treasure house has much potential for bioprospecting. Towards this end, biotechnology is seen as an area which could enable sustainable use of this rich biological endowment.

National Policy and Regulation

General Policy

Biotechnology is one of five key enabling technologies identified by the government in its Seventh Malaysia Plan (1996-2000) for priority action. The others include advanced materials, information technology, electronics and advanced manufacturing technology. The Plan envisages that the greater use of technology and the development of a strong domestic capability are major factors in meeting the objectives of productivity-driven growth and industrial competitiveness. Therefore, increased funding for R&D will be made available to strengthen domestic capability in the above technologies. The Plan also calls for the establishment of centres of excellence within and among universities and research institutions as well as efforts to commercialize R&D activities.

In an effort to strengthen and coordinate the numerous activities in biotechnology among the scientific community in Malaysia, and eventually with a view to commercializing

R&D products of biotechnology, the government, through the National Council for Scientific Research and Development (MPKSN) established the National Biotechnology Directorate (NBD) in April 1995. The linkages between NBD and other agencies are shown in Figure 4.1. The NBD's mission statement is 'to spearhead the development of biotechnology in Malaysia through research and related activities directed at the commercialization of biotechnology and to establish Malaysia as the leading centre for biotechnology industry'. Its goals include strengthening research capacity in biotechnology, commercializing government-funded biotechnology research, and facilitating the development of biotechnology-based industry.

The NBD envisages several functions for itself, namely:

- To assess national and global development in biotechnology with a view to identifying opportunities that could be harnessed by Malaysian industry.
- To identify focal points for specific areas of biotechnology.
- To promote collaboration among institutes of higher education.
- To enhance national capabilities through training of scientists, technologists and other related personnel.
- To identify and recommend funds for scaling up R&D projects.
- To provide common facilities for scale-up and prototyping where necessary.
- To promote biotechnology within the business sector with the aim of commercializing R&D outputs and creating spin-off commercial activities.
- To establish and maintain a national directorate of scientists and companies involved in biotechnology.
- To conduct an audit of technology achievements.

The strategies to be adopted by the NBD include:

- The rapid translation of applied research into commercial application through cooperative activities among government agencies, industry and universities.
- Enhanced support of research in molecular biology.
- Promotion of the development of the essential infrastructure for biotechnology research.
- Enhancement of education and training programmes to ensure the human resources needed are adequate for the development of biotechnology R&D.

To achieve the above goals, functions and strategies the NBD will carry out three main programmes, namely:

- The Research, Development and Commercialization Programme.
- The Corporate Services Programme.
- The Communication Programme.

Funding

The main source of government funding for R&D, including in particular biotechnology, is from a financial mechanism instituted by the Government of Malaysia in 1987. The Intensified Research in Priority Areas (IRPA) initiative has been an important enabling strategy to encourage the development of science and technology in the country, with the long-range objective of turning Malaysia into an advanced industrial society by the year 2020. The IRPA programme was initiated in 1988 during the middle of the Fifth Malaysia Plan (1985-1990) and a sum of US $152 million was set aside for R&D. This amount was raised to US $235 million for the Sixth Malaysia Plan (1991-1995). For the current Seventh Malaysia Plan (1996-2000), the allocation for R&D under IRPA has been pegged at US $480 million. This is double the amount allocated in the previous Plan, but still short of

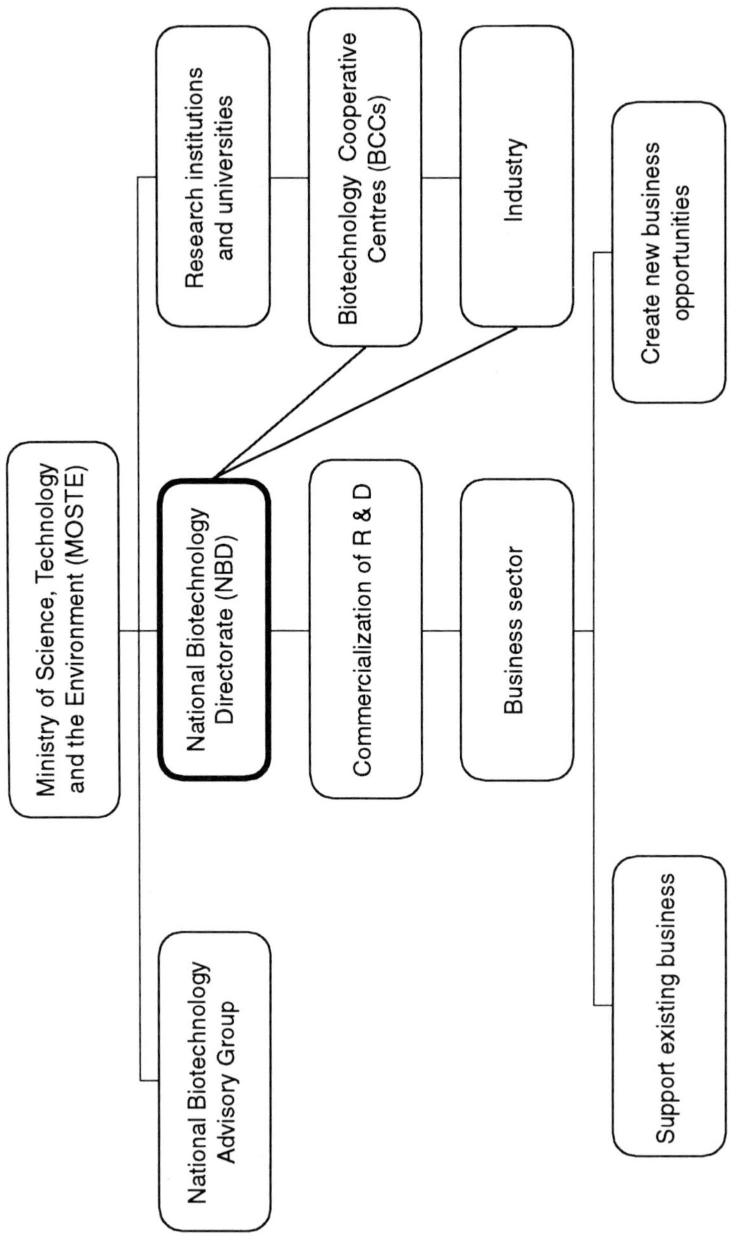

Figure 4.1. Linkages between Malaysian National Biotechnology Directorate (NBD) and other agencies.

the 1.5 % of the GDP target which the government has set itself.

The aims of the IRPA programme can be summarized as follows:

- The development of indigenous research capacity for industrial development, especially in new and emerging technologies.
- Guidance of the overall direction of research programmes of public sector research institutions.
- The creation of linkages with postgraduate education in areas of national need.
- The promotion of closer linkages between public sector research and industrial organizations.

In the Fifth and Sixth Malaysia Plans, the R&D sectors under IRPA were agriculture, industry, medical, strategic and social science. An analysis of IRPA-funded R&D under the Sixth Malaysia Plan revealed that the agricultural sector received the largest percentage of funds (46.55%), followed by industrial manufacturing-oriented research (28.19%) (Table 4.14). The strategic sector which comprises new and emerging technologies accounts for 13.36% of the allocation, and of this amount two-thirds went to biotechnology.

The large amount of funding for agriculture and/or agriculture-related disciplines does not reflect the actual intention of the government, which is to reduce the emphasis given to agriculture in favour of industrialization, but rather reflects the low absorption capacity of many of the areas, in particular those which are new and emerging. Agricultural-oriented research and agricultural-oriented institutions are able to absorb new research funds more effectively. However, the low absorptive capacity of engineering and technology is due to the low levels of research capability in these two areas across sectors and institutions.

Many of the research institutions and universities that received IRPA funding are involved in biotechnology activities. In four of the five sectors (social science is the exception), a total of US $20 million for 67 projects was handed out during the Sixth Malaysia Plan. This works out to an average of about US $304,000 per project (Table 4.15).

Apart from the IRPA funding, biotechnology activities in the Sixth Malaysia Plan are supported through the National Biotechnology Programme administered by the MOSTE. The Programme has the following objectives:

- To improve the scientific manpower base in all aspects of biotechnology through education and training.
- To stimulate and fund basic and applied research in selected fields of biotechnology.
- To encourage and facilitate the application and commercialization of research results.
- To strengthen the existing network of interaction and collaboration between local universities and institutions, the private sector and selected centres overseas.

There are six sub-programmes in the Plan, namely, molecular biology and genetic engineering, plant biotechnology, animal biotechnology, industrial and environmental biotechnology, medical biotechnology and commercialization. In the Sixth Malaysia Plan,

Table 4.14. R&D allocation: Sixth Malaysia Plan (US $ millions).

Sector	Allocation	%
Agriculture	109.5	46.5
Industry	66.3	28.2
Medical	23.9	10.2
Strategic	31.4	13.4
Social science	4.0	1.7
Total	235.1	100.0

approximately US $2.4 million was allocated to this programme to support basic and applied research in the above areas deemed relevant to the Malaysian economy.

Regulatory Structure

Biosafety

In 1996, the government endorsed the establishment of the Genetic Modification Advisory Committee (GMAC) with a mandate to deal with all issues pertaining to genetically modified organisms (GMOs), namely the use, safe transfer and handling of GMOs. The GMAC will assume an advisory role and enforcement will have to be effected through mechanisms already established such as the Plant Quarantine Act 1976 and the Food Act 1983. However, the Committee will have oversight of risk assessment and risk management and is responsible for coordinating the relevant enforcement agencies and the institutional biosafety committees to be set up by universities, research institutions and the private sector doing work on GMOs. The membership of GMAC comprises scientists, policy-makers, legal experts and NGOs. As a first step the Committee brought out a publication entitled, 'National Guidelines on the Release of Genetically Modified Organisms (GMOs) into the Environment' in January 1997. This document is based on the experience of other countries and documents published by international organizations such as the 1996 UNEP International Technical Guidelines on Safety in Biotechnology and the 1991 UNIDO Voluntary Code of Conduct for Release of Organisms into the Environment, as well as those published by national organizations such as the Australian GMAC's 1993 Guidelines for Planned Release of Genetically Manipulated Organisms.

The Committee is in the process of drafting national legislation to regulate the transfer, use and handling of GMOs in the country and this should be ready in 1997. Malaysia is a strong advocate of a legally-binding international instrument to regulate GMOs and the country is very supportive of the move to formulate a Biosafety Protocol under the Convention on Biological Diversity.

Biodiversity

As mentioned above, Malaysia is one of 12 megadiversity countries of the world. Its gene-rich status has prompted the government to take several measures to conserve biological diversity and promote sustainable use of the country's biological resources. At present, about 58% of the land area is still covered with natural forest, and there is existing legislation which regulates *in situ* conservation efforts such as national parks, virgin jungle reserves, permanent forest estates and marine parks as well as the protection of fauna such as the Wildlife Act 1972 and the Fisheries Act 1985. *Ex situ* conservation in the form of arboreta, field collections of important agricultural species, seed banks and *in vitro* gene banks are satisfactorily maintained by various government agencies, research institutions and universities.

Malaysia signed the Convention on Biological Diversity (CBD) during the Earth Summit held in Rio de Janeiro in June 1992, and subsequently ratified the treaty a year later. Malaysia is committed to its three-pronged objectives, namely conservation and sustainable use of biological diversity, and equitable sharing of benefits arising from the utilization of genetic resources. In fulfilling its obligations, the country has undertaken a national study on the status of biological diversity in order to assess and identify the gaps in conservation and management of biological diversity. As a follow-up action to this study, a National Policy on Biological Diversity has been drawn up and will be launched

Table 4.15. Distribution of IRPA grants for biotechnology research in the Sixth Malaysia Plan (1990-1995) (US $ thousands).

	Agriculture		Industry		Medicine		Strategic		Total	
Institution[a]	Number	Amount	Number	Amount	Number	Amount	Number	Amount	Number	Amount
PORIM	2	1,810	1	360					3	2,170
RRIM	2	59					1	360	3	419
SIRIM			1	1,643					1	1,643
IMR					3	232			3	232
MARDI							5	3,639	5	3,639
UM	6	648	2	840	10	1,258	5	400	23	3,146
UKM	4	2,690			1[b]		2	838	7	3,528
UPM	8	1,494			4	461	3	520	15	2,475
USM	1	1,480			3	920			4	2,400
UTM	1[b]								1	
IPH	2	753							2	753
Total	26	8,934	4	2,843	21	2,871	16	5,757	67	20,405

[a] Full names of institutions: PORIM, Palm Oil Research Institute of Malaysia; RRIM, Rubber Research Institute of Malaysia; SIRIM, Standards and Industrial Research Institute of Malaysia; IMR, Institute of Medical Research; IPH, Institut Penyelidikan Haiwan; MARDI, Malaysian Agricultural Research and Development Institute; UM, Universiti Malaya; UKM, Universiti Kebangsaan Malaysia; UPM, Universiti Putra Malaysia; USM, Universiti Sains Malaysia; UTM, Universiti Teknologi Malaysia.

[b] Amount of grant less than US $1000.

in early 1997. In this policy document, there is a reiteration of the national commitment to conserve biological diversity and to promote its sustainable use.

Whether for conservation or for sustainable use of biodiversity, biotechnology is seen as an important tool to achieve the objectives set in the national policy. Apart from being used for *ex situ* conservation efforts, biotechnology is to be used for the following:

- To assess levels of genetic variation in natural populations of forest species or germplasm collections of agricultural crops.
- To modify existing crops to include new and novel agronomic traits.
- To evaluate and utilize little-known native flora and fauna for possible biopharmaceutical products.

A number of strategic alliances are being contemplated between Malaysian universities and research institutions and foreign organizations, such as those initiated in late 1996 with the New Energy Development Organization (NEDO) and Japan Bioindustry Association (JBA) on conservation and sustainable utilization of biological diversity.

Prior to ratification of the CBD, the issue of access to genetic resources had not been fully addressed by many countries including Malaysia. There is existing legislation that regulates the removal and/or export of biological material from its *in situ* location, such as the Protection of Wildlife Act 1972, the Plant Quarantine Act 1976, the National Forestry Act 1984 and the Fisheries Act 1985. In addition, foreign researchers who intend to conduct research in Malaysia and use local resources have to obtain permission and adhere to certain conditions stipulated by the government.

However, none of the above regulations includes access to genetic resources along the lines stipulated in Article 15 of the CBD, namely that access should be on mutually agreed terms, subject to prior informed consent of the party providing such resources and that the benefits arising from use of the genetic resources should be shared on a fair and equitable basis. The Government of Malaysia is now taking the necessary steps to incorporate such elements by way of amendments of the existing legislation pertaining to biological/genetic resources.

Research, Development and Marketing

Major Projects and Institutes

Research in the field of molecular biology is rapidly developing in a number of institutes and universities. There are ongoing attempts to increase the number of research workers in molecular biology, which is currently low by international standards. A typical example would be only four or five molecular biologists working in a major research institution such as the Palm Oil Research Institute of Malaysia (PORIM), and less than this in other similar research institutions such as the Rubber Research Institute of Malaysia (RRIM) and the Malaysian Agricultural Research and Development Institute (MARDI). The facilities in these centres, however, are up-to-date, including all the modern DNA-related analytical and preparatory equipment. A greater number of researchers are found in each of the public universities carrying out research in molecular biology, namely the Universiti Kebangsaan Malaysia (UKM), the Universiti Malaya (UM), the Universiti Putra Malaysia (UPM), the Universiti Malaysia Sarawak (UNIMAS) and the Universiti Sains Malaysia (USM).

Tissue Culture research is being conducted in several universities as well as MARDI, PORIM, RRIM and the Forest Research Institute of Malaysia (FRIM). Work is focused on micropropagation of the oil palm, ornamentals and forest plantation species.

Diagnostics and Vaccine Development are being given some emphasis in areas of regional infectious diseases, particularly dengue fever, malaria, Japanese encephalitis, typhoid and hepatitis. Several of the universities and the Institute of Medical Research (IMR) are involved in this work. Research at Universiti Putra Malaysia has resulted in the development of a vaccine against Newcastle disease for chickens. The Veterinary Research Institute in Ipoh also has a wide-ranging programme on vaccine development.

Chemistry of Natural Products and their Development is being researched by a number of institutions, often in multidisciplinary groups at FRIM, MARDI, Universiti Kebangsaan Malaysia, Universiti Malaya, Universiti Putra Malaysia and Universiti Sains Malaysia. This is an area of great promise in the light of bioprospecting and taking into consideration the large resources of plants in Malaysian forests and the considerable expertise in botany, plant chemistry and drug screening, in addition to the increasing expertise in molecular biology.

Biotechnology Cooperative Centres

In August 1996 the MPKSN endorsed the proposal of the NBD to set up the Biotechnology Cooperative Centre (BCC) Programme with the following objectives:

- To manage and coordinate R&D and training programmes more efficiently by combining existing research groups from various institutions in a particular field into one strong team.
- To enhance commercialization of research by strengthening linkages between universities, research institutions and industry.
- To promote cooperation among researchers through efficient utilization of resources and to strengthen research networks.

Under the programme seven BCCs have been established in the following institutions:

1. Universiti Kebangsaan Malaysia and Universiti Malaya Molecular Biology Cooperative Centre.
2. MARDI Plant Biotechnology Cooperative Centre.
3. Universiti Putra Malaysia Animal Biotechnology Cooperative Centre.
4. Universiti Malaysia Sarawak and IMR Medical Biotechnology Research Centre.
5. Standards and Industrial Research Institute of Malaysia Environmental/Industrial Biotechnology Cooperative Centre.
6. Universiti Sains Malaysia Biopharmacy Cooperative Centre.
7. Universiti Putra and MARDI Food Biotechnology Cooperative Centre.

The emphasis of the programme is on high quality applied research relevant to the Malaysian economy although basic research such as recombinant proteins and metabolic pathway engineering will also be undertaken and strengthened. Initially, most centres will be established on existing research activities whilst new areas are being developed. One of the important facets of the BCCs is education and training in molecular biology. Each BCC is also encouraged to develop linkages with centres of excellence overseas. The NBD, as the secretariat for the BCC, is currently finalizing arrangements to establish the Malaysia-Australia Cooperative Programme in Biotechnology, with the collaboration of the Cooperative Research Centre for Biopharmaceutical Research Pty. Ltd., Australia.

In a recent development, the NBD also initiated and acts as a clearing house for the joint programme between the NEDO, the Japan Bioindustry Association (JBA) of Japan and a number of Malaysian universities and research institutions. This collaborative research venture on biological diversity comprises five areas of work, namely: biodiversity databases and gene banks; evaluation and monitoring of marine communities; screening and isolation of bioactive compounds in microorganisms and higher plants; ecosystem evaluation and inventory through advanced technology; and evaluation of therapeutic and toxic potential of natural products. The Malaysian institutions involved are Universiti Kebangsaan Malaysia, Universiti Malaysia, Universiti Putra Malaysia, Universiti Malaysia Sarawak, MARDI and FRIM while the Japanese counterparts are Ehime University, Matsuyama College of Agriculture, Rikkyo University, Kyoto University, Hokkaido University, Tokyo University, Kinki University and the Marine Biotechnology Institute, Tokyo.

Major Research Institutes

Major research institutes include:

- National Biotechnology Directorate, Ministry of Science, Technology and the Environment, Kuala Lumpur.
- Centre for Gene Analysis and Technology (CGAT), Faculty of Life Sciences, University of Kebangsaan Malaysia, Bangi, Selangor *(for further details of this centre see Appendix I).*
- Biotechnology Research Centre, MARDI, Kuala Lumpur *(for further details of this centre see Appendix I).*
- Institute of Medical Research, Jalan Pahang, Kuala Lumpur.
- PORIM, Selangor.
- RRIM, Kuala Lumpur.

Private Sector Engagement

At present, biotechnology-based industry is not an important sector in the Malaysian economy. The involvement of the private sector in R&D, in particular in genetic engineering, is negligible. However, in a recent survey five areas of relevance to biotechnology were identified. These are food processing, agriculture, aquaculture, pharmaceuticals and waste utilization.

In food processing, the activities range from traditional fermentation in brewery, bakery, soy sauce, dairy products and traditional fermented products to the production of flavour enhancer, yeast and sugars such as high fructose corn syrup, starch hydrolysate and glucose syrup. The food industry is characterized by a few big concerns, including transnational corporations, and many small and medium-size companies. Despite the fact that processing activities are on the rise, the country is very much dependent on imports which total in excess of US $1.6 billion a year with more than 70% of raw materials being imported. The industry relies heavily on foreign technology. Small and medium-size companies receive assistance from the public R&D organizations, in particular, MARDI. Large, sophisticated companies are practically on their own, with no local R&D involvement but well supported by their parent company laboratories located overseas.

In agriculture and forestry, two plantation crops still dominate the landscape; palm oil and natural rubber. The average price of crude palm oil (CPO) in 1996 was US $500 a tonne, down from the 1995 figure of US $580, while that for Standard Malaysian Rubber (SMR20) was 330 cents per kilogram compared to 350 cents before. Production of CPO is expected to remain above eight million tonnes. In 1996, production was an estimated 8.04 million tonnes, up from 7.8 million tonnes in 1995. In the case of natural rubber, although 1995's historic performance was not matched in 1996, prices were still maintained at a favourable level. Palm oil is the world's second most important vegetable oil after soybean oil, and for Malaysia it provides a livelihood to 200,000 rural families and employment to 120,000 workers on estates.

In the 1970s and 1980s there had been a lot of interest in the potential of clonal oil palm to propagate elite planting material. However, the industry has suffered setbacks in this area since abnormalities have been detected in fruits of oil palm derived from tissue culture. This is unfortunate both in the short term and long term since the industry will have to embark on a massive replanting scheme and the demand for selected planting material will be high. Apart from elite planting material, R&D institutions such as PORIM are involved in strategic biotechnology research to genetically engineer the palm oil for specific uses, namely tailoring its eating qualities to resemble those of olive oil, and also to extend its uses for industrial purposes. PORIM has strong links with the oil palm industry in Malaysia, such as Guthrie, Sime Darby, United Plantation, Golden Hope and others.

Rubber can easily be cloned by bud-grafting, so there is no impetus to embark on micropropagation for mass production of planting material. Research by conventional plant breeding has resulted in the release of a number of high-yielding rubber clones over the years. Research in rubber biotechnology in RRIM has focused on the molecular biology of the plant itself, in particular the use of RFLPs to study genetic diversity in *Hevea brasiliensis, H. camargoana* and *H. pauciflora*.

With respect to planting materials for fruit orchards and ornamental farms, their production is normally undertaken by many small nurseries which are not bound by any guidelines to maintain standards. There appears to be neither order in the market nor financial or technical barriers to entry. The use of tissue culture techniques to produce quality seedlings is still restricted to certain ornamentals (such as orchids) or fruits (such as banana). Several large plantation companies traditionally involved in the main commodity crops appear to be putting some investment into R&D of non-traditional crops, such as papaya, guava, banana and ornamentals. Apart from the local market, which is comparatively small (about US $1 million per year) and highly fragmented and disorganized, several large companies are examining the possibility of overseas marketing of superior clones of plants produced through tissue culture. The use of molecular biology and genetic engineering to modify traits in plants has yet to be attempted on a commercial scale in Malaysia. Another newcomer on the local scene is TropBio Research Sdn. Bhd. Malaysia, a pioneer biotechnology company established in October 1995 (*see Appendix I for more information*). This contract research company aims to make biotechnology expertise available for the development of Malaysia's biodiversity resources. TropBio is a joint venture with ForBio Ltd., an Australian forest biotechnology company. TropBio plans, through collaboration with Malaysian institutions and the ForBio Group, to develop technologies for identifying elite plants, creating new hybrids, introducing novel and foreign genes, and propagating plants on a massive scale using tissue culture and robotics. The species targeted by the company include rubber, teak, *Acacia*, indigenous hardwood, oil palm and other tropical plant species.

The prospect of using biotechnological tools to improve aquaculture production in Malaysia looks favourable. Currently, slightly more than 25,000 operators with a total area of 5500 ha are involved in the culture of fish and shellfish (prawns, cockles and mussels). The annual output is valued at more than US $40 million which constitutes approximately 5% of the national fishery output.

Related to aquaculture is the culture of ornamental fish. The world market for these in 1987 alone was valued at US $700 million and it is estimated that the market size is expanding at a rate of 10-15% per year. The Malaysian output was valued at US $8.4 million in 1990. Currently, the industry mainly comprises small enterprises, but a number of large operators have recently entered the scene. These large companies are attempting to penetrate the Asian market, specifically Korea, Japan and Taiwan.

The potential for expansion of the ornamental fish industry in Malaysia is high. Peninsular Malaysia has more than 250 species of freshwater fish and about 55 species of aquatic plants. Coupled with some innovation in genetic manipulation, a variety of ornamental fish could possibly be bred and introduced to the market. However, some basic practical problems need to be solved before contemplating the export market, such as reduction of the mortality rate of the fish consignment and development of improved means of packaging.

The market value of the medical pharmaceutical industry in Malaysia is estimated at US $160 million and is growing by 10% per year. The products include vaccines, antibiotics, biodiagnostic kits and hormones. Local production is limited to formulation, pelleting or repackaging while the active ingredients used are invariably imported. There are about 40 companies, a large number of which are representatives of Western drug companies. Vaccine and antibiotic production for veterinary use are estimated to be worth US $4 million and US $16 million, respectively. Hence there is scope for commercialization in these areas given the importance of veterinary services in the country.

The processing of rubber, oil palm, cocoa and coconut inevitably generates large quantities of waste materials. Emission regulations have resulted in efforts by major plantation companies to treat such waste by microbial means instead of incineration. Lately, there has been growing interest in more innovative and profitable utilization of wastes by the production of value-added products such as biofertilizer and biofeed via solid substrate fermentation, novel products, commodity chemicals and certain plant metabolites.

Constraints

In the Malaysian scenario the commercialization of biotechnology is hampered by the lack of products ready to be commercialized. Therefore, it is imperative that research in biotechnology be strengthened. It is therefore necessary to upgrade competence in biotechnology in each research institution. Ongoing research in biotechnology needs to be strengthened and intensified. It has also been recommended that the government fully support the establishment of centres of excellence in molecular biology in *all* the local universities. These centres should also serve as a platform to link the universities with industry. This will ensure the sustainability of the biotechnology industry in this country. The preceding recommendations, made several years ago, have now been adopted and elaborated with the establishment of the NBD and the launching of its BCC programme described earlier. It is expected that these new mechanisms will ensure that Malaysia's interest in biotechnology will make an important impact on the national economy.

Future Prospects

An attractiveness-strength matrix analysis was constructed based on a recent industry survey to evaluate areas of potential for biotechnology in the country. Results from the analysis are summarized in Table 4.16. It is anticipated that products acting 'external' to plants or animals are likely to be more easily produced, in the short term, than biotechnological products involving DNA manipulation which are 'internal' to plants or animals. Continuation of biotechnological research with plant, animal and human disease diagnostic techniques and the associated health enhancement treatments (vaccines for animals and humans, and micropropagated virus and disease-free plants) will lead to the production of many new or improved commercial products. Action needs to be initiated, or continued, to gain benefits in the near future from genetic resources through breeding better food crops using biotechnological tools, and from bioprocessing to give new products, create products of improved quality, and to convert wastes into safe and/or value-added products.

Areas of potential for commercialization of biotechnology in Malaysia are in:

- Agro-based industry.
- Aquaculture and ornamental fish (marine, fresh/brackish water).
- Animal and human health.

Agro-based Industry

In agro-based industry the priority areas identified for commercialization include production of elite propagation materials and breeds, enhancement of crop yield and animal productivity, bioprocessing and bioconversion of waste. *In vitro* technology, which encompasses micropropagation and production of artificial seeds, is considered relevant to production of propagation material on a commercial scale. Potential clones that are envisaged to have immediate commercial impact are oil palm, ornamentals with emphasis on native and hybrid orchids, and fruits such as papaya, banana and pineapple. Except for oil palm which already has a high internal and in-house market, ornamentals and fruits can be targeted for the export market. Development of transgenics is considered imperative as a long-term strategy to meet future challenges in this sector.

In line with the expansion of the livestock industry to meet local consumption, increased production of superior breeds of cattle, sheep and goat by means of embryo transfer technology is commercially viable.

Malaysia has to manage its existing commodity and abundant agricultural waste effectively and profitably through product enhancement as well as diversification. Current efforts to convert agricultural waste into biofertilizer and biofeed via solid substrate fermentation is deemed viable for immediate commercialization. Long-term efforts in the utilization of other agricultural residues should be channelled into spin-off of novel products, commodity chemicals and plant metabolites.

Aquaculture and Ornamental Fish

Aquaculture is becoming an increasingly profitable venture in Malaysia both for local and export market. At present marine and fresh water breeding programmes using conventional methods are adequate to meet current demand, and are profitable. However, in the long term, novel approaches utilizing biotechnology in the breeding programmes are needed to make this industry more competitive.

The ornamental fish industry is also viewed as an attractive venture grossing over US $21 million in 1989 when 80% of production was exported. This rapidly expanding industry is envisaged to be more profitable via development of 'designer fish' through biotechnology methods.

Animal and Human Health

In the field of human and animal health, the main areas of interest are therapeutics and prophylactics. All too often, diseases which are confined to the tropical regions are considered less profitable by multinational companies. There is, therefore, need to give more emphasis to diseases or threats to health which are endemic to local and regional areas. In evaluating targets for commercialization, diagnostic methods utilizing molecular

Table 4.16. Attractiveness-strength matrix of biotechnology industry in Malaysia.

Competitive position	Local strength: Strong	Local strength: Average	Local strength: Weak
High	Micropropagation: oil palm fruit spp. (papaya, banana) ornamentals forest spp.	Animal diagnostics Biofeed Biofertilizer Secondary products Vaccines Ornamental fish and aquaculture	Biotherapeutics Human diagnostics
Medium	Embryo transfer	Biopesticides Growth hormones	Novel compounds Bioconversion of wastes Plant diagnostics Transgenic plants
Low		Microbial fermentation	Transgenic animals

biology techniques and recombinant DNA should be seriously considered; these include immunological reagents, DNA probes and monoclonal antibodies.

The development of new therapeutics for animal or human health is known to be costly and requires stringent regulatory approval. It is, therefore, desirable that production of therapeutics for local consumption be made through joint venture arrangements involving established international corporations.

Current efforts in the production of whole cell vaccines for livestock should be commercialized in an effort to meet the local demand. In the long term, emphasis needs to be shifted towards construction of a polyvalent vaccine that provides immunity to a variety of pathogens. At present, vaccines for human health are all imported. As a long-term strategy, it would be better if vaccines were produced for local and regional markets through alliances such as licensing and marketing arrangements. Such ventures would need to be endorsed by, and utilized in, public health programmes in the country.

Acknowledgements

In preparing this article, the author wishes to express his appreciation to the following colleagues: Prof Abdul Latif Ibrahim, Managing Director, NBD; Prof Mohamad Noor Embi and Prof Nor Muhammad Mahadi of the Faculty of Life Sciences, UKM; Dr Jalani Sukaimi of PORIM; and Mr Mohd. Zahir Kechot of the Faculty of Economics, UKM.

Mexico

José Luis Solleiro

Introduction

Development of biotechnology in Mexico has taken place in a changing economic context. Prior to 1984, Mexico had a model of economic development driven by the objective of import substitution. Under this model, strong protection of domestic enterprises was granted in the form of high tariffs for imported goods, important incentives and subsidies, and strong State intervention in the economy. In 1984, Mexico began a drastic transformation, discarding the import-substitution model and the protected economy in favour of export promotion and participation in the global market. An important part of the new economic strategy has been based on free trade, with the expectation that market forces will foster the competitiveness of Mexican industry. This export-promotion policy was successful in the 1990s. Total exports reached US $42.6 billion in 1991, US $46.2 billion in 1992, US $51.9 billion in 1993, US $60.9 in 1994 and US $78 billion in 1995.

Nevertheless, the Mexican economy still performs in a very heterogeneous way. Of a total of almost 1.5 million companies in Mexico, 700 contribute to 80% of total exports. In general, Mexican industry lacks sound innovation and marketing strategies to be competitive in a global economy.

Mexican science and technology policy in the 1990s, after the sharp fall in expenditure on research during the 1980s, has aimed at promoting scientific achievements that meet international standards. In technology, the aim is to increase productivity, strengthen competitiveness, and favour technology that will conserve the environment and make optimum use of natural resources. With this policy, it has become apparent that the government is supporting scientific development and leaving responsibility for technological development to the private sector. This view is reinforced by the National Programme of Industrial Policy and Foreign Trade, which sets a framework of objectives for industrial development, emphasizing three strategies:

- To create favourable conditions for exports.
- To develop the domestic market as a base for import substitution and for the integration of Mexican industries in the international economy.
- To promote the integration of clusters of highly competitive industrial groups with participation of smaller companies.

Little attention is being paid to the creation of specific incentives for investment in innovative projects with the unfortunate result that only a few firms have reacted positively. The policy-makers' belief that expectations of commercial rewards would be sufficient incentive to stimulate the innovative spirit has not been substantiated.

National Policy and Regulation

General Policy

Biotechnology research in Mexico has its origins in the late 1960s and early 1970s, when important investments in development of infrastructure and research personnel were made at the main public universities. This established a solid foundation for the creation of research capacity in some public institutions. Since then, important academic achievements have been made[a] and the number of research groups dealing with biotechnology has shown steady growth (see Table 4.17). Some of these groups have international prestige and form part of global collaborative networks.

Recent government scientific policy has given strong stimulus to scientific research, training and the creation of a scientific research infrastructure. The National Science and Technology Council (CONACYT) has established a series of mechanisms based on peer review to promote high academic and technical standards in the activities it supports.

The establishment of priority areas is not contemplated and the criterion of excellence is the only one being followed for supporting projects. Supply and demand mechanisms, combined with academic evaluations (both of the research group and the proposal), have been established as the main means for selection in basic and applied research (Eastmond, *et al.*, 1992). Although the idea of developing a National Biotechnology Plan has been examined from a scientific, political and industrial standpoint, it has not led to the definition of a policy or programme specifically dedicated to the promotion of biotechnology.

Funding

Support for biotechnology is granted according to scientific demand. About a third of the members of the National System of Researchers[b] belong to the subject areas of biological, biomedical or chemical sciences. This means that 1874 researchers in these fields demand support for research. According to CONACYT's figures, Mexico's global expenditures in research and development reached US $1.95 billion in 1993. In 1995, 2.99% was allocated for agricultural research, 2.13% for research on natural resources and environment, 5.13% for research on health and 19.3% for research programmes sponsored by CONACYT (mostly basic research). These figures give only an idea of the size of the resources available for biotechnology research, given that no official numbers reflecting the exact expenditures are published. The national expenditures for biotechnology research could be estimated considering that 15%-20% of the resources managed by the above mentioned sectors are budgeted for biotechnology projects (between US $88 million and $118 million annually).

Regulatory Structure

Mexico has made major efforts to establish a suitable regulatory framework in areas affecting biotechnology development and applications. For example, it now has modern laws governing pollution control, sanitary regulation and regulation of medical research.

Biosafety

Following international recommendations, and making use of sources of international cooperation, in 1988 Mexico created a National Agricultural Biosafety Committee with the aim of supporting decision-making by the Ministry of Agriculture regarding the release of genetically modified plants in the environment. The Committee is comprised of leading scientists, government officials and representatives of producers. It works on a case-by-case basis, focusing on the characteristics and possible risks of the product, rather than the

---

[a] Mexican researchers of different academic institutions have obtained international research awards. Some others have completed successful transfers of biotechnologies like fermentation processes for the production of cattle feed, equipment for biological waste water treatment and filtration of industrial gases, and diagnostic kits for tropical diseases.

[b] The National System of Researchers is a federal programme that stimulates scientific and technological development in Mexico. It confers an increase in researchers' salaries (CONACYT, 1996).

Table 4.17. Main biotechnology research institutions in Mexico.

Institution	Main research disciplines
Centro de Ciencias de Sinaloa	Agricultural biotechnology and agroindustry; microbiology; aquaculture; soil science; ecology
Universidad de Guanajuato	Molecular biology; biochemistry; genetic engineering; microbiology; biological control of pests; agricultural biotechnology and agroindustry; tissue culture
Instituto Politécnico Nacional: Centro de Desarrollo de Productos Bióticos	Agricultural biotechnology and agroindustry; bioprocessing; biological control of pests; agroindustries
Escuela Superior de Ingenieria Quimica e Industrias Extractivas	Environmental protection
Escuela Nacional de Ciencias Biológicas	Food processing; enzyme technology
Universidad Autónoma de Nuevo León	Molecular biology; biochemistry; genetic engineering; epidemiology and diagnostics
Universidad Autónoma de Chihuahua	Agricultural biotechnology and agroindustry; cattle technology; bioprocessing; bioremediation; human health
Centro de Investigación y Estudios Avanzados (CINVESTAV):	
Mexico City Unit	Molecular biology; genetic engineering; diagnostics; parasitology; microbiology; biochemistry; neurochemistry; immunology; bioprocessing; tissue culture; agroindustries; bioremediation
Irapuato Unit	Plant molecular biology; biochemistry; plant genetic engineering; microbiology; tissue culture
Mérida Unit	Food technology; animal health; aquaculture; marine ecology; parasitology
Instituto Mexicano del Petróleo	Bioprocessing; bioremediation; genetic engineering
Colegio de Posgraduados en Ciencias Agrícolas	Agricultural biotechnology and agroindustry; microbiology; tissue culture; natural production; genetic engineering
Instituto Tecnológico de Durango	Food technology; agroindustries; mineralogy
Instituto Tecnológico de Veracruz	Enzyme technology; genetic engineering; microbiology; biological control of pests

Table 4.17. Continued.

Institution	Main research disciplines
Universidad Autónoma de Aguascalientes	Bioprocessing; biochemistry; genetic engineering; tissue culture
Instituto Tecnológico de Tehuacán	Environmental protection; bioremediation
Universidad Nacional Autónoma de México (UNAM):	
Instituto de Biotecnología	Molecular biology; bioprocessing; enzyme technology; genetic engineering; microbiology; plant biotechnology; human health
Instituto de Biología	Plant biotechnology and tissue culture
Instituto de Investigaciones Biomédicas	Bioprocessing; genetic engineering; human health
Instituto de Fisiología Celular	Molecular and cell biology; immunology; human health
Facultad de Medicina	Molecular biology; genetic engineering; immunology; gene therapy
Facultad de Química	Food technology; microbiology; tissue culture; mycology
Instituto Tecnológico y de Estudios Superiores de Monterrey-Centro de Biotecnología	Food technology; molecular biology; environmental protection; toxicology; bioremediation
Instituto Tecnológico Agropecuario	Agricultural biotechnology and agroindustry; tissue culture
Instituto Tecnológico de Tepic	Agroindustries
Instituto Tecnológico de Culiacán	Food technology; agricultural biotechnology
Universidad de Sonora	Aquaculture; soil science; waste treatment and bioremediation
Universidad Autónoma Metropolitana	Food technology; microbiology; natural production; environmental protection; bioremediation; agroindustries
Universidad Autónoma del Estado de México	Food technology and bioprocessing
Instituto Tecnológico de Celaya	Bioprocessing; process engineering
Instituto de Ecología	Agroindustries; animal production; microbiology; environmental protection; plant biotechnology; mycology; biological control of pests

Table 4.17. Continued.

Institution	Main research disciplines
Centro de Investigaciones Científicas y Educación Superior de Ensenada	Ecology; aquaculture
Centro de Investigaciones Biológicas del Noroeste	Marine biology; ecology; aquaculture; plant biotechnology; tissue culture
Centro de Investigaciones Científicas de Yucatán	Plant biotechnology; tissue culture; agroindustries
Instituto Mexicano del Seguro Social	Cell biology; molecular biology; genetic engineering; immunology; human health; tissue culture
National Institutes of Health (Secretaría de Salud)	Human health; molecular biology; immunology; diagnostics; immunology; virology; nutrition; bioprocessing; enzyme technology
Centro de Investigación en Alimentos y Desarrollo	Molecular biology; biochemistry; food technology
Centro de Investigación y Asistencia Tecnológica de Jalisco Centro Internacional para el Mejoramiento del Maíz y el Trigo	Food technology; plant biotechnology; tissue culture
Universidad Autónoma de San Luis Potosí	Biological control of pests
Instituto Nacional de Investigaciones Forestales, Agrícolas y Pecuarias	Plant biotechnology; tissue culture; molecular biology; genetic engineering; biological control of pests; animal nutrition and health; cattle technology; food technology
Centro Internacional para el Mejoramiento del Maíz y el Trigo (CIMMYT)	Molecular biology; genetic engineering; plant biotechnology; biological control of pests

process by which it was created. After gaining experience with over 40 cases, an Official Standard was issued in 1996 to regulate agricultural biosafety. Biosafety in research is also regulated by specific by-laws of the General Law of Health. These by-laws establish responsibilities and obligations to those conducting genetic engineering research. The Mexican biosafety framework still lacks provisions for the release of genetically modified microorganisms into the environment, as well as measures regarding the operation of large-scale industrial facilities using recombinant organisms.

Mexico has also developed a modern policy framework for environmental protection and sustainable development which creates important opportunities for the application of agricultural and environmental biotechnologies. A National Programme of the Environment and reforms to the General Law for Ecological Equilibrium and Environmental Protection were approved in 1996, establishing a very complete framework as well as legal and economic instruments for protecting the environment and promoting private sector participation in environmental management activities. However, economic, operative and cultural problems will have to be overcome before this framework can be applied.

An amendment to the General Law of Health made in May 1997 established the need for creation of new advisory committees on food safety and biosafety for the release of microorganisms into the environment. These committees are in the process of being formed.

Intellectual Property Rights

Protecting intellectual property rights (IPR) for biotechnology innovations has become pivotal in the promotion of research, technology transfer and investment. In June 1991, a new Mexican law for the Promotion and Protection of Industrial Property[c] was approved. It opened up the possibility of patenting most inventions related to biotechnology. In 1994, to fulfil IPR standards established by the TRIPS agreement and those of the NAFTA (North American Free Trade Agreement), the Mexican legislation was reformed. Currently, biotechnology products and processes, microorganisms, and transgenic animals are patentable. Patent protection is also available for plants containing a novel gene. In the case of new plant varieties, Plant Breeders' Rights are granted to plant breeders under the Law of Plant Varieties approved in October 1996. This Law is compatible with the 1978 Act of the International Union for the Protection of New Plant Varieties (UPOV) which Mexico joined in December 1995. Important steps have been taken for the enforcement of IPR legislations too, particularly by strengthening the infrastructure and personnel of the institutions responsible for intellectual property management.

Research, Development and Marketing

Major Projects and Institutes

The major institutes carrying out biotechnology research in Mexico and the projects under way are shown in Table 4.17. A directory of biotechnology in Mexico has been published by the Mexican Society for Biotechnology and Bioengineering (Sociedad Mexicana de Biotecnología y Bioingeniería, 1996).

Private Sector Engagement

There are no specific programmes for supporting biotechnology innovation, while support for firms wanting to develop technologies is limited to credit lines from the FIDETEC (Fideicomiso de Desarrollo Tecnologico) Trust, managed by CONACYT, and the National Financial Institution (Nacional Financiera-NAFIN) and NAFIN's own Programme for Technological Development. These credit lines do not provide special interest rates, but they support the firm's guarantees. However, the two financial programmes must channel their resources through private banks. This makes negotiating credit a real nightmare. Very few firms can satisfy the requirements imposed by the banks. Even though FIDETEC and NAFIN provide guarantees, the banks increase their demands to safeguard against loss and this drives firms away. There are no special programmes for financing industrial biotechnology projects or for providing venture capital. Furthermore, there are no tax incentives for biotechnology enterprises. In this environment, only a handful of biotechnology-based companies have been successful in developing and commercializing their products internationally. The prominent examples of such companies belong to the

---

[c] In August 1994, the Law was reformed and its name was changed to the 'Industrial Property Law'.

agro-industrial sector, where the biotechnology strategy of the PULSAR group must be underscored. A company of this Group, Agroindustrias Moderna, has implemented aggressive mechanisms to access new biotechnologies to compete in the production of vegetables, tropical fruits and seeds. The strategy combines mergers and acquisitions[d] with a small in-house R&D capacity.

From this, we may conclude that Mexican biotechnology is driven by scientific research. While an important number of scientific achievements can be shown, development of a biotechnology industry in Mexico has to follow a path full of obstacles. Thus, the number of biotechnology firms is still very small[e] having shown a very slow rate of growth over the last five years. On the other hand, three important companies recently suspended their biotechnology operations in the country (Quintero, 1994). In sharp contrast to the scientific policy, Mexican industrial policy has not considered development of bioindustry as a serious objective. Consistent with this, is the indifference of financial institutions towards the creation of biotechnology-based enterprises.

Morocco

Arifi Abdelaziz

National Policy and Regulation

Funding

The biotechnology research programme of Morocco has received funding through collaborations with a number of national and international organizations such as UNDP, FAO, USAID, Deutsche Gesellschaft für Technische Zusammenarbeit, Banque Islamique pour le Développement, International Fund for Agricultural Development, EC, Fondation Internationale pour la Science, the Third World Academy of Sciences and Agence Francophone pour l'Enseignement et la Recherche (AUPELF) as well as the public and private sector.

Regulation

Considerable progress has to be made in the area of biosafety regulation and consumer protection before a fully-integrated biotechnology programme, encompassing the entire process from laboratory research to commercial end product, can be realized.

Research, Development and Marketing

Major Projects

The research topics in biotechnology being pursued by Moroccan scientists are presented in Table 4.18. In most institutions, plant tissue culture and related aspects of biotechnology

[d] PULSAR acquired the following companies in the last three years: Asgrow Seeds, Peto Seed, Royal Sluis and DNA Plant Technologies.

[e] A recent survey conducted by CamBioTec has shown there are about 100 Mexican firms to some extent using modern biotechnology.

are the predominant areas. A few laboratories deal with molecular biology and some have just started work on genetic transformation and transgenesis.

Research focuses on more than 30 plant species, but horticultural crops are prevalent. Since 1982, research programmes have been carried out on tissue culture of banana, date palm, citrus fruits, strawberry, olive, potato, tomato and pistachio. Recently, tissue culture studies have been conducted on food legumes, medicinal plants and some arid land species.

Research in molecular biology on the date palm has recently been initiated at the National Institute of Agronomic Research (INRA) Regional Centre and the Faculty of Sciences of Marrakech University. However, advanced techniques in molecular biology and genetic engineering have yet to be employed in the improvement of the main crops in Morocco. Cereals have been the subject of intensive research programmes in several laboratories for many years, but few of these incorporate biotechnology into their programmes. A joint project under the AUPELF-UREF (Université de Reseaux d'Expression Française) programme implementing androgenesis was recently launched at IAV (Institut Agronomique et Veterinaire) Hassan II in collaboration with the Faculty of Science in Rabat University.

Biotechnology applications in industry are limited to isolated examples. At the National Centre for Sugar Cane in Kénitra, research has been carried out on the clonal selection and micropropagation of local and newly-introduced varieties of sugar cane since 1985. At the Faculties of Science in Fez and Meknès, biotechnology research on sunflower has been recently initiated.

Research in forest biotechnology is restricted to a few laboratories. Tissue culture of eucalyptus and cork oak is carried out at the Centre National de la Recherche Forestière (CNRF) in Rabat and at the Laboratory of Plant Biotechnology and Physiology at the Horticultural Complex of IAV Hassan II in Agadir a research programme on argan (*Argana spinosa*) has been underway since 1993. There is also some work being done on cedar and holm-oak.

In the area of medical biotechnology, veterinary departments at the IAV Hassan II are conducting studies in the isolation, purification and characterization of antigens as possible candidates for vaccines against leishmaniasis and hyaloma. These departments are also involved in the production of recombinant vaccines for equine plague and diagnostic tests for mycobacteria.

The Food Microbiology and Biotechnology Department at the IAV Hassan II is carrying studies on the production of industrial enzymes from hyperthermophilic bacteria.

Major Institutes

Biotechnology has emerged only recently in Morocco, so research programmes were launched first on plant tissue culture technologies and later on various aspects of animal and food biotechnology (molecular biology, genetic engineering and research work with transgenic organisms). A recent survey conducted by the National Centre for Scientific and Technical Planning indicated that a large number of research units and laboratories are carrying out projects in biotechnology. Most biotechnology R&D projects involve collaboration between at least two research institutions. Institutions involved in agricultural biotechnology come under the authority of the Ministry of Agriculture and Agricultural development (INRA, IAV Hassan II, École Nationale d'Agriculture, CNRF, Biopharma, Societé de Développement Agricole, Societé de Gestion des Terres Agricoles, Centre Technique de la Canne à Sucre) and the Ministry of Higher Education (universities). In the case of certain crops, private companies are building up their own research groups.

Table 4.18. Biotechnology research at public sector institutions in Morocco.

Institution/ laboratory	Area of research/ techniques	Species
INRA:		
Citrus Programme, Kénitra	*In vitro* selection, isozymes, molecular diagnostics, protoplast culture, protoplast fusion, somatic embryogenesis, somatic hybridization	Citrus fruits
Plant Biotechnology Research Laboratory, Marrakech	*In vitro* selection, micropropagation, protoplast culture, protoplast fusion, somatic embryogenesis	Date palm
Plant Pathology Laboratory, Marrakech	Molecular diagnostics, polymorphisms	Date palm
Cereal Programme, Rabat	Dihaploid production, molecular diagnostics	Cereals
Biotechnology Laboratory, Settat	Genetic engineering, molecular diagnostics, molecular markers, molecular diagnostics	Cereals, chickpea
Institut Agronomique et Veterinaire (IAV) Hassan II:		
Plant Biotechnology Laboratory, Horticultural Department, Rabat	*In vitro* selection, protoplast culture, protoplast fusion, somatic embryogenesis	Artichoke, eucalyptus, fig, olive, pistachio, potato
Plant Biotechnology and Physiology Laboratory, Agadir	Micropropagation, protoplast culture, somatic embryogenesis	Argan, banana, caper, carnation, potato, rose, strawberry
Molecular Biology and Genetic Engineering Laboratory, Plant Ecology Department	Molecular biology, molecular markers, transgenesis	Potato, sunflower, wheat
Plant Tissue Culture and Cytogenesis Laboratory, Agronomy and Plant Breeding Department	Dihaploid production, micropropagation, somatic embryogenesis	Cereals
École Nationale d'Agriculture, Meknès:		
Animal Production Department	Genetic engineering, immunological diagnostics, molecular biology, molecular markers	Livestock
Pomology Department	Bioenergy, isozymes, molecular diagnostics	Fig, olive

Table 4.18. Continued.

Institution/ laboratory	Area of research/ techniques	Species
Faculté des Sciences de Marrakech, Marrakech:		
Algology and Hydrophyte Laboratory	Biogas production, bioenergy, microbiology	
Plant Physiology Laboratory	*In vitro* selection, micropropagation, somatic embryogenesis	Arid zone plants, date palm
Office Regional de Mise en Valeur Agricole, Kénitra	Micropropagation	Sugar cane
Water and Forest Administration Quartier Administratif, Rabat, Chellah:		
Micropropagation Laboratory, Sidi Amira	Micropropagation	Eucalyptus, oak
National Centre of Hydrobiology	Artificial insemination, embryonic transplantation	Black bass, carp, pike, trout
Societé de Gestion des Terres Agricoles, Rabat	Micropropagation	Potato

Private Sector Engagement

Since biotechnology research is at present mainly a public sector activity carried out at universities and research institutions, linking up with the private and industrial sectors promises to be one of the most important challenges lying ahead. For the time being, the private sector relies mainly on importation of existing technologies and components (e.g. varieties of transgenic plants).

National and International Collaboration Initiatives

National and international collaboration consists mainly of training in research and development. Many of the research workers currently employed in biotechnology were trained through international collaborative research programmes. For example, research workers at IAV Hassan II were trained through a cooperation programme with the US Agency for International Development (USAID) and workers at INRA through the Mid-America International Agricultural Consortium programme.

Joint collaboration has been developed with international agencies and networks such as the ICGEB, the International Plant Biotech (IPB) Network, the African Plant Biotechnology Network, the Pan African Biotechnology Association, AUPELF-UREF and the Date Palm Network.

Future Prospects

In Morocco, as is the case for many developing countries, opportunities for application of biotechnology are addressed in terms of increased agricultural production and yield stability. Thus, future work in biotechnology will focus on the use of transgenic technology to obtain plants with desired traits such as biotic and abiotic stress tolerance, pest and disease resistance and insecticide tolerance. In parallel to these developments, integrated commodity research programmes have a favourable perspective as they could maximize the benefits of research.

Nigeria

Nduka Okafor

National Policy and Regulation

Over the past 30 years or so, there have been fitful efforts to achieve central coordination of science and technology in Nigeria, beginning with the formation of the National Council for Science and Technology in 1969, and followed by the establishment of the National Science and Technology Development Agency in 1975, the Federal Ministry of Science and Technology (FMST) in 1980, the merger of the FMST and the Ministry of Education in 1984, the re-establishment of the FMST in 1985, the 'replacement' of the FMST by the National Agency for Science and Engineering Infrastructure (NASENI) in 1992, and the final re-establishment of the FMST in 1993. The FMST as presently constituted has the following functions and mandates *inter alia*:

- Formulation, monitoring and review of the National Policy on Science and Technology.
- Promotion and administration of technology transfer programmes.
- Promotion and coordination of scientific and technological research and development activities and technological innovation in agriculture, industry, communications, electronics, medicine, pharmaceuticals, roads and construction, energy research and the basic sciences.
- Promotion and coordination of scientific and technological innovation, development, assessment, adaptation and production, including the development of scientific and technological infrastructure.
- Coordination and supervision of the research and development efforts of all the national research institutes and agencies.
- Establishing, promoting and maintaining federal-state linkages in science and technology.
- Establishing and maintaining relations with scientific, technical and technological bodies and agencies of the Organization for African Unity (OAU), the United Nations, the Economic Community of West African States, as well as other regional bodies and bilateral and multilateral arrangements, and maintaining active contributions to, and national representation in, these bodies, agencies and arrangements.

In spite of the above mandate, the FMST has not yet brought its influence to bear in the area of science coordination. This is almost certainly a result of its chequered existence and the short period of time since its most recent re-establishment in 1993. There is as yet no evidence of coordination in biotechnology, or the regulation of its practice, documented or otherwise. Nevertheless, as shown below, there has been some movement towards a national policy on biotechnology.

Although there is as yet no Nigerian government policy on biotechnology, Nigerian scientists have been working towards elaborating such a policy. Thus the 'Workshop on Biotechnology in National Development,' sponsored by the National Agency for Science and Engineering Infrastructure (NASENI) in April 1993, and attended by some 150 scientists from academia, research institutes and the private sector, made the following observations and recommendations regarding the direction of biotechnology in Nigeria:

Observations by the 'Workshop on Biotechnology for National Development'

1. Manpower in biotechnology is extremely limited in Nigeria.
2. Research activities are conducted in universities and research institutes mainly on an individual basis and without adequate resources.
3. There has been instability over the years (i.e. frequent changes) in institutional structure and policy instruments on science and technology development in the country.
4. Biotechnology is of importance in, *inter alia*, food/agriculture, the pharmaceutical and health sector, the manufacturing industries and the environment.
5. There is lack of adequate infrastructure for the National Biotechnology Programme.
6. Biotechnology is multidisciplinary in nature and also requires specialized training.
7. There is insufficient interaction between scientists, policy-makers and the private sector.
8. The government has committed 1% of the Federal Account as annual subvention to NASENI for funding its science and engineering programme.

Considering the above observations, the following resolutions were made at the Workshop:

Recommendations by the 'Workshop on Biotechnology for National Development'

1. For better coordination of the existing manpower in achieving defined national goals, there is a need for a directory of available manpower in biotechnology.
2. Since biotechnology is a modern and dynamic discipline, there is the need for specialized training of manpower required for the National Biotechnology Programme.
3. There is the need for the establishment of a national data bank and a national gene bank.
4. Adequate equipment and computer facilities should be procured for the National Biotechnology Programme. However, on a long-term basis, the workshop recommends the evolution of endogenous capability for the development, production and maintenance of equipment.
5. There should be a National Biotechnology Centre responsible for the coordination of biotechnology activities in the country, as well as regional centres of excellence in biotechnology.
6. The existing framework, i.e. NASENI, for the development of science and engineering infrastructure in the country should be allowed to become stabilized.
7. There is a need to establish biosafety regulations in biotechnology R&D and in the testing and release of GMOs to the environment. A National Biosafety Committee should be appointed to prepare and administer the regulations.
8. There is a need to form policies and regulations to protect intellectual property rights.
9. The opinion of indigenous scientific experts should be taken into consideration in decision taking.
10. There should be active interaction between the universities, research institutes and industry.

11. There should be a team approach and effective networking of research activities in order to optimize the use of funds, facilities and manpower.
12. The funding of programmes in biotechnology should be multilateral, that is by the government, the private sector, individuals and international agencies. However, the main source should be the Federal Government.
13. In order for biotechnology to enhance socioeconomic development of the country in the aforementioned areas, adequate and continuous funding would be necessary.
14. A strategic Master Plan on Biotechnology should be developed.

Funding

Since biotechnology has not been singled out for special government attention, no special budget exists for it.

Regulatory Structure

There is no special regulation on biotechnology, and thus no regulations or legislation exist at present concerning safety in biotechnology, preservation of biological diversity, or trade issues as these relate to biotechnology. However, the issues of safety and intellectual property rights are addressed in the recommendations of the Workshop mentioned above (see Recommendations 7 and 8 above). Further to this, a biosafety meeting was convened at the International Institute of Tropical Agriculture (IITA), Ibadan, in 1993 and the report of this meeting is likely to form the backbone of any safety legislation in Nigeria.

Research, Development and Marketing

Major Projects

Medical diagnostic kits, e.g. for HIV and hepatitis B, are available in the market as imports from industrialized countries. Biotechnologically modified plant materials do not seem as yet to be on the market.

Major Institutes

Institutions engaged in biotechnology research and development in Nigeria comprise:

Organizations Funded by the Nigerian Government
Several of the 25 Federally-governed, and 11 State-governed universities.
Sixteen of the 18 agricultural research institutes under the Federal Ministry of Agriculture.
Eight of the 18 research institutes under the Federal Ministry of Science and Technology.

International Intergovernmental Organizations
The International Institute of Tropical Agriculture (IITA), Ibadan. At present this is the best equipped research establishment for biotechnology in Nigeria.

Non-Governmental Organizations (NGOs)
The Foundation for African Development through International Biotechnology (FADIB).

Table 4.19. Work in biotechnology in Nigerian universities.

University	Department	Type of work/project
University of Ibadan, Ibadan	University College Hospital, Department of Zoology	Malarial parasite, *Plasmodium falciparum*
	Faculty of Veterinary Medicine	Monoclonal antibodies, PCR-related techniques
	Department of Zoology	Trypanosomes, *Trypanosoma brucei*
University of Benin, Benin City	Department of Pharmacy	Bacterial plasmids
University of Nigeria, Nsukka	Department of Botany	Plant tissue culture
Nnamdi Azikiwe University, Awka	Biological Sciences Biotechnology Unit	Plant tissue culture, DNA-related work on *Rhizobia*
Obafemi Awolowo University, Ile-Ife	Department of Chemical Engineering/ Biochemistry	Production of enzymes for industrial use

The work in biotechnology currently being carried out in the above public sector establishments is presented in Tables 4.19 to 4.22.

Biotechnology Research in Nigerian Universities
Table 4.19 shows some of the work being carried out in Nigerian universities.

Biotechnology in Research Institutes of the Federal Ministry of Agriculture
Plant tissue culture is carried out at many of the research institutes of the Federal Ministry of Agriculture whose work centres around plants (Table 4.20). At the Root Crops Research Station, radiation-induced mutagenesis of plant cell cultures is performed. Plant tissue culture work is also practised at the Cocoa Research Institute of Nigeria. The Nigerian Institute for Veterinary Research carries out work with animal tissue cultures.

Research Institutes in the Federal Ministry of Science and Technology
The biology-based research institutes in the Federal Ministry of Science and Technology are strongly biotechnology-oriented (Table 4.21). The National Institute for Medical Research in Lagos is nearly as well equipped as the IITA (see below) in terms of hardware and trained personnel for recombinant DNA research. The National Agency for Science and Engineering Infrastructure conducts a programme of workshops on biotechnology. At the Sheda Science Complex in Abuja, about half of the budget and research efforts are devoted to projects in biotechnology (the other half being given over to work in physical sciences).

Research at the International Institute of Tropical Agriculture (IITA)
The IITA is the leading establishment in biotechnology research and development in West Africa. It is an intergovernmental international organization and owing to its superior

Table 4.20. Research institutes of the Nigerian Ministry of Agriculture and their mandates.

Name of Ministry	Mandate
Cocoa Research Institute of Nigeria	Research on cocoa, kola, cashew, coffee and tea
Forestry Research Institute of Nigeria, Ibadan	Research on natural forests, plantations, wood products and wildlife
Institute for Agricultural Research, Zaria	Genetic improvement of sorghum, groundnut, cowpea, cotton and sunflower Farming systems research and extension covering north-west Nigeria
Institute of Agricultural Research and Training, Ibadan	Research on maize, kenaf, jute Soil sciences
National Institute for Freshwater Fisheries Research (formerly Kainji Lake Research Institute), New Bussa	Research on freshwater fisheries and other aquatic resources in rivers and natural and man-made lakes
Lake Chad Research Institute, Maiduguri	Variety improvement of wheat and barley Improvement of farming systems and agricultural crops in Bornu and Gongola regions
National Animal Production Research Institute, Zaria	Research on cattle, goats, sheep and poultry as well as exotic breeds and animal feeds
National Cereals Research Station, Badeggi, Niger State	Research on rice, soybeans, sesame seed and sugarcane
National Horticultural Research Institute, Ibadan	Research on fruits, vegetables, their processing and preservation Development of indigenous ornamental plants
National Root Crops Research Station, Umudike	Research on yams, cassava, cocoyams, Irish potatoes, sweet potatoes and ginger Farming systems research and extension covering the south-eastern agricultural zone
National Veterinary Research Institute, Jos	Research on livestock diseases and their control including the production of vaccines and sera
National Institute for Oceanography and Marine Research, Lagos	Research on the geography of the Nigerian ocean bed and contiguous land mass Research on marine livestock and fisheries
Nigerian Institute for Trypanosomiasis Research, Kaduna	Research on tsetse and simulium and diagnostic methods for the control of onchoceriasis and trypanosomiasis

Table 4.20. Continued.

Name of Ministry	Mandate
Nigerian Institute for Palm Oil Research, Benin City	Research on coconut, date, raffia, and ornamental palms
Nigerian Institute for Stored Products Research, Ilorin	Research on storage and preservation systems
Rubber Research Institute of Nigeria, Benin City	Research on natural rubber and its by-products

financial position compared to the public sector educational and research institutions, it is endowed with high quality equipment, library facilities and trained personnel.

The goal of the IITA is to increase the productivity of key food crops and to develop sustainable agricultural systems that can replace the traditional bush fallow, and bush-burning system of cultivation in the humid and sub-humid tropics of Africa. Crop improvement programmes focus on cassava, cowpeas, maize, plantain/banana, soybean and yams. Research findings are shared through international cooperation activities which include training and exchange of information and germplasm.

The IITA established its Biotechnology Unit in 1987 to complement ongoing research and to tackle research problems which could not be solved by more conventional methods. The biotechnology research activities of the IITA are presented in Table 4.22.

Table 4.21. Research institutes in the Federal Ministry of Science and Technology of Nigeria dealing with biotechnology.

Research institute	Mandate
Federal Institute of Industrial Research, Oshodi	Food processing, agro-allied products, textiles, pulp and paper design, fabrication of prototypes
National Centre for Genetic Engineering and Biotechnology , Ibadan	Husbanding of plant and animal genetic resources; Developmental research in genetic engineering and biotechnology
National Institute for Medical Research, Lagos	Medical research on communicable diseases, e.g. malaria, human parasites Nutritional defect problems, genetic diseases, public health
National Institute for Pharmaceutical Research and Production	Research on medicinal plants, herbs Drug development and formulation
National Agency for Science and Engineering Infrastructure (NASENI), Lagos	Planning for research and development of engineering and scientific infrastructure development complexes, including biotechnology
Sheda Science Complex, Abuja	Research in biotechnology

Table 4.22. Research activities of the International Institute of Tropical Agriculture (IITA) concerning biotechnology.

Area	Project
Molecular genetics	Molecular taxonomy/phylogeny of *Dioscorea, Manihot, Musa* and *Vigna* Molecular mapping/genomic library of *Dioscorea, Manihot, Musa* and *Vigna* RFLP and RAPD-PCR analyses to detect germplasm duplicates and interspecific crosses as well as to determine the genetic diversity of *Dioscorea, Manihot, Musa* and *Vigna*
Non-sexual gene transfer	Somatic embryo genesis and plant regeneration in cowpea, cassava, yams and plantain/banana Protoplast transformation by electroporation and polyethylene glycol-mediated methods in cowpea Gene transformation by DNA gun and *Agrobacterium* in cowpea and plantain
Cytogenetics	DNA flow cytometry in cassava, yams and plantain/banana for ploidy determination
Molecular diagnostics	Indexing of viruses, fungi and bacteria in yam, cassava, cowpea and plantain using RAPD-PCR, universal polyvirus monoclonal and polyclonal antiserum
Tissue culture	*In vitro* germplasm preservation for cassava, yams and plantain/banana Micro propagation *in vitro* for cassava, yams and plantain/banana
Wide crosses - embryo culture and rescue	Wide crosses in cassava, yams and cowpea for the introduction of novel traits Embryo culture and rescue in plantain/banana, cassava, yams and cowpea

Biotechnology at the Foundation for African Development through International Biotechnology (FADIB)

FADIB is a non-governmental organization founded in 1992 to train Africans in biotechnology, particularly genetic engineering, and to carry out research in the area. FADIB was formed in recognition of the role biotechnology, especially genetic engineering, can play in the development of African countries through improvements in agriculture, health, agro-based industries and the environment. Its aim is to have trained 1000 Africans by the year 2002. FADIB organizes short training courses in genetic engineering with funds obtained from the international community, and had organized seven workshops and symposia at the time of writing (November 1996). The introduction of desirable properties into crops by genetic engineering is a major research initiative of the Foundation.

FADIB has members in about 34 countries, ten of which are non-African. The Foundation's headquarters are located in Awka in Anambra State, at the planned location of its laboratories and conference centre, but it currently operates from Enugu.

Private Sector Engagement

Private sector involvement in biotechnology may come from the food processing, brewing and pharmaceutical industries. The major food processing industries and breweries have foreign counterparts; in the past, all R&D work was conducted in the foreign laboratories of these companies. Recently, however, following public demand that these firms should carry out R&D activities in Nigeria, many food processing companies and breweries have now started local R&D efforts. However, recombinant technologies do not feature in the work of these companies, even though scope for its use exists. In the case of the pharmaceutical industry, local companies are simply engaged in the packaging of basic drugs manufactured outside the country and therefore do not engage in biotechnology.

Constraints

Problems for the rapid development and exploitation of biotechnology in Nigeria have been identified by the Workshop on Biotechnology for National Development as recorded above. These problems include policy instability in the governmental organization of science, under-funding, lack of trained manpower and lack of coordination and direction from the government, among others.

Pakistan

S. Riazuddin

National Policy and Regulation

Pakistan has a slowly developing industrial sector and therefore relies mostly on an agricultural economy. The government encourages cutting edge research in biotechnology, though the country's financial situation is such that it cannot afford to spend the required amount on biotechnology or any R&D initiative. Nevertheless, some institutes do receive adequate funding from governmental and non-governmental sources and it is these that are making significant progress in biotechnology research and product development. Notable among these are the National Centre of Excellence in Molecular Biology (NCEMB) established in 1985 and the National Institute of Biotechnology and Genetic Engineering (NIBGE), established in 1992. NCEMB's main strength is plant transformation for the development of transgenic crop varieties, whereas NIBGE specializes in biofertilizers, biomass, bioleaching and nitrogen fixation.

Regulation

The regulatory situation for biotechnology R&D is in its infancy in Pakistan. Legal protection for intellectual property is rather weak. Pakistani research institutes do follow international biosafety guidelines such as those passed by the US National Institutes of Health, National Biosafety Guidelines. Rules to govern and supervise research or products involving genetically modified organisms are in preparation. The government has formulated a high-level federal committee entrusted with the job of framing a national set of guidelines for biosafety.

Research, Development and Marketing

Major Projects

Currently promising investment opportunities in biotechnology-related products in Pakistan include the following:

- Biofertilizer and biopesticide production.
- Transgenic plants of agricultural importance harbouring genes conferring resistance to various crop pests or bacterial, fungal or viral infection.
- Microbial fermentation of lignocellulosic biomass for production of fuel (ethanol) using genetically engineered microbes.
- Genetically engineered microbes or transgenic plants which can biodegrade pollutants in soil, natural water reserves, sewage and industrial effluents.

Major Institutes

The major government initiatives in biotechnology research in Pakistan include the National Centre of Excellence in Molecular Biology, Lahore, the National Institute of Biotechnology and Genetic Engineering, Faisalabad and the Biomedical and Genetic Engineering Division, A.Q. Khan Labs, Islamabad.

National Centre of Excellence in Molecular Biology (NCEMB) (see also Appendix I)
NCEMB is housed on 58 acres of suburban land in Lahore, Pakistan. It was founded in 1985 and now comprises state-of-the-art labs and modern equipment for carrying out cutting-edge research in molecular biology and biotechnology. The premises consist of four major research wings, an administration block, a seminar room, a library and a guest house for visiting scientists, besides acres of land for field testing of transgenic crop varieties. A staff of 106 persons including 31 senior scientists, 41 MPhil and PhD research scholars, nine technicians and 66 administrative and technical staff are employed. It has a total annual budget of US $300,000.

The major fields of interest at NCEMB are plant transformation, use of *Bacillus thuringiensis* crystal proteins against common crop pests, plant pathogenic fungi and their control, restriction endonucleases, DNA-based diagnosis of infectious and heritable diseases, hepatitis B immunology and forensic DNA typing. The major research projects being studied at NCEMB include plant transformation for the development of transgenic Basmati rice varieties, genetically engineered cotton cultivars resistant to lepidopteral insects, fungus and pest resistant transgenic chickpea, search for new strains of *B. thuringiensis* with novel pesticidal activities, Bt-resistance mechanism in pest populations, Bt-formulation as a pesticidal spray, prenatal diagnosis of ß-thalassaemia, diagnosis of tuberculosis pathogen, therapeutic monoclonal antibodies against Asian isolates of hepatitis B virus (for treatment of neonates born to infected mothers), establishment of a database of allele frequencies of some forensically important genetic markers in four sub-populations of metropolitan Lahore, search for novel restriction endonucleases and large-scale production of commonly used restriction endonucleases, DNA vectors and Taq DNA polymerase. NCEMB also offers customized synthesis of oligodeoxyribonucleotides (oligos), recombinant Taq DNA polymerase, T7 phage DNA, lambda DNA and various cloning vectors.

NCEMB offers state-of-the-art biotechnology and molecular biology research facilities including a biolistic gun for plant transformation, plant tissue culture facilities, transgenic plant testing growth chambers, greenhouses and testing fields, a DNA synthesizer, several programmable thermal cyclers, several bench fermentors, biohazard safety cabinets, medium and ultracentrifuges, HPLC, FPLC and conventional ion exchange and affinity column chromatography equipment, the latest two-dimensional PAGE and isoelectric focusing equipment, manual and automatic DNA sequencing apparatus, gyratory shakers, incubators as well as computerized bibliographic databases (Reference Update and MEDLINE), subscription to 30 major scientific journals by its library and remote access searching of GenBank and Genome Data Base.

Within Pakistan, NCEMB has set up a research collaboration with the Biomedical and Genetic Engineering Division of the A.Q. Khan Labs in Islamabad to pursue mapping of recessive gene loci affecting hereditary deafness in humans via short tandem repeat genotype analysis of large pedigrees. The Centre is also collaborating with the Nuclear Institute for Agriculture and Biology, Faisalabad and the Central Cotton Research Institute, Multan for the large-scale field testing and breeding of transgenic plants.

On the international scene, NCEMB has ongoing intensive collaboration with the University of Washington, Seattle, USA, and New England Biolabs, Boston, USA; Laboratorium Genetika, University of Ghent, Belgium; Instituto di Genetica Biochimica ed Evoluzionistica, Pavia, Italy; and Dipartimento di Genetica e Microbiologia Universita di Pavia, Italy. Programmes at the Centre are supported by the Rockefeller Foundation, the International Atomic Energy Agency and the Asian Development Bank.

The NCEMB publishes research articles in such major scientific journals as *Nucleic Acids Research, Plant Tissue Culture, Plant Pathology, Plant Cell Reports* and *International Chickpea Newsletter.*

A.Q. Khan Research Labs, Biomedical and Genetic Engineering Division

The Biomedical and Genetic Engineering Division of A.Q. Khan Research Laboratories is a government funded group of laboratories, and there is no contribution towards the R&D efforts of this division by the private sector.

The following research interests and projects are being pursued at the centre:

- Human eye diseases.
- Investigation of the genetic basis of autosomal recessive non-syndromic deafness.
- Linkage analysis of genes involved in mental retardation, infertility, cataract, Ushers syndrome, glaucoma and other inherited diseases.
- PCR-based diagnosis of viral and bacterial diseases in Pakistan.
- Use of a combination of HLA typing and microsatellite mapping to determine the genetic diversity amongst the different ethnic populations of Pakistan.
- Participation in the HLA workshops. The centre is involved in typing 100-150 individuals from selected populations of Pakistan for the HLA Class 1 antigens.
- Participation in the Human Genome Diversity project. The centre is studying the genetic diversity of the various unique populations within Pakistan by the application of microsatellite mapping.
- Typing of male individuals from selected Pakistani sub-populations for Y-chromosome polymorphisms.
- Production of monoclonal antibodies of clinical importance against human peripheral

blood mononuclear cells surface antigens and identification of antibodies against markers unique to the Pakistani population. This project is being conducted in collaboration with Prof. Leonard Herzenberg of the Dept. of Genetics, Beckman Centre, Stanford University, Stanford, California, USA.
- Cytotoxicity screening of different biological products with an aim to isolate biologically active compounds for their antineoplastic activity.

National Institute of Biotechnology and Genetic Engineering (NIBGE) (see also Appendix I)

This institute was set up in 1992 and formally inaugurated by the President of Pakistan in 1994. The six main research divisions of the NIBGE are:

1. Plant biotechnology.
2. Biofertilizers.
3. Biofuels.
4. Basic biology (for health-related research).
5. Environmental biotechnology.
6. Biotechnology of minerals and fossil fuels.

In the plant biotechnology division, the major emphasis has been on the cotton crop. One research group in the plant biotechnology division is engaged in the development of virus resistant cotton through genetic engineering. In addition, a cotton genome project has been initiated with the objective of identifying various genes of agronomic importance. The work on cotton biotechnology is funded by the Ministry of Food and Agriculture through a loan from the Asian Development Bank. In addition, a tripartite project with the John Innes Centre, Norwich, UK and the University of Arizona, USA has also been awarded by the Common Fund for Commodities through the International Cotton Advisory Committee, Washington, USA.

The research being conducted on biofertilizers has reached the commercialization stage. The inoculum production technology has been optimized and is ready for scaling up. Work in the biofuels division has also reached a stage where scaling up of the process for microbial conversion of biomass into methane and alcohol has become feasible. Selection of potent strains capable of hyperproduction of cellulase, hemicellulase, amylase and citric acid has been carried out with concomitant studies of enzyme kinetics.

In the basic biology division, the health-related biotechnology being conducted has revolved around PCR and DNA probe-based diagnostic methods for the detection of infectious diseases. Such diagnostic tests for tuberculosis and hepatitis C have now been developed and will shortly be available to physicians. Facilities and expertise are also available for prenatal diagnosis of genetic disorders. Another project in progress in this division is one concerning optimization of expression of a bacterial gene conferring salt tolerance in transgenic alfalfa plants.

In environmental biotechnology, processes based on microbial detoxification of industrial effluents have been developed and an active collaboration with the private sector has been established. Presently, the emphasis is on textile and leather industry effluents. In addition, contract research is being carried out for a pharmaceutical company for the removal of phenol toxicity from their effluent. NIBGE has developed a battery of microbiological tests for determining the toxicity and mutagenicity of a range of effluents.

In mineral and fossil fuel biotechnology, research work on the recovery of uranium by microbial leaching of low grade uranium ores is considerably advanced. Similarly, work on copper bioleaching has been carried out under contract from the Saindhak project.

Another viable project is microbial desulphurization of indigenous coal which is of high sulphur content. In laboratory experiments, 80-90% sulphur has been removed by microbial methods.

In order to further the commercial exploitation of technologies developed at NIBGE, a biotechnology company named Pakistan Innovative Biotechnology Services (PIBS) has been established as a subsidiary of the Al-Technique Corporation of the Pakistan Atomic Energy Commission. PIBS is entrusted with the task of marketing the technology and services/expertise available at NIBGE.

The NIBGE laboratories have state-of-the-art equipment including a capillary electrophoresis based automated DNA sequencer (Perkin Elmer), fermentors, a DNA synthesizer, FPLC, HPLC, high speed centrifuges, gel imaging system, etc. The library subscribes to about 30 scientific journals as well as Current Contents on diskette. A local area network is being established. E-mail facilities and Internet access are available. NIBGE is an affiliate centre of the International Centre for Genetic Engineering and Biotechnology. NIBGE is formally associated with the Quaid-e-Azam University, Islamabad, Pakistan for the conferring of MPhil and PhD degrees in Biotechnology.

Private Sector Engagement

Lack of political support and the presence of a unique socioeconomic set-up in the country has resulted in very little investment in biotechnology joint ventures by partnership between government biotechnology institutes and the private sector. However, there has been some notable progress. The NCEMB has established successful collaboration and cooperation with New England Biolabs, USA, and is in the process of establishing a private company, Molecular Technologies, that will market products developed by NCEMB. The initial focus will be on the production of DNA-based kits for the diagnosis of highly infectious or heritable diseases (tuberculosis, hepatitis, and ß-thalassaemia). The National Institute for Biotechnology and Genetic Engineering, Faisalabad, has also recently created a commercial venture for marketing biotechnology products.

Constraints

The major constraints on progress in biotechnology in Pakistan are as follows:

- Insufficient government funds.
- Insufficient funding from sources other than own government.
- Difficulty in obtaining necessary equipment.
- Difficulty in obtaining necessary chemicals, reagents.
- Poor maintenance or after-sales service of laboratory equipment.
- Lack of biotechnology-related academic courses in order to develop human resources in biotechnology.
- Lack of personnel specifically trained in laboratory techniques.
- Unique socioeconomic structure leading to lack of investment by the private sector.

Philippines

Virginia G. Novenario-Enriquez

National Policy and Regulation

General Policy

In the Philippines, the crucial role of biotechnology for the global competitiveness and sustainable development of the country has been recognized by government agencies at the highest level, and special reference has been made to biotechnology under the Medium-Term Plan designed by the National Economic Development Authority.

The Office of the President has moved to promote biotechnology as a major sector in the country's economic, scientific and technological development. Upon Presidential instruction, an Export Strategy for Biotechnology was drawn up in 1995 as a national blueprint to harness the commercial potential of biotechnology. Intending initially to focus on the local market, the plan aims to build the Philippine biotechnology industry as an export winner, with target revenues of not less than US $100 million by the year 2000. Among the cornerstones of this national export strategy are the forging of strategic alliances with multinational companies, and the institution of a comprehensive programme of fiscal incentives. The Export Development Council included biotechnology in the country's Omnibus Investments Incentives Act.

Furthermore, in order to rationalize and strengthen the biotechnology research and development (R&D) infrastructure, Presidential Proclamation No. 526 (signed 30 January 1995) constitutes a Network of National Institutes of Biotechnology (NNIB) from four units of the University of the Philippines System. The Network is expected to focus on agricultural, medical, industrial, and environmental applications of biotechnology. A key NNIB member is the National Institute of Molecular Biology and Biotechnology (BIOTECH), first established in 1979 as a dedicated R&D institution based at the University of the Philippines in Los Baños, Laguna. Proclamation No. 526 also directs the Department of Science and Technology (DOST) and the National Economic Development Authority to mobilize donations and external funding for the Network.

As the central agency mandated to direct and supervise national science and technology efforts, DOST, an executive body directly under the Philippine President, has been at the forefront of developing and harnessing national potential in biotechnology. In policy-making and programme planning, DOST is guided by the Science and Technology Coordinating Council, a multi-sectoral body chaired by the DOST Secretary.

DOST has identified biotechnology (along with information technology and materials science) as a strategic tool for global competitiveness, commanding top priority for DOST investments in human resources development, infrastructure improvement, and R&D funding.

Funding

Lack of adequate and/or sustained funding has historically plagued the Philippines' R&D sector, biotechnology included. In 1996, for instance, the DOST's institutional R&D budget totalled US $19.63 million and has averaged a mere 0.20% of GNP in the last five years, despite policy-makers' avowed support for science and technology. This low level of R&D funding is provided on a annual basis, subject to approval of the General Appropriations Act, and released according to a quarterly schedule, a situation which is not conducive to effective R&D planning and implementation.

Given the financial constraints of developing countries, international/multilateral sources present themselves as alternative funding sources. In the case of the Philippines, however, foreign funding has not been significant. This may be due to the fact that the priorities and interests of foreign and local parties often do not coincide. Furthermore, support from multilateral agencies usually necessitates some counterpart funding, which may not be within the capacity of the requesting party at the time of need.

R&D funding for biotechnology and other S&T (science and technology) fields is provided and managed mainly by the Department of Science and Technology, although the Departments of Agriculture and Health also provide some funding. Although these and other chartered public institutions exercise fiscal autonomy to fund and undertake R&D projects, the DOST has been granted authority to review R&D budgets since 1995 as provided in the country's General Appropriations Act. This R&D budget review function is performed by DOST through an Inter-Council Committee composed of representatives from its five Sectoral Councils: The Philippine Council for Agriculture, Forestry and Natural Resources Research and Development; the Philippine Council for Aquaculture and Marine Resources Research and Development; the Philippine Council for Health Research and Development (PCHRD); the Philippine Council for Industry and Energy Research and Development; and the Philippine Council for Advanced Science and Technology Research and Development (PCASTRD). The Councils fund biotechnology R&D projects consistent with national priorities, according to their institutional mandates.

Overall, however, the Philippine Council for Advanced Science and Technology Research and Development (PCASTRD), created in 1987 by Executive Order No. 128, serves as the country's leading agency in biotechnology. Its mandate is 'to develop, integrate, and coordinate the national research systems for advanced S&T and related fields', including biotechnology. The PCASTRD hosts the Biotechnology Technical Panel of the Science and Technology Coordinating Council, which is responsible for periodically developing a Biotechnology Sectoral Plan. In pursuit of its mandate, PCASTRD administers, coordinates, and monitors the national R&D programme in biotechnology as outlined in the Biotechnology Sectoral Plan.

Exact figures for national biotechnology R&D expenditure are hard to evaluate, given the absence of a comprehensive R&D information system and the diversity of autonomous funding sources. DOST Councils are major funding sources; however, R&D institutes also receive direct R&D funding through their institutional budgets provided in the General Appropriations Act. Furthermore, these various government agencies can tap both foreign and private-sector support.

Data from key institutions nevertheless indicate that R&D expenditure in 1996 totalled not less than US \$3.431 million, of which US \$0.719 million came from international research or development agencies. These foreign sources included the Asian Vegetable Research Centre, the Australian Centre for International Agriculture, Queensland Institute of Medical Research, the Rockefeller Institute, the International Atomic Energy Agency, the United States Agency for International Aid, and the World Health Organization.

Regulatory Framework

Biosafety

The Philippines was one of the first Asian countries to implement biosafety guidelines. Originally drafted in 1987, the guidelines underwent a series of consultations with the private and public sector before their adoption in 1991. The guidelines cover activities, such as importation, research, development, introduction, field release, and production or manufacture of biological materials, particularly where genetic manipulation is involved, or where there is importation of exotic microorganisms, plants, or animals.

Implementation of the guidelines is overseen by the National Committee on Biosafety of the Philippines (NCBP) which is chaired by the DOST Undersecretary for Research and Development. As provided by Executive Order (EO) 430 of 1990 creating the NCBP, its membership consists of the following: a biologist, an environmental scientist, a physical scientist, a social scientist, two respected members of the community, and a representative each from the departments of Agriculture, Health, and Environment and Natural Resources. The NCBP guidelines provide for the creation of an Institutional Biosafety Committee, and the appointment of a Biological Safety Officer, in institutions engaged in research or production activities involving genetic engineering. Under EO 430, the NCBP is likewise empowered to coordinate with national regulatory bodies over any violation of its guidelines. The NCBP is mandated to meet once a month, or as decided by the chair on urgent matters.

Currently, the 1991 biosafety guidelines are being revised. The approved 'Guidelines on Planned Release of Genetically Modified Organisms (GMOs) and Potentially Harmful Exotic Species (PHES)' were published in May 1998.

Intellectual Property Rights (IPR)

A new Intellectual Property Rights Code of the Philippines was enacted by Congress on 6 June 1996 and signed by the President. The law is in compliance with the Philippines' commitment to the minimum standards set by the Agreement on Trade Related Aspects of Intellectual Property Rights (TRIPS). The Code adopted *in toto* the TRIPS definition of intellectual property rights. For patents, the first-to-invent rule was abandoned in favour of the first-to-file rule with patent protection extending 20 years from the date of filing of patent application.

Under Section 22.4 of the new IPR Code, patent protection of life-forms is now allowed but is limited to 'microorganisms and non-biological and microbiological processes'. Excluded are 'plant varieties or animal breeds or essentially biological processes for the production of plants or animals'. However, the section expressly declares that the exclusion of plant varieties or animal breeds from patent protection does not preclude the enactment by Congress of a law providing *sui generis* protection for these biological resources. Moreover, Section 22.4 refers to the enactment of a law on 'community intellectual rights protection' as a measure to respond to the equity demands of source communities, particularly indigenous communities. The Philippines subscribes to the Budapest Treaty on the International Recognition of the Deposit of Microorganisms for the Purposes of Patent Procedure.

The Intellectual Property Office under the Department of Trade and Industry is the regulatory body on IPR in the Philippines.

Biodiversity

As with biosafety guidelines, the Philippines preceded other countries in having formulated a regulatory framework on bioprospecting. Executive Order 247 'Prescribing Guidelines and Establishing a Regulatory Framework for the Prospecting of Biological and Genetic Resources, Their By-products and Derivatives For Scientific and Commercial Purposes, and for Other Purposes' was approved by President Ramos on 18 May 1995. Its implementing rules and regulations were signed by the Secretary of the Department of Environment and Natural Resources (DENR) on 21 June 1996.

Executive Order 247 (EO 247) covers prospecting of all biological and genetic resources in the public domain, including natural organisms growing on private land, intended to be utilized by both foreign and local entities, whether government or private. It also encompasses all activities aimed at discovering, exploring, or using these resources for pharmaceutical development and agricultural and commercial applications, excluding traditional uses.

Under EO 247, sites for bioprospecting can include protected areas and ancestral lands, domain and local communities (with prior informed consent). The terms for bioprospecting activities are to be defined in either an Academic Research or Commercial Research Agreement drawn up with the Philippine government.

The implementing body for EO 247 is comprised of an Inter-Agency Committee on Biological and Genetic Resources (IACBGR) whose Technical Secretariat is chaired by the director of DENR's Protected Areas and Wildlife Bureau. Members of IACBGR include representatives of the line departments of Agriculture, Foreign Affairs, Health, Science and Technology; the Philippine National Museum; the Philippine scientific community (academia); and non-governmental/people's organizations.

Research, Development and Marketing

Biotechnology Infrastructure

The Philippines' infrastructure for biotechnology is public-sector based. It comprises mainly the newly-created Network of National Institutes of Biotechnology (NNIB) based at the University of the Philippines (UP) and the agencies of the Department of Science and Technology. A number of R&D centres affiliated with the departments of Agriculture and of Health, and a few other state-run universities, also engage in biotechnology R&D.

The UP-based NNIB provides the primary backbone for biotechnology training and R&D. Aside from undertaking R&D, UP Diliman in Quezon City conducts baccalaureate and graduate (MSc and PhD) degree programmes in molecular biology and biotechnology. There are other UP Diliman units implementing biotechnology R&D projects at the Marine Science Institute and the Natural Science Research Institute. At UP Los Banos in Laguna province, the BIOTECH Institute serves as a core R&D centre undertaking agro-industrial, plant and animal health, and environmental biotechnology R&D. Other UP Los Banos units significantly involved in biotechnology R&D are the Institute of Plant Breeding, the National Crop Protection Centre, the Institute of Biological Sciences, and the College of Forestry. Meanwhile, UP Manila concentrates on health biotechnology R&D. UP Visayas located in the island province of Iloilo, though still to become fully operational as a biotechnology R&D centre, is programmed for aquatic and marine biotechnology R&D.

Other public-sector R&D centres with significant involvement in biotechnology are the Research Institute for Tropical Medicine under the Department of Health, the Department of Agriculture's Bureau of Plant Industry, the Philippine Coconut Authority, and state universities like Benguet State University and Central Luzon State University.

Biotechnology Sectoral Plan

The Biotechnology Sectoral Plan (1995-2000), the second biotechnology midterm plan, underscores the national need for world-class products and services which can compete as export winners and/or which meet basic domestic needs. It aims to increase national productivity and global competitiveness, protect and conserve the environment and biodiversity, accelerate human resource and infrastructure development for biotechnology, and promote the role of science in national development.

Three main programmes are considered basic to the attainment of the objectives of the Biotechnology Sectoral Plan: capability-building in terms of acquisition of new technologies and human resource development, research and development, and public education and advocacy.

Attention is focused on selected areas of application through R&D programmes in leading institutions and the advocacy of so-called 'vanguard projects' under the Biotechnology Sectoral Plan. Under the new Biotechnology Sectoral Plan, R&D for agricultural, industrial, and health applications are to be prioritized as vanguard projects for attaining global competitiveness.

Agricultural Biotechnology

The development through genetic engineering of superior trees and crops has been identified as midterm R&D priority in agriculture. A project on the development of genome maps and genetic markers in coconut and mango using molecular marker technologies is also being funded under this R&D programme. The programme is currently implemented by PCASTRD through the UP Institute of Plant Breeding. In addition to this, the agricultural biotechnology R&D initiative is substantiated through the institutional programmes of several UP Los Banos-based institutes (mainly the National Institute of Molecular Biology and Biotechnology (BIOTECH), the Institute of Plant Breeding, and the National Crop Protection Centre).

Four R&D programmes are based at BIOTECH, which was set up in 1979. These are Biotechnology for Agriculture and Forestry; the Foods, Feeds and Speciality Products Biotechnology Programme; the Environmental Biotechnology Programme; and Biotechnology for Health (Plant and Animal).

The Agriculture and Forestry Programme focuses on:

- Development of mixed inoculant and biofertilizer technology for selected crops like field, forage and pasture legumes.
- Evaluation of microbial-based fertilizers for rice, corn, fruit crops, and forest species.
- Novel propagation methods for coconut, ornamentals, medicinal plants, and selected trees.
- Bio-utilization of coconut water.
- Crop tolerance to environmental stress.

In the Foods, Feeds and Speciality Products Programme, efforts centre on:

- Development of PCR-based detection kits and monoclonal antibody-based technique for selected microbial contaminants/toxins (including red tide toxin) in foods, feeds and water.
- Production of selected enzymes using local microbial strains.
- Production of fats from tropical oils, and other industrially-useful bioproducts like rennet and vitamin B12.
- Bioprocessing of agro-industrial wastes for feeds.

The Environmental Programme concentrates on:

- Development of biotraps and microbial mats for bioremediation of heavy metals and recalcitrant organics.
- Deodourization and decolouration of agro-industrial wastes.
- Studies of microorganisms tolerant to metals and industrial wastes.
- Utilization of agro-industrial waste products (for example, for the production of biodegradable plastics).

The Plant and Animal Health Programme explores:

- Prevention and control of pests and diseases using vaccines and monoclonal antibodies in economically important crops like mango, banana and papaya and in fowl and livestock.
- Production of secondary metabolites as feed additives or plant growth enhancers.

BIOTECH's institutional budget for R&D totalled some US $577,000 in 1996, and was supplemented a further US $385,000 by external funding from DOST and foreign sources.

The Plant Biotechnology Programme of the Institute of Plant Breeding (IPB)(UP Los Banos), established in 1975, is concentrated on important dry-land crops, such as fruits, vegetables, and cereals. Research areas include development and application of *in vitro* culture technology; genome mapping and identification and use of molecular markers for crops and pests for marker-assisted selection, determination of genetic variability, and identification of crop varieties; cloning of important genes; and other genetic engineering applications. Ongoing R&D projects concern crops of economic value such as vegetables (e.g. mungbean and other legumes, cabbage, pechay, shallot, cucurbits), fruits (e.g. tomato, banana, papaya), coconut, corn, abaca, bamboo, ornamentals, and root crops. In 1996, IPB's local R&D funding amounted to US $170,000, while foreign support totalled US $172,000.

At the National Crop Protection Centre, set up in 1982, R&D programmes are directed toward the biological control of pests and diseases of rice, corn, vegetables, fibre crops and ornamentals. In 1996, the Centre's R&D expenditures amounted to US $313,000.

Industrial Biotechnology

In industry, the focus is the development of industrial chemicals over a six-year period. One of the lead agencies in this area is DOST's Industrial Technology Development Institute (ITDI), which has been undertaking biotechnology R&D using protoplast fusion since 1982 and genetic engineering work since 1988. ITDI's institutional research programmes have been in the following areas; cassava-based agro-industrial chemicals, mushroom culture, microbial degradation and remediation (pollution control), strain improvement, and food and health care. In 1997, an R&D programme on biodegradable plastics was started.

The Institute's agrochemicals programme is working on pilot-scale production of cyclodextrin, citric acid, high-fructose syrup, lactic acid, starch-hydrolysing enzymes, and microfungal mycelium (using agro-industrial wastes). The mushroom programme is aimed at pilot-scale production of heat-tolerant *Agaricus* species and production in submerged culture. The microbial degradation and remediation programme is targeted at solid municipal wastes (organic wastes) including liquid effluents. In the area of strain improvement, research is focused on mushrooms, ornamentals, and microorganisms for the pulp and paper industry. The food and health care research programme is directed towards bench-scale production of food colours. ITDI's R&D spending in 1996 for maintenance and operating expenses alone totalled US $248,000.

Health Biotechnology

In the health sector, the emphasis is on development of diagnostic methods for tropical diseases. The health biotechnology R&D programme, administered by DOST's Philippine Council for Health Research and Development (PCHRD), is aimed at capacity-building for production of vaccines and other biological products, and basic research and clinical immunological studies. One major project (initially granted US $1.35 million by DOST and the Department of Health) is the establishment of current good manufacturing practice and pilot plant facilities for R&D of recombinant vaccines and other biologicals. A related PCHRD project implemented by the Research Institute for Tropical Medicine and the National Institute of Health (UP Manila) concerns development of a vaccine against *Schistosoma japonicum*. These major R&D projects are complemented by a host of other studies for the development of diagnostic tools/reagents and other biological products for effective prevention and control of major public-health diseases. At the National Institute of Health (NIH) (UP Manila), R&D is geared towards development and/or improvement of safer and more effective vaccines for prevention of leading infectious diseases in the Philippines; simple, rapid, highly sensitive and specific immunodiagnostic tests for field use; biological control methods of animal vectors of disease; and production of antisera for diseases requiring immunotherapy. The NIH's annual R&D budget in 1995 amounted to US $192,000 which is supplemented by DOST grants and foreign funding. The PCHRD is also administering a project for development of a DNA analysis laboratory in forensics, and significant government investments have been channelled into training of health-related S&T personnel through graduate scholarships or short-term training abroad.

Other Areas

Aside from these priority areas, certain DOST institutes conduct research on biotechnology applications in forestry management and forest products processing, food processing, fibre/textile processing and assessment of nuclear techniques. The UP Marine Science Institute also undertakes research on marine natural products and their biochemistry, mariculture, marine environment biology and restoration ecology.

Major Institutes

Biotechnology R&D is becoming increasingly dynamic and is conducted mostly in public-sector research centres and state-run colleges and universities. These currently host more than 302 PhD and 346 MSc level personnel specialized in the biosciences alone. Greater numbers of highly-trained S&T workers are employed in the public sector, such as those with expertise in chemistry or engineering. Many of these individuals are equally qualified for the interdisciplinary field of biotechnology R&D.

In the Philippines, the following count among the major biotechnology R&D centres:

- National Institute of Molecular Biology and Biotechnology (BIOTECH) at University of the Philippines (UP) Los Banos.
- National Institute of Health at UP Manila.
- National Institute of Molecular Biology and Biotechnology at UP Diliman.
- Institute of Plant Breeding, UP Los Banos.
- National Crop Protection Centre, UP Los Banos.
- Industrial Technology Development Institute, Department of Science and Technology.

- Research Institute for Tropical Medicine, Department of Health.
- Natural Science Research Institute, UP Diliman.

Private Sector Involvement

As a matter of policy and procedure, DOST involves the private sector in science and technology planning and R&D evaluation. The Science and Technology Coordinating Council, the highest policy-making body in science and technology, has business/industry leaders as members. Private sector representatives likewise serve on the DOST Councils' various Technical Panels. These spearhead the formulation of sectoral plans, and evaluate R&D project proposals/outputs on a regular basis. Similarly, private sector involvement is possible through R&D co-funding, contract research, or technology transfer agreements.

Despite these various opportunities, private sector involvement in the Philippines in biotechnology R&D or commercialization has been minimal. At present, a local company is involved in the commercialization of an antibiotic sensitivity disc developed through a project with DOST's Philippine Council for Health Research and Development. Likewise, three small-to-medium size enterprises have entered into tie-ups with BIOTECH-UP Los Banos for the commercialization of four areas of agricultural biotechnology. The four areas concern animal health and nutrition: L-lysine as feed additive, probiotics production and evaluation, subunit vaccine for haemorrhagic septicaemia in cattle and carabao, and protein-enriched pineapple pulp and peeling.

In September 1996, private sector involvement in biotechnology took a new turn with the organization of the Biotechnology Association of the Philippines Inc. (BAPI), which enjoyed the active support of the UNIDO Philippine Country Office, DOST's Philippine Council for Advanced Science and Technology Research and Development, BIOTECH of UP Los Banos, and the Department of Trade and Industry. With the country's goal of making biotechnology an export winner by the year 2000, BAPI is expected to take the lead role in this endeavour.

In Asia, a 1994 study of the Asian Productivity Organization has indicated that the Philippines stands at a comparative advantage to its neighbours. This position of strength has been attributed to several factors, including availability of highly qualified personnel, adequate education and training, and relatively well-developed R&D infrastructure, aside from other advantages such as the country's rich diversity of flora and fauna and its huge annual accumulation of biomass. Furthermore, the Philippines' biotechnology-related policies are well in place. On the other hand, certain drawbacks have also been recognized in these and other areas, including funding, regulations, private-sector involvement, international/multilateral linkages and technology transfer.

Constraints

Among developing countries, the Philippines has distinguished itself by the high-level advocacy for biotechnology development. Morever, the country is amongst the first to promulgate up-to-date policies and laws on biosafety, bioprospecting, and intellectual property rights. However, these policies have often met with obstacles in the implementation phase, due to operational constraints and lack of coordination. This is particularly true in R&D, where critical-level funding is not always assured or where efforts are often duplicated. It is also true in biosafety and IPR, where line personnel are inadequately trained or where institutions are not fully equipped to substantiate the spirit of the guidelines.

Furthermore, biotechnology policies still need to be expanded and coordinated to maximize benefits in terms of human resources and investment. For example, institutional mandates need to be reviewed and revised to allow certain agencies to really regulate or else implement biotechnology projects extending to commercialization. At the same time, harmonization is necessary so that public institutions do not work at cross-purposes. A comprehensive approach to policy formulation and implementation needs to be pursued.

Academic and Technical Infrastructure

The country's academic and technical infrastructure has been judged as generally adequate. Training and education in biology and biotechnology up to the PhD level are up to par, particularly with the recent upgrading of human and laboratory resources in centres of excellence under DOST's engineering and science education project. However, some institutional deficiencies are felt in certain basic sciences like immunology, virology, cell physiology and taxonomy and application-specific areas, such as in drugs and vaccine development, production and quality control, biodiversity assessment and management, and bioengineering. The recent rationalization, by presidential fiat, of biotechnology R&D institutes into the National Network of Biotechnology Institutes should further enhance the academic and technical infrastructure in biotechnology.

However, the regulatory agencies, and the associated support agencies of the National Committee on Biosafety of the Philippines (Quarantine Office, Bureau of Food and Drugs, institutional biosafety committees) and the Intellectual Property Office are lagging behind in capability and resources. Since biotechnology is a very dynamic field, R&D institutions, educational centres, and policy-making bodies need to be constantly updated on developments in the discipline. This poses a big challenge to a struggling economy like the Philippines.

Human Resources

Data compiled by DOST's Philippine Council for Advanced Science and Technology Research and Development (from its network of 16 academic and R&D institutions) show that, as of March 1998, there were 302 PhD and 346 MSc degree holders in the biosciences, representing more than the critical number of qualified personnel for needed biotechnology research, development and education. Among these, many have been exposed to advanced techniques abroad, particularly with DOST's current overseas education and training support programme.

Although these personnel represent a broad base of expertise, their optimal use as a critical resource is not realized. In most instances, their attention is divided between teaching and research. Meanwhile, many researchers still have to learn to conceptualize and undertake market-driven research projects rather than those according to their area of interest or subject specialization. Moreover, it is desirable for them to develop a culture of interdisciplinary collaboration. In addition to technical areas of biotechnology, training of personnel has to be undertaken in biotechnology management, bioinformatics, bioethics, science and technology (S&T) studies, impact assessment, etc.

With the national leadership's advocacy of biotechnology development as an export industry, the availability of well-trained S&T personnel, and the extent of existing R&D infrastructure for various biotechnology applications, Philippine biotechnology has excellent prospects as an investment area.

Through DOST's implementation in 1992 of the US $85 million Engineering Science Education Project under the auspices of the World Bank and the Overseas Economic Cooperation

Fund, the Philippines' personnel capability and infrastructure in the advanced S&T fields, including biotechnology, are being strengthened to world-class standards. Major seeding investments are continually being achieved, for example in overseas and local scholarships (PhD and MSc level), laboratory upgrading, and journal and book acquisitions under DOST's priority-setting scheme for vanguard areas towards global competitiveness. Even trade promotion and investment policies and programmes have been tailored to prime the biotechnology sector. An Omnibus Investment Code, passed in 1991, provides fiscal incentives for R&D projects. These incentives consist of income tax holidays (for a maximum of eight years), duty-free importation of breeding stocks and genetic materials, tax credits on domestic breeding stocks and genetic materials, employment of foreign nationals, etc.

Future Prospects

In the study of biotechnology in developing countries made by the Tokyo-based Asian Productivity Organization in 1994, the Philippines was ranked third in research, particularly in the areas of agriculture/livestock and food/food processing (topping the list were China and Korea). Microbial fertilizers are specifically identified as the Philippines' area of superiority, with additional strengths in biopesticides, animal vaccines, diagnostic tools for plant and animal diseases, and food fermentation. Commercialization of agricultural biotechnology, such as biofertilizers and biocontrol agents for various pests and diseases, is also a very promising area. Backed by the Philippines' internationally-renowned expertise in agricultural research, such technology has the advantage of being tailored for use in the wide Asian market of small farmers and producers. This market is now growing conscious of environmental concerns and the challenges of a globalizing economy.

More areas of biotechnology, mostly for agricultural applications, are ripe for wide-scale commercialization. From BIOTECH-UP Los Banos, the list includes four biofertilizers recommended for various crops and trees and a growth hormone for ornamentals and vegetables. More projects need only to be taken beyond the pilot scale through infusion of adequate post-R&D funding for wider trials or test production. These include industrial enzymes like amylases, proteases, and ligninase; diagnostic kits for some plant and animal diseases; biocides/larvicides; and bioremediation technologies for livestock and industrial wastes. Fromthe National Crop Protection Centre, nuclear polyhedrosis viruses and Metarhizium, Trichogramma and Diadegma technologies for pest/insect control in vegetables are ready for either commercial-scale testing or production.

Given the government's recent seeding investments for a vaccine production project, coupled with its institutional health biotechnology R&D and human resource development programmes, medical biotechnology may prove to be an attractive investment area, with the booming population of the tropical/developing world as a target market.

Another possible investment area, largely unexplored but nevertheless highly attractive, is marine biotechnology. The Philippines has a wealth of marine resources with immense potential for the development of both natural and biotechnological products. While institutional foundations (e.g. the UP Marine Science Institute, the UP Visayas) and highly qualified core R&D personnel exist, a critical level of financing is needed to systematically and effectively exploit these natural resources.

Bioprospecting in general offers similar enticements to investors. Strategically located in the equatorial belt where East meets West, the Philippines is blessed with a rich diversity of tropical flora and fauna, assets for development of both natural and biotechnology products. In this area, government resources have been directed towards systematic collection and cataloguing of genetic resources. The establishment of a national strain bank has been recently funded by DOST's Philippine Council for Advanced Science and Technology Research, and is intended to expand culture collection centres in the country as well as systematize existing microbial collections held at BIOTECH in UP Los Banos. The National Museum and the Institute of Plant Breeding, the National Herbarium and the Plant Genetic Resources Laboratory provide facilities which, although not yet adequate, represent good starting collections for bioprospecting.

Poland

Tomasz Twardowski, Andrzej B. Legocki and Andrzej Zabza

National Policy and Regulation

General Policy

In 1989, Poland started its transformation to a free market economy. Political and bureaucratic upheavals are a dominant feature of Poland's political landscape today: they affect all aspects of everyday life. Poland is undergoing a period of transition, and society, as well Government, is dominated by very important problems such as privatization, inflation and unemployment, OECD and NATO memberships, application to EU, etc. These problems are widely considered top priorities for the country. In terms of the speed of restructuring processes, Poland (together with Hungary and Czech Republic) is considered the leader in Central Europe. Biotechnology as a modern discipline started in Poland in the late 1980s. The Biotechnology Committee at the Polish Academy of Sciences was established in 1987. Transition from a centrally planned to a free market economy has dominated the government agenda since 1989. As a result, debate on a national biotechnology policy receives little attention.

Funding

Investment in science is insufficient, amounting to just 0.5% of GNP on an annual basis; three persons in a thousand are involved in research and development (data for 1997).

Regulatory Structure

Before 1980, it was only the scientific establishment that registered an interest in legislation related to biotechnology. Prior to 1990, regulations were adopted on the American model, though these were not legally enforceable. After 1990, Poland's membership in the EU required acceptance of European legislation (with regard to biotechnology, EC Directives 90/219 and 90/220). Since 1990, international conventions have influenced national legislation strongly - in particular, the Convention on Biological Diversity (1995), membership of OECD (1996), the Budapest Treaty, and protection of intellectual property rights (1993).

In the 1990s, the first legal acts related to biotechnology were passed, with ratification of the Convention on Biodiversity (1995) and introduction of patent law (1993) being the most significant.

Over the last three years, there has been an intention to implement a new Polish law, which would be a 'national copy' of the EU Directives, drafted with the assistance of international organizations (OECD, UNEP, UNIDO). In November 1997, a team of experts submitted a proposal for a Polish 'Gene Law' to the government; the proposal is based on Directives 90/219 and 90/220.

The first field experiments with transgenic plants (potato, corn and beet) were performed in 1997, in Poland. In 1998, more than 20 field experiments were carried out. All the field experiments are performed with the permission of the Minister of Agriculture, and under the strict supervision of experts. A new environmental protection law, complying with the Biodiversity Convention, was passed by Parliament and signed by the President (September 1997). The Minister of Agriculture, in cooperation with the Minister of Environment Protection, the State Committee for Scientific Research and the Minister of Health, appointed an Expert Committee for Genetically Modified Organisms (GMOs) (July 1996) and a procedure for applications concerning the introduction of GMOs to the environment (February 1997). The aims of the Committee are:

- To draw up national regulations.
- To evaluate applications concerning the release of GMOs.

A new law concerning the protection of the environment was passed by Parliament on 1 January 1998. In the new law, a separate paragraph is dedicated to biotechnology (establishment of the competent authority for granting permits for GMO releases, trade and GMO labelling).

Polish patent law is compatible with the rules of the European Union. The issue of intellectual property rights over natural genetic resources is yet to be decided.

Research, Development and Marketing

Major Projects

At present Poland does not have a modern biotechnology industry, although 'traditional' biotechnology industries (antibiotics and fermentation, for example) are well organized. Modern biotechnology is still too poorly established to perform independent research. The Polish Academy of Sciences has set up several scientific committees, including the Biotechnology Committee. The committee has set priorities in the fields of agrobiotechnology, medical diagnostics and pharmaceutical products.

Major Institutes

The State Committee for Scientific Research has issued a list ranking the best institutes. In the case of biotechnology, the following research institutes of the Polish Academy of Sciences are considered leaders: The Institutes of Experimental Biotechnology, and of Biochemistry and Biophysics (Warsaw), the Institute of Bioorganic Chemistry (Poznañ), the Centre for Molecular and Macromolecular Research (Lodz), and the Institute of Immunology (Wroclaw) (*see Appendix I for further information on these centres*).

As far as education in biotechnology at university level is concerned, several programmes at the largest Polish universities were introduced, and biotechnology faculties established (in Poznañ, Lublin, Lodz, Gdañsk, Wroclaw, Warsaw). It is estimated that about 500 students graduate in biotechnology (including food and agrobiotechnology) as a first degree every year. About 1000 scientists and researchers are engaged in biotechnology-oriented studies.

Future Prospects

Future prospects for biotechnology in Poland are connected with agrobiotechnology, health industry and services. These major directions of development are directly related with the sustainable development of biotechnology as covered by the Convention on Biological Diversity, with non-governmental organizations emphasizing consideration of socioeconomic factors prior to the development of new biotechnologies.

Russian Federation

K.G. Skryabin and L.P. Matyash

National Policy and Regulation

General Policy

The Government of the Russian Federation recognizes the strategic role of biotechnology in achieving economic growth. A federal programme, 'Living Systems Technologies', was approved by the Government in 1996. The following subsidiary R&D programmes are closely related to biology and biotechnology:

The subsidiary programme *Advanced Biological Engineering Methods* is aimed at accelerated development and effective utilization of advanced bioengineering methods in various areas of health care, agriculture and industry. Genetic engineering will be used to produce substances involved in regulatory and protective functions in humans and animals, such as interferons, insulin, interleukins, vaccines, growth hormones, and neuropeptides, and to promote the development of new and effective methods for diagnosis, treatment and prevention of the most widespread diseases. The programme aims at the development of novel forms of transgenic high-yield crops resistant to various diseases or transgenic animals of high productivity, as well as the use of transgenic plants and animals as 'live bioreactors'. Enzyme and protein engineering research focuses on the development of new biocatalytic processes for production of chemicals, while new bioengineering methods will be applied in mineral and trace element leaching, in biosensors and in environmental protection.

The subsidiary project *Human Genome* aims at the complete decoding of the molecular structure of the human genome and the creation of new methods for diagnosis, prevention and treatment of various human diseases. Within the framework of the joint Human Genome Programme, with the collaboration of the Engelhardt Institute of Molecular Biology, Russian Academy of Sciences (RAS) and Argonne National Laboratory (USA), a novel technology for production and application of biological microchips is being developed. Oligonucleotide microchips have already been applied, for performing a simple, fast and reliable DNA sequence analysis, for diagnostics of some genetic diseases, in gene polymorphism studies, and for the identification of microorganisms in environmental samples.

The subsidiary programmes *High-efficiency Food Production Processes* and *Perspective Processes of Agricultural Production* are aimed at the development and introduction of new ecologically safe technologies for agriculture and food production. The programmes include such aspects as increasing soil fertility, development of integrated plant protection, breeding of improved animal stocks, development of new technologies for production of food and fodder proteins with utilization of non-traditional kinds of vegetative materials, and new methods and equipment to reduce losses of agricultural produce in transportation and long-term storage.

The subsidiary programmes dealing with the development of new drugs and biomedical and healthcare applications include R&D in recombinant and hybridoma technologies.

The aim of these programmes is to develop and introduce new effective methods of prevention, diagnosis and treatment of the most widespread and intractable diseases such as atherosclerosis, cancer, influenza, hepatitis, diabetes, alcoholism, narcotics abuse and AIDS.

There is a series of culture collections: the All-Russian Collection of Microorganisms at the Institute of Biochemistry and Physiology of Microorganisms (Pushchino, Moscow), the All-Russian Collection of Industrial Microorganisms at the Institute of Genetics and Selection of Industrial Microorganisms (Moscow), the Culture Collection of the Biotechnological Centre, St. Petersburg University and other culture collections (including animal and plant cell lines) at various institutes.

Basic studies in molecular biology, biochemistry and biotechnology are supported by the Russian Foundation for Basic Research. The Ministry of Sciences and Technology of the Russian Federation includes a Department of Life Sciences which coordinates funding for the promotion of life science research and biotechnology in particular.

Other ministries supporting biotechnology-related work are:

- The Ministry of Health Care.
- The Ministry of Agriculture and Food.
- The State Committee on Protection of the Environment.

Regulatory Structure

The first guidelines for research involving recombinant DNA molecules were elaborated in the former USSR in 1978 when recombinant DNA technology was basically a research tool. Ten years later these guidelines were revised and adopted in their new form. At the present time, the focus of public concern has shifted from regulation of recombinant DNA research to the nature and scope of the Government's role in regulating industrial development to minimize environmental damage and risk to humans. In 1993, the Ministry of Science and Technology of the Russian Federation established a National Committee for the Elaboration of Legislation for Work with Genetically Modified Microorganisms. This Committee included representatives of various interests (scientists, senior officials of the Ministries of Agriculture and Ecology). In order to develop regulation in cooperation with the international community and compatible with international practice, UNIDO, at Russia's request, organized an International Meeting in October 1993 in Moscow, where biotechnology safety and regulatory issues were discussed with a group of international experts. The Informational Centre of COBIOTECH, housed at the Centre of Bioengineering of the Russian Academy of Sciences, was subsequently designated as the official node of the UNIDO Biosafety Information Network and Advisory Service (BINAS).

The Federal Act of the Russian Federation on State Regulation of Genetic Engineering Activity was finally adopted by the State Duma in June 1996 and signed by the President. This Federal Act establishes regulations governing utilization of genetic resources, protection of the environment and assurance of environmental safety arising in connection with genetic engineering activity. The Act does not consider the application of genetic engineering techniques to human beings or to tissues or cells in the human organism. In March 1997, the Government of the Russian Federation established the Inter-Agency Commission on Genetic Engineering Activity.

Research, Development and Marketing

Russia has considerable potential for creation of a pharmaceutical industry based on modern biotechnology. Unfortunately, the recent economic crisis has significantly affected Russian industry, the pharmaceutical industry in particular: drastic budget cuts, lack of funds, the need to re-equip and other problems put obstacles in the way of the transfer of technology from laboratory to industry.

Despite severe financial constraints, a number of significant results have been achieved. There are a few genetically engineered drugs on the Russian pharmaceutical market:

- Reaferon: Genetically engineered human α-2-interferon was approved by Pharmcommittee in 1990 for treatment of diseases related to human disorders (for example, auto immune disease and cancer) and different viral diseases.
- Interleukin-1-β (betaleukin): A recombinant human interleukin-1-β, this is a universal immunomodulator for the correction of various immunopathologies.
- Hepatitis B vaccine: A recombinant yeast hepatitis B vaccine.

In addition, the following genetically engineered potential medicines are under clinical or preclinical trials:

- Somatogen: A genetically engineered human somatotropic growth hormone aimed at pituitary dysfunction treatment.
- Gammaferon: A genetically engineered human g-interferon, an immunomodulator with antiviral and anticancer effects.
- Betaferon: A recombinant human β-interferon - a broad spectrum antiviral drug - appears to protect AIDS patients against opportunistic infections. Betaferon appears to be well tolerated in patients.
- Refnolin: A tumour necrosis factor - genetically engineered immune-mediator of potential use against cancer.

Still under development are genetically engineered human proteins to be used as potential medicines: interleukin-1 and -2, human granulocyte colony stimulating factor, erythropoietin, thymosin-1-α, immunotoxins and a series of vaccines (hepatitis A, C; tick-borne encephalitis, diphtheria, whooping-cough).

The expansion of modern biotechnology into the field of large-scale chemical production is proceeding slowly. One successful example appears to be the process of biocatalytic hydrolysis of acrylonitrile to acrylamide. The advantages of the process are the simplicity of equipment, energy efficiency, and comparative environmental friendliness. In 1995, the process was implemented in three plants within the Russian Federation.

In agriculture, development of commercial transgenic crop varieties has received a boost from the adoption and implementation of the State Regulation of Genetic Engineering Activity. The first transgenic crops including potato, developed by Monsanto for resistance to the Colorado beetle, and the RAS Bioengineering Centre for resistance to potato virus Y, are in the final stage of registration in the All-Russian State Commission for Testing and Protection of Selection Achievements. Monsanto and AgrEvo also perform field trials on six transgenic crop plants including potato, soybean, beet and maize.

Another area of quite successful application of biotechnology in Russian industry is biogeotechnology. The development of this new technology is based on the use of microorganisms which enable enhanced extraction of oil from wells. Application of this technology has allowed the pumping out of more than 200,000 tonnes of additional oil over the last few years. New biotechnological methods have been successfully applied for leaching low-grade ores and concentrating such metals as copper and gold.

Efforts in environmental biotechnology focus on the selection or construction of the most effective strains of microorganisms - biodegraders - for remediation of soils and water reservoirs polluted with oil or different toxic substances, for purification of sewage, etc. In line with the International Chemical Weapons Convention (1993), the development of environmentally safe technology for the destruction of chemical weapons has become a pressing problem. One successful example is the ecologically safe destruction of lewisite and mustard gas, which has been achieved by electrochemical pretreatment of their neutralization products followed by biological oxidation of the effluent.

Private Sector Engagement

There are almost 100 pharmaceutical plants in Russia of which 49 plants were privatized by the beginning of 1996. The main producer of pharmaceuticals in Russia is the joint-stock company 'Biopreparat'. This company has under development (preclinical and clinical trials) more than 100 new processes for the production of recombinant interferons, interleukins, erythropoietin, hormones, vaccines and new diagnostics. However, funds for commercial development are insufficient, and the following are prioritized: production of genetically engineered human insulin, genetically engineered hepatitis B vaccine, peroral forms of immunostimulators based on genetically engineered interferons and cytokines.

Major Institutes

Basic research in biology and biotechnology is carried out in such institutes as:

Russian Academy of Sciences (RAS)

- The Institute of Bioorganic Chemistry, Moscow.
- The Centre of Bioengineering, Moscow.
- The Institute of Molecular Biology, Moscow.
- The Institute of Gene Biology, Moscow.
- The Institute of Molecular Genetics, Moscow.
- The Institute of Microbiology, Moscow.
- The Institute of Plant Physiology, Moscow.
- The Institute of Biochemistry and Physiology of Microorganisms, Pushchino.
- The Institute of Protein Research, Pushchino.
- The Institute of Cytology and Genetics, Novosibirsk.
- The Novosibirsk Institute of Bioorganic Chemistry, Novosibirsk.

Within the Russian Academy of Sciences, the Scientific Council for Biotechnology was set up in 1980 and the Interdepartmental Commission on Recombinant DNA in 1988.

Russian Academy of Agricultural Sciences

- The Institute of Agricultural Biotechnology, Moscow.
- The Institute of Agricultural Microbiology, St. Petersburg.
- The Institute of Phytopathology, Odintsovo.
- The Scientific Industrial Biotechnology Centre in Animal Husbandry, Dubrovitsy.

The Russian Academy of Medical Sciences
- The Institute of Biomedical Chemistry, Moscow.
- The Institute of Nutrition, Moscow.
- The Institute of Experimental Medicine, St. Petersburg.

The Ministry of Health Care
- The State Research Centre of Virology and Biotechnology 'Vector', Koltsovo, Novosibirsk region.
- The State Research Centre of Genetics and Selection of Industrial Microorganisms, Moscow.
- The State Research Centre for Applied Microbiology, Obolensk.
- The Research Centre of Molecular Diagnostics and Therapy, Moscow.

Ministry of Education
- Moscow State University/Department of Biology, Department of Chemistry, International Biotechnological Centre, Immunotek, Moscow.
- The Biotechnological Centre of St. Petersburg State University, St. Petersburg.
- The Moscow State Academy of Applied Biotechnology, Moscow.
- The Moscow State Academy of Fine Chemical Technology, Moscow.

Future Prospects and Constraints

The future of biotechnology in Russia depends foremost on the economic situation of the country. Having a considerable potential for biological and genetic research leading to the creation of a new bioindustry, Russia still has opportunities to overcome the severe constraints on biotechnology development.

The realization of the federal programmes needs much more financial support and investment than the Government has budgeted for over the last few years. The economic crisis in Russia has brought about a reduction in the budget for all scientific activities including biotechnology. The Russian Foundation for Basic Research can support only a limited number of research groups because of shortage of funds. Many young and well-trained scientists leave the country looking for better conditions for research in the USA and Western Europe. Many laboratories need re-equipping in order to introduce new techniques.

South Africa

Muffy Koch and Jocelyn Webster

National Policy and Regulation

General Policy

Given the diversity of South African society, extensive negotiation is required in order to find common ground and working solutions for nation building between the developed and the underdeveloped. At present, policies are being addressed and the role of science and technology is just one of many areas of discussion. In all policies, the guiding principle is the government's Reconstruction and Development Programme, which focuses on the basic needs of people and the redressing of past inequalities.

The distribution of the benefits of science, engineering and technology (SET) in South Africa has been the subject of much controversy in the past. The current need is to mobilize the power of SET to generate the innovations needed to bridge the development gap and improve quality of life for all South Africans. To this end, the establishment of a government ministry of Arts, Culture, Science and Technology is a promising sign of commitment to technology. Science, engineering and technology hold the key to meeting basic needs (health, adequate nutrition, housing, a strong economy) and it is in this framework that biotechnology must function and grow in South Africa.

Biotechnology will need to concentrate its role on raising the living standards of developing communities, while also stimulating the worldwide competitiveness of South African industries. To achieve this, situation and needs analyses are necessary to ensure a focused strategy for biotechnology innovation. The biotechnology community has initiated this process (CSIR-FRD AMI, 1993; Directory of Biotechnology in South Africa, 1995; Webster and Koch, 1995). Implementation requires strong government policies to create a climate in which biotechnology business can grow.

Funding

The R&D expenditure in South Africa is comparatively low. There has been a decline in government R&D spending from 1.04% of the GDP in 1987 to 0.75% in 1993 and to 0.68% in 1995. This trend does not bode well for South Africa's future world competitiveness. Spending on non-medical biotechnology R&D was approximately US $4 million in 1994. The largest sectors in non-medical biotechnology are agriculture and food, but most of the food research is second generation biotechnology.

Regulatory Structure

Biotechnology developments in South Africa are supported by the existence of legal instruments covering patent rights and biosafety, which are being updated and improved as the technology develops. In recognition of the potential importance of biotechnology to the economy of the country, it is intended that legislation should provide a constructive, rather than excessively restrictive, framework for the safe application of the technology.

Biosafety

Modern biotechnology was introduced into South Africa in the 1970s, when the technology was still in its infancy. In 1978, the South African Committee for Genetic Experimentation (SAGENE) was created as an advisory body concerned with laboratory containment of recombinant DNA experiments, approval of projects, training courses and laboratory standards. Since 1989, SAGENE has been mandated to act as an advisory body to government departments, scientists, industry and the public on issues of environmental release of genetically modified organisms (GMOs). SAGENE deals with genetic modification of microorganisms, plants and animals. From 1990, SAGENE has reviewed 100 applications resulting in over 200 field trials on 12 crops. During this period four scale-up trials, three industrial and one medical, were also applied for. Although constituted by a notice in the Government Gazette, SAGENE has no legal authority and all applications are submitted on a voluntary basis. Import of GMOs is covered through a permit system operated by the government's Department of Agriculture, which seeks SAGENE's recommendations on all applications to import GMOs. Containment conditions attached to the import permit are legally binding.

A number of activities relating to GMOs are already controlled in terms of existing legislation. In recognition of the need for additional integrated legislation which applies specifically to GMOs, a government interdepartmental Working Group drafted legislation by means of which a new statutory committee with executive powers will be formed. The new committee will act as the Competent Authority in all matters concerning the use and release of GMOs. The new legislation deals specifically with the development, production, use and application of GMOs and includes applications of human gene therapy, whilst harmonizing with existing legislation. The GMO Act (No. 15 of 1997) was passed on 23 May 1997 and will be implemented in 1999. The Act will be administered by a Registrar for GMOs, whose duties will include:

- Registration of all national facilities working with GMOs.
- Ensuring review of all applications received for trial and commercial release of GMOs.
- Monitoring all imports and exports of GMOs.
- Inspection of GMO facilities and trials.
- Negotiation of South Africa's position on regional and international initiatives regarding GMOs.
- Review and updating of the regulations and guidelines supporting the GMO Act.

Review of applications and regulations will be carried out by a Scientific Advisory Committee, made up of not more than ten scientists from various specified fields.

Intellectual Property Rights

Patent rights in South Africa are protected by the Patents Act No. 57 of 1978. The Act specifically provides that a patent shall not be granted for any variety of animal or plant or any essentially biological process for the production of animals or plants, not being a microbiological process or the product of such a process. The South African courts have not yet pronounced upon this section of the Act and thus there is no case law in South Africa to guide a patentee as to whether products such as genetically engineered plants and animals are patentable or not. However, experts in patent law believe that such products, as well as truly microbiological products, are patentable in South Africa, and international companies believe that with South Africa's proven patent law process, their biotechnology-related property rights will be protected.

Biodiversity

South Africa has signed the Convention on Biological Diversity and is actively contributing to the development of the Biosafety Protocol. Transboundary movement of GMOs is already monitored and South Africa is a member of the Regional Biosafety Focal Point that aims to build biosafety capacity and cooperation between the 13 member countries in east and southern Africa. (These member countries are Botswana, Kenya, Lesotho, Malawi, Mauritius, Mozambique, Namibia, South Africa, Swaziland, Tanzania, Uganda, Zambia and Zimbabwe. Ethiopia participates, but has not officially joined.) The Dutch funding for the RBFP has finished, and the future of this valuable, active and effective regional biosafety focal point looks uncertain.

Research, Development and Marketing

Major Projects and Institutes

South Africa has three large national research institutes with established biotechnology R&D:

- The Agricultural Research Council (ARC).
- The CSIR (previously the Council for Scientific and Industrial Research).
- The Medical Research Council (MRC).

The ARC is primarily responsible for applications addressing crop improvement, animal production and veterinary needs. Current expenditure is concentrated on addressing small and rural farmer needs, with local and international seed companies providing most of the funding for conventional crop improvement.

The CSIR is a national industrial research institute using biotechnology in food security, water and environmental applications. Government funding at the CSIR is targeted at raising the living standards of developing communities, but this organization obtains nearly 60% of its funding from industrial partners.

The MRC focuses its biotechnology research on addressing local disease problems and providing improved diagnostics and treatment.

The Universities of Cape Town, the Free State, Natal, Pretoria, Stellenbosch and the Witwatersrand are the major tertiary institutes actively applying biotechnology in R&D projects. Each of these universities has active medical and life science programmes, but there is a growing concern that many graduates are leaving the country to escape escalating violent crime, reverse discrimination and frustration resulting from affirmative action programmes. All of these institutes have actively sought to increase their intake of students from disadvantaged communities and it is hoped that this will address the alarming loss of skilled professionals through emigration.

Private Sector Engagement

The major biotechnology investment in the country is the African Explosives and Chemicals Industry (AECI) commercialized lysine plant aimed primarily at export and/or import replacement. This company focuses its biotechnology R&D on industrial fermentations and crop improvement. Ten local seed companies and research organizations are carrying out field trials with genetically modified crops, three of these arising from local research, the rest using imported technology from international biotechnology and seed companies in America and Europe.

Three new biotechnology industries (fermentation, biomining and vaccines) and a dozen or so products have evolved in the field of biotechnology in this country. The 1998 Directory of Biotechnology in South Africa lists 46 private companies involved in biotechnology. Of these, 40 produce or use biotechnology products and six are consultancies or services. The biotechnology products produced or in use in South Africa cover a wide range of applications including industrial enzymes, diagnostics, vaccines, plants, pharmaceuticals, chemicals and environmental applications. The first conditional general release of a genetically engineered crop was approved in 1996. The first commercial genetically modified crops (maize and cotton) were planted in 1998.

Development of products targeted at international niche markets include the new lysine fermentation plant built by AECI. This facility went into production in late 1996 and is based on local R&D. Other examples of successful competitiveness in niche export markets are the use of local strains of pyrite-oxidizing bacteria selected for the leaching of metals from arsenopyrite ores and the production of vaccines targeting control of African diseases for both humans and animals.

Constraints

The major constraints upon developing local biotechnology businesses are poor communication between academics, research organizations and industry; a lack of product development research leading to innovative applications; a perception of inaccessible venture capital funding; inadequate national policy for generating new business; high cost of entry and current lack of legislation to control use of GMOs. Universities are beginning to address these problems, for example the 1994 launch of the Education and Innovation Foundation at the University of Natal to protect and exploit its intellectual property in Africa. The 1996 government paper on Science and Technology has identified better communication between academics, research organizations and business as a priority.

Another major drawback in developing a biotechnology industry in South Africa is the poor availability of skilled human resources. Frank Fisher, head of R&D at AECI, has stated that availability of future human resources in biotechnology is a key concern. He believes the State is not spending nearly enough on developing top class scientists in the area of microbiology and adds that AECI is spending more every year on equipment, training and research than all South Africa's universities combined (these have to share an annual budget of about US $1.25 million for equipment). Although tertiary education should be increasing the enrolment and degrees awarded in science and technology, there has been a clear shift towards the arts in the last ten years. The constraints faced by the new government are that out of 21 million adults, about eight million are illiterate and another seven to eight million are only functionally literate. The government is already spending approximately 22% of the national budget on education and additional expenditure is not likely to be forthcoming. The solution will have to lie in better use of the money available for education.

Future Prospects

Food security, adequate health care and a manufacturing export economy have been identified as major aims for the new government. Applications of biotechnology to crop improvement and the establishment of a thriving indigenous foods industry that exports into Africa are the primary agricultural opportunities. Coupled to this, is the opportunity to apply biotechnology to developing a sustainable market in medicinal plant uses and to the medical needs of the country (primarily tuberculosis, HIV and malaria). Both food security and medical applications of biotechnology have been identified as major investment opportunities. Locally suited biotechnology products are frequently applicable to neighbouring countries and collaborative projects are strongly encouraged to reduce duplication and enable more efficient spending of the available funds in the region.

Recent EU funding has stimulated major bi- and multilateral R&D projects that address needs in South Africa and collaborating African countries with the assistance of EU expertise. Although most of these projects are in the early stage of needs and situation analyses, some of this funding will be channelled to biotechnology products in the future, once the needs have been clearly identified. Areas of common interest include environmental protection, food security, veterinary vaccination and medical care.

Acknowledgements
This information has been taken from a survey funded by the Foundation for Research Development, which relied on contributions by Dr E.J. Morris, Prof D.E. Rawlings, Dr P.G. Cilliers, Prof P.D. van Helden, Dr C. Donninger, Dr D.K. Berger, Dr D.W. Verwoerd, Mr N.D. Nel and Dr H.C. Kasan. Our thanks to Janice Dewar and Michael Koch for editing and input.

Thailand

Sutat Sriwatanapongse, Morakot Tanticharoen and Sakarindr Bhumiratana

National Policy and Regulation

General Policy

Recognizing the potential of biotechnology to affect a broad spectrum of industries, Thailand has placed increased emphasis on biotechnology over the last decade. This is reflected in the establishment of the National Centre for Genetic Engineering and Biotechnology (BIOTEC) in 1983 and the launching of the Science and Technology Development Project (STDB) in 1985. During the first eight years of its existence, BIOTEC was under the jurisdiction of the Ministry of Science, Technology and Energy (currently, the Ministry of Science, Technology and the Environment) and was in charge of coordinating biotechnology research and development across the country. At the end of 1991, the activities of BIOTEC and STDB were merged as part of the National Science and Technology Development Agency (NSTDA), an autonomous organization with the mandate to be the main 'driving force' for rapid science and technology development. Opportunities to apply biotechnology in the public and business sectors have been expanding, and the government, through the NSTDA, has enhanced not only national R&D and training but also collaboration with the private sector.

Funding

In the present national economic and social plan, the Thai government anticipates increasing the R&D and training budget to 0.5% of its annual budget. The private sector, through various incentive and collaborative schemes, will allocate up to 0.25% of sales in R,D&E investment. The annual operating budget allocated to BIOTEC alone has grown several fold from US $3.5 million in 1993 to over US $10 million in 1996. Similar operating budgets have been allocated to other government agencies and universities.

In terms of investment, government support is focused on three areas: infrastructure, human resource development and information service.

Infrastructure
Investment in new BIOTEC office and laboratory facilities is estimated to be around US $5 million. Construction of a BIOTEC building which will house offices and laboratories

at the Science and Technology Park is underway, with estimated costs of US $10 million. According to the five-year plan of the NSTDA (1997-2002), BIOTEC will receive a budget of US $33.5 million in total for the purchase of equipment. The facilities at the NSTDA Yothi Building and at the Science and Technology Park (laboratory equipment, pilot plant, incubator unit and greenhouses) add up to about US $50 million in total investment in biotechnology capacity building under NSTDA. Among public institutes, the Ministry of Agriculture and Cooperatives is investing about US $10 million in biotechnology capacity building within the Ministry. Several universities such as Mahidol, Kasetsart, Chulalongkorn and Chiang Mai, are supporting biotechnology capacity building by establishing their own biotechnology centres.

At the NSTDA Yothi Building, BIOTEC has established national laboratories in order to carry out in-house research projects. These laboratories are equipped to meet BIOTEC's research needs in the following fields:

- Food biotechnology.
- Environmental biotechnology.
- Bioresources utilization and culture collection.
- Fermentation technology and biochemical engineering.
- Monoclonal antibody and hybridoma technology.
- Molecular biology and biochemistry.
- Protein engineering.

Human Resource Development

During the first phase (1990-1994), 200 students were sent to study abroad in various areas of biotechnology and related fields. The cost of investment is estimated to be about US $32 million. The same number of students are being sent out in the second phase (1995-1999).

Information Service

BIOTEC supports technical information services with a special focus on research work undertaken in the ASEAN region. These services support over 2000 subscribers worldwide and are available on the Internet (http://www.biotec.or.th). In addition, BIOTEC issues a regular newsletter, BIOTEC News, in both Thai and English, as well as supporting a number of monographs, proceedings and occasional publications.

Regulatory Structure

Biosafety

Biosafety issues are only now starting to be debated in Thailand and the underlying concepts are unfamiliar even amongst academics and certain regulatory agencies. The development of biosafety guidelines in Thailand commenced with a feasibility study commissioned by BIOTEC in 1990. This followed a review of the status report on the prospects of biotechnology in agriculture commissioned in 1986. As a result, the Biosafety Subcommittee was established under BIOTEC in November 1990 by the Science and Technology Development Executive Board. After the reorganization of BIOTEC under the National Science and Technology Development Agency (NSTDA), the *ad hoc* Biosafety Subcommittee was appointed in April 1992 with the sole responsibility of drafting the Biosafety Guidelines. The draft was completed in June 1992.

Essentially the Biosafety Guidelines encompass principles for good laboratory practice and the release of GMOs into the environment. A National Biosafety Committee (NBC) was established in January 1993. The establishment of institutional biosafety committees at various universities, government departments, research institutes, regulatory agencies and private agencies was also strongly recommended, and in many cases these recommendations have been implemented.

Prior to the formulation of biosafety guidelines, a number of rDNA-related research projects had already been undertaken at several universities in Thailand, utilizing locally-generated genetic material. Although most of the material thus investigated can be classified under 'Generally Recognized as Safe', the principal investigators usually adhere to safety precautions and biosafety guidelines observed in other countries where similar work is carried out.

Applications or requests for information regarding field trials of transgenic material have been received from many transnational companies. The following cases have received permission to perform field testing:

- An application for field trials for the production of seeds from genetically modified tomato plants, the Flavr Savr tomato produced by the company Calgene, was submitted for consideration in 1993. The application was approved by the NBC since it was the general opinion of the scientific community that the work had no risk or very low risk. The field trials were performed by the local company, Thaipan, in early 1994, under close supervision of the NBC. It was the first case of GMO field testing in the country and it served as a model for both public and private agency personnel to practise the biosafety monitoring process. Field testing of this Flavr Savr tomato was repeated in 1996 in the Northeastern province.
- A field trial of Monsanto Bt cotton was carried out under restricted containment in a netted house in 1996. The field test site was in the district of Pakchong, Nakhon Ratchasima. The field testings were carried out in the following year in open fields. Results from these field trials indicated that there was no risk in using Bt cotton and it is now under consideration for commercial release.
- In 1997, Bt corn field trials were approved to be carried out by Novartis at their experimental station under netted screenhouses.

Intellectual Property Rights

The Thailand Patent Act of 1979 and its Amendment of 1992 has the following key elements:

- The patentable material must be the product of 'human invention', not that of discovery of natural products.
- The patentable material or invention must be a new or novel product for which no disclosure has been made.
- The invention involves innovative steps which add value, or lead to technological development.
- The invention must be industrially applicable leading to commercialization or its use in the agriculture industry.

Under the Thailand Patent Act, there is some limitation in patenting living organisms. Article 9(1) states that microorganisms and parts thereof, as well as animals and plants collected from nature, cannot be patented. So it does not protect traditional plant varieties and animal breeds. Where microorganisms are concerned, individual cases must be considered separately for protection. In general, parts of microorganisms cannot be patented, although man-made parts may be considered (for example, modified Bt gene, isolated strains and purified strains).

The Plant Variety Act of 1965 and its Amendment of 1992 under the responsibility of the Ministry of Agriculture and Cooperatives, protects the conservation, quality and import-export of plant varieties, but does not protect new plant varieties; neither does the Act address Plant Breeders' Rights or refer to the rights of owners of varieties. It is clear that the present Patent Act under the responsibility of the Ministry of Commerce does not protect new varieties of plants (or animals).

The Ministry of Agriculture and Cooperatives and the Ministry of Commerce have drafted the Plant Variety Protection Act. This will protect new varieties (under the UPOV system), traditional varieties and wild plants. The draft has been considered in various group meetings and it is now under consideration in the Parliament for improvement and approval.

Biodiversity

Thailand has joined other countries in signing the Convention on Biological Diversity in 1992, but has not yet ratified the Convention. However, the government has been considering proceeding with ratification which is planned soon. Although Thailand has not yet ratified the Convention, many activities which comply with the Convention guidelines have been carried out. One important activity concerns access to Thailand's biological resources. The draft of the national regulatory framework has been completed and it is now under the review of the National Environment Committee chaired by the Prime Minister. It is anticipated that the draft, if approved, will be announced as the 'Prime Minister's Declaration on the Regulation of the Access to Biological Resources'.

In the 'Prime Minister's Declaration', the implementing agency is proposed to be the 'National Biodiversity Centre'. The new centre will be soon established under the umbrella of the National Science and Technology Development Agency (NSTDA) with the responsibility of coordinating biodiversity projects with both public and private agencies, besides the regulatory work on the access to biological resources.

Research, Development and Marketing

Major Projects

For more effective management of the wide scope of biotechnology, R&D activities are grouped into six major project areas as follows:

- Plant and Plant Product Improvement.
- Animal and Animal Product Improvement.
- Rural Development and Small Farmers.
- Sustainable Development.
- Health Improvement.
- Novel Products and Process Improvement.

The above six mission areas are recognized as priorities for the development and application of biotechnology in Thailand. A multifaceted approach is taken within each area and this can include upgrading the human resources necessary to support study and management of the area, support for research, support for private sector development, training and public awareness. Research and development activities in these six mission areas can be highlighted as follows:

Plants, Animals and their Product Improvement

This area incorporates most of the classical biotechnology approaches to the improvement of agriculture, including tissue culture and embryo transplantation. Innovation is being attempted in fields of particular importance to the Thai economy, notably rice and cassava and starch conversion technology. The focus on agriculture also benefits the poorer sector of the economy, namely farmers, as well as being of potential benefit to many small to medium range enterprises, such as cut-flower producers and fruit orchard growers. BIOTEC's programme incorporates research on improvement in seed technology in addition to alternative natural fertilizers and biotechnology for the control of insect pests. Product development is undertaken involving the private sector at an early stage, often co-funding the research. BIOTEC involvement often continues right through to the pilot plant, or pre-production stage. The programme also incorporates an innovative focus on small farmers, providing improved off-season crops as well as learning from the farmer's traditional use of technology, proven strains and the interrelationship between the farmer and the surrounding ecology.

Environmental Biotechnology
The country's rapid industrialization is providing new challenges for waste treatment. In addition to promoting improved waste management by working closely with industrial groups, BIOTEC is at the forefront of research in the country on treatment of waste waters, particularly from the food-related industries, as well as waste recovery and developing useful products from these waste streams. The application of technology appropriate to Thailand for the treatment of and recovery from solid and hazardous wastes is also undertaken.

Aquaculture
As Thailand is now one of the largest prawn producers and exporters in the world, BIOTEC pays special attention to this field which lends itself very well to modern biotechnology through breeding, feed development and supplementation, disease control, quality assurance and waste treatment and recovery. As worldwide stocks diminish, the country is turning more to farming the sea, and biotechnology has a major role to play. Breakthroughs have already been made in the area of diseases of major economic species of prawns that are proving a boon for producers nationwide. It should be noted, however, that large-scale prawn farming in Thailand has been promoted at heavy cost to traditional fishing families and the coastal ecology, leading to mangrove destruction and pollution of waters.

Medical Biotechnology
Biotechnology has already proven itself an invaluable tool in the field of medical and pharmacological advancement. BIOTEC is naturally focusing on tropical disease research, although there is also considerable emphasis on diagnostics. Much innovative research is also being undertaken in the area of medicinal plants and traditional medicine, and many native species are being examined for their potential medicinal value. The use of plants and herbs as traditional curatives is also being re-examined to determine what role biotechnology might play in their enhanced application and possible commercialization.

Food Biotechnology
In this mission area, the focus is on two major streams, one being fermentation technology for food products, the other the enhancement of the quality of Thai food products, especially those for export. With the international demand for higher quality food products, BIOTEC has been conducting an intensive campaign directly with industry, introducing such techniques as HACCP (Hazard Analysis Critical Control Point), to keep Thailand competitive in the world food market. More and more food producers are realizing the need for better in-house quality control and research and development, and BIOTEC

training programmes are helping to this end. Since food plays such an important role in the Thai economy, the BIOTEC food technology programme is likely to become the country's primary focus for research, quality control, inspection, testing, training and information services.

Biodiversity

In common with her neighbours in South East Asia, Thailand has suffered heavily from deforestation and environmental degradation as a direct result of industrial development. In response, the first step has been taken to allocate areas as national parks or forest reserves. However, very little work has been undertaken to date on identifying and measuring the biological resources existing in these remaining areas and their interrelationship with humans. BIOTEC has embarked on an ambitious survey campaign to measure these resources and to develop the human resource skills base, to build up more accurate inventories. Besides training and the development of field manuals, bibliographies and monographs, BIOTEC will also place strong emphasis on public awareness of what these biological resources constitute, their usefulness and conservation. Basic sciences are also being promoted, to create both interest and potential employment opportunities for young graduates and technicians. The support for sophisticated laboratories also ensures that the economic potential of bioresources is not overlooked and that the programme provides unique opportunities for scientists to work alongside NGOs and minority groups who have traditional knowledge to offer in the management of these resources.

In order to encourage the above activities, BIOTEC has supported a number of specialized laboratories around the country with research institutes which are regarded as the leaders in their respective fields. Such institutes also form part of the nationwide Biotechnology Network. They are:

- The Plant Genetic Engineering Unit at Kasetsart University, Kamphaengsaen Campus.
- The Microbial Genetic Engineering Unit at Mahidol University.
- The Marine Biotechnology Laboratory at Chulalongkorn University.
- The Biochemical Engineering and Pilot Plant Development and Training Institute at King Mongkut's Institute of Technology.
- The Medical Biotechnology Unit at Siriraj Hospital, Mahidol University.
- The Plant DNA Fingerprinting Unit at Kasetsart University, Kamphaengsaen Campus.
- The Cassava and Starch Conversion Technology Unit at Kasetsart University.

Major Institutes

In order to promote enhanced cooperation amongst the various research institutes nationwide, BIOTEC has formed a consortium known as the Biotechnology Network. The Centre services the Network by involving members in Centre activities and providing them with information services. Current members of the Biotechnology Network in Thailand include the following institutes, listed in order of size:

- Mahidol University, Bangkok.
- Kasetsart University, Bangkok.
- Chulalongkorn University, Thonburi, Bangkok.
- King Mongkut's Institute of Technology, Bangkok.
- Chiang Mai University, Chiang Mai.
- Khon Kaen University, Khon Kaen.

- Prince of Songkla University, Hat Yai Campus, Songkla.
- Srinakharinwirot University, Prasarnmit Campus, Bangkok.
- King Mongkut's Institute of Technology, Lat Krabang, Bangkok.
- King Mongkut's Institute of Technology, North Bangkok, Bangkok.
- Thammasat University, Bangkok.
- Silpakorn University, Bangkok.
- Burapa University, Chonburi.
- Maejo University, Chiang Mai.
- Asian Institute of Technology, Bangkok.
- Thailand Institute of Scientific and Technological Research, Bangkok.
- Department of Agriculture, Ministry of Agriculture and Cooperatives, Bangkok.
- Department of Fisheries, Ministry of Agriculture and Cooperatives, Bangkok.
- Royal Forest Department, Ministry of Agriculture and Cooperatives, Bangkok.
- Department of Livestock, Ministry of Agriculture and Cooperatives, Bangkok.
- The Food and Drug Administration, Bangkok.
- Thai Industrial Standards Institute, Bangkok.
- Suranari University of Technology, Nakorn Ratchasima.

BIOTEC is quite active in the international research collaboration arena, having worked with institutions in the ASEAN countries, and also in Australia, Belgium, Canada, the United Kingdom and the United States of America. BIOTEC was until recently the Secretariat for the National Chairman of the ASEAN Sub Committee on Biotechnology and the Chairman of the ASEAN Sub Committee on Food Science and Technology. BIOTEC serves as the Secretariat for the Asia-Oceania Network of Biological Sciences in collaboration with the Committee on Science and Technology in Developing Countries and the International Biosciences Network.

The Centre is also an active participant in UN programmes such as the FAO/UNIDO FARM (Farmer Centred Agricultural Resource Management) Programme and the UNIDO BINAS (Biosafety Information Network and Advisory Service) Programme. BIOTEC is at present acting as the BINAS mirror node in the Asian region. In addition, BIOTEC coordinates biotechnology activities with IRRI (International Rice Research Institute), UNEP, the Rockefeller Foundation and other international agencies.

Private Sector Engagement

Since the NCGEB is an autonomous government agency, it has more flexibility in dealing with the private sector. A good example of close public-private collaboration is provided by the consortium research and development, with universities and private companies targeted at solving the yellow headed disease of shrimp. The activities have led to the establishment of a joint venture company, the Shrimp Culture Research and Development. BIOTEC and a group of companies have agreed that the company will carry out research work leading to the production of the specific pathogen free stock. In the long run, this will be a good base for selective breeding and the production of hybrid shrimp.

The government recognizes the importance of public-private participation in R&D and training. Some incentives have been used to stimulate more private involvement in research. Tax incentives, for example, are expected to help promote more private participation in R&D and training. The government has recently reduced import duties on research equipment and materials from the range 30-50% to that of 5-10%. Moreover, deduction of R,D&E costs at the rate of 150% before taxation is allowed.

At present, strategic industrial sectors are surcharged a small amount on their export products. It has been suggested that the funds from the surcharge be used to partially support research at consortium research centres under the direction of the contributors. As a result, a consortium research on tapioca has been developed in the joint BIOTEC-Thai Tapioca Development Institute (TTDI), with the TTDI agreeing to allocate about 10 million baht (at the time, nearly US $400,000) to support this research. BIOTEC has established a specialized laboratory, the Cassava and Starch Conversion Technology Unit, at Kasetsart University in support of the activities. BIOTEC has also been involved in the establishment of the National Food Institute in collaboration with the Ministries of Industry, Commerce, Public Health and Agriculture. It is envisaged that there will be increasingly more of this public-private collaboration in the future.

Commercialization

BIOTEC has been in operation for more than ten years and there are many research products ready for commercialization. Efforts have been made during the past five years to commercialize these products. A working group has been established comprising four to five experts from the private sector, and two to three from BIOTEC. The research results are prioritized according to their marketing potential, and the highest priority product is chosen for consideration.

There have been a few cases of success where private companies have agreed to commercialize a product. The first of these involved the production of spirulina using waste water from starch factories. The process has resulted in waste water treatment and the simultaneous production of a high protein source for food and feed products. In another case, *Trichoderma* was produced as a biofungicide for the treatment of fruit tree root rot. A third case concerned the production of Nuclear Polyhedrosis Virus as a bioinsecticide for the control of cotton bollworm.

Constraints

Despite a considerable amount of funding for research from the government, Thailand still faces a number of problems in increasing its global competitiveness in biotechnology. The government infrastructure and capacity building within the NSTDA and through BIOTEC alone is insufficient for successful biotechnology development in the country. There is still a need to improve the infrastructure and capacity in other government agencies and universities in order that they can effectively collaborate with BIOTEC researchers.

Thailand faces a shortage of skilled personnel in biotechnology as well as in other science and technology areas. Research manpower is estimated to be only about 1.4 per 10,000 population, compared to 27 in Korea and 42 in Taiwan. There is a need for a critical mass of scientists to carry out research work and the only solution is through more private sector involvement in R&D activities. The public sector alone is not able to increase the number of well qualified research personnel to the required level. Another limitation concerns financial support from both public and private sectors which needs to be increased as stated above. There is little chance for Thailand to receive research grants from international funding agencies owing to its status as a more advanced developing country. Any such international funding is likely to be in the form of a partnership programme with co-funding.

Ukraine

L.N. Shynkarenko, O.V. Novozhilov and Ya. B. Blume

National Policy and Regulation

General Policy

The conditions that led biotechnology to flourish in the USA and some countries of Western Europe did not exist in Ukraine at the end of the 1980s, nor indeed in the rest of the former Soviet Union. Despite a long tradition of research and development in the biological sciences, especially in the Academy of Sciences of Ukraine, the Ukrainian scientific community lagged years behind those of the USA and Western Europe in terms of commercializing research efforts. Ukrainian academics, as a rule, had ineffective contacts with the industry and functioned in isolation, under their own set of rules. These rules were the result of a publicly funded academic research structure under the close regulatory control of the state.

It was under such circumstances that the Soviet Government took an initiative at the beginning of the 1980s to speed up the development of the biotechnology sector. With the cooperation of institutions of the former Soviet Union, Ukrainian scientific institutions took part in the implementation of a programme of biotechnology research and development in member-countries of the Eastern block, intended to last until the year 2000. Furthermore, a number of agricultural biotechnology centres were created, including one in Ukraine. The Ukrainian Agricultural Biotechnological Centre was founded to exploit the new techniques of genetic and cell engineering, and to provide vocational training. Two regional centres were also set up, in Odessa, and in L'viv.

At the same time, the Government of the USSR created the Joint Scientific and Technical Complex, 'Biogen', which aimed to accelerate the transfer of know-how from biotechnological research establishments to industry, under the conditions of a highly centralized industrial market. It should be pointed out that the projects of Ukrainian scientists took a prominent place in the framework of 'Biogen', especially in the field of plant biotechnology. Since the time of disintegration of the USSR in 1991, the Ukrainian Agricultural Biotechnological Centre has become the state institution that coordinates the development of biotechnology in Ukraine. However, due to a lack of state support, and the rapidly deteriorating financial situation in the country, the Centre's activity was practically paralysed.

Since 1991, the State Committee on Science and Technology, which became the Ukrainian Ministry for Science and Technology in 1996, played an active role in maintaining scientific potential in the field of biotechnology. The Committee continued to support 'Biogen' from the state budget, as an independent international programme. In 1994, 'Biogen' was transformed into a new state programme, under the name 'Perspective Biotechnologies'. This is still in operation.

Increasing interest in the development of biotechnology was also shown by the National Academy of Sciences of Ukraine (NASU). In 1993, the Presidium of NASU established the 'Bioengineering' Scientific Council, and initiated an interdisciplinary research programme, 'Principles of Bioengineering'. However, this did not receive state financial support.

Ukrainian independence was followed by the establishment of new state institutions, promoting development of biotechnology in their particular areas of competence. The

State Committee of Medical and Microbiological Industry began to play an active role in the practical application of biotechnological 'know-how'. The sphere of its interests ranged from drugs to some types of veterinary products that had been developed on the basis of novel biotechnologies. Expertise in the fields of plant biotechnology, immunobiotechnology, and reproductive biotechnology was available in the institutions of the Ukrainian Academy of Agricultural Sciences (UAAS).

At the beginning of 1997, a new Department of Agriculture, Medicine and Biotechnology was set up within the Ministry of Science and Technology. The aim of this Department has been the coordination of research, development and investment activities in the field of biotechnology. Later the same year, the Joint Committee on Biotechnology was also established under the auspices of this Ministry. The Committee's staff include representatives of the Government responsible for the development of biotechnology, prominent scientists, and specialists from industry. There are four sections in the Committee - biotechnology of microbial synthesis, plant biotechnology, biotechnology of animal cells and immunobiotechnology. The Committee's objectives are the analysis of the latest achievements, assessment of prospects for the development of biotechnology in Ukraine, and the administration of R&D programmes in the area.

The Parliament has recognized biotechnology as one of the strategic R&D priorities for Ukraine, and has given the Ministry of Science and Technology responsibility for:

- Directing pure and applied research towards the development of up-to-date biotechnology to meet long-term industrial aims, through immediate financial support.
- Concentrating intellectual potential, material, technical and financial resources to speed up innovation in biotechnology.
- Developing vocational training programmes in the area of biotechnology.
- Promoting the domestic biotechnological industry and domestic market development.
- Developing an adequate legal environment for the development of the biotechnology industry.

Funding

State investment in biotechnology R&D is significantly less than in industrialized countries. By and large, biotechnology is financed by the state, and carried out in the research institutions of NASU (13 institutes), UAAS (20 institutes), some institutes under the auspices of the Ministry of Health, the Academy of Medical Sciences and the Ministry of Food Industry, as well as in some state universities. Additionally, the Ministry of Science and Technology supports all these institutes, and other non-state research establishments through research grants. Grants for state research programmes were approximately US $1.5 million in 1996, and for innovation projects about US $10 million over the last two years.

High priority biotechnology research lies within the framework of the 'Perspective Biotechnologies' programme, consisting of 20 projects with a total budget of US $300,000 in 1997. There is no private investment in biotechnological research as yet, except for some projects in traditional biotechnology (in the fields of wine and cheese making or the preservation of vegetables, for example).

Regulatory Structure

Regulations governing biotechnology in Ukraine are typical of those in operation throughout the former USSR. Most of the legal instruments covering IPRs and trade issues have remained unchanged. Over the last few years, however, the regulatory framework for biotechnology has been widened significantly to include biosafety and biological diversity. The Committee on Safety Control and Regulation of Work with Genetically Modified Organisms under the State Committee on Science and Technology started to operate as early as 1992. Later on, this Committee was transformed into the Committee on Biosafety affiliated to the Ministry of Science and Technology. The main goals of the Committee on Biosafety are to:

- Ensure the establishment and improvement of regulatory oversights in biosafety.
- Elaborate the rules for safe production, use and environmental release of modified organisms.
- Create and maintain central databases in genetic engineering and biosafety.
- Coordinate the elaboration and implementation of an application/notification system based on risk assessment/risk management procedures.
- Coordinate activities on harmonization of the national biosafety system with extant regional and international mechanisms.

The Government will soon consider the issue of granting interdepartmental status to the Committee. If granted, the Committee will be authorized to permit environmental release of genetically modified organisms (GMOs). In 1997, the Committee presented a Bill to Parliament on the state regulation of genetic engineering, although it is likely to take between one and two years for this to become law.

This law is based on the corresponding Federal Law of Russia, though it is more detailed, and consists of many sections, including general principles and ideas, areas of application, tasks for the state government, main trends of state policy, licensing, the system of safety, the legislative status of institutions carrying out biotechnology R&D, approval for activities in genetic engineering, certification of production, liability, possible mechanisms for providing information to the public, international cooperation, and the funding of regulatory agencies.

The law envisages:

- State regulation of genetic engineering will not deal with food labelling. However, it will define the classes of GMOs which represent a possible risk in the food chain.
- The State Committee on Biosafety will be established by the Cabinet of the Ministry of Science and Technology.
- Decisions issued by this Committee will be available to the public, and will be mandatory on the territory of Ukraine.
- The Committee will register all users handling GMOs. This registration will be issued only if the facilities and personnel guarantee the safe use of GMOs.
- Based on previous experience with GMOs, it is suggested that principles of Good Microbiology Practice are sufficient to ensure the safety of contained uses of GMOs.
- Use of all transgenic crops should be regulated through authorization for environmental release. In some cases the Committee should be granted the authority to withdraw certain GMOs for safety reasons.
- The law should provide a means to recognize (through bilateral agreements) certain documents issued by other countries, approving the release of transgenic organisms.
- The law will request both the user and the Committee to publicize information of general concern.

In connection with Ukraine joining the Convention on Biological Diversity, a regulation has defined which resources are considered Ukraine's national property (Governmental Decision 1174, 18.02.97). These include unique collections of microorganisms, fungi, hydrobionts, plant and animal cells and genes.

Research, Development and Marketing

Major Projects

As mentioned above, a number of projects are being carried out in the framework of the State Scientific and Technical Programme 'Perspective Biotechnologies'. The main projects, defining the general directions of this Programme are:

- Development of a new generation of biosensors for the analysis and control of biotechnological processes (Institute of Molecular Biology and Genetics of NASU).
- Creation of new cell cultivars of Arnebia as a source of shikonin (Institute of Molecular Biology and Genetics of NASU).
- Analysis of sugar metabolism in peas, and investigations of transgenic expression (Institute of Cell Biology and Genetic Engineering of NASU).
- Construction and selection of a *Streptomyces peucetius* strain - used for production of the anticancer drug rubomycin (State University of L'viv).
- Manufacture of immunobiological substances with direct application for treatment of cytomegaloviral and adenoviral infections (Zabolotny Institute of Microbiology and Virology of NASU).
- Creation of new cell cultivars for production of dioscorea, ginseng, levsea and mandragora (Institute of Food Chemistry and Technology of NASU).

Some original biotechnological research is also being carried out under the auspices of specific programmes on human health, environmental protection and agriculture:

- Development of technologies for obtaining and using chemokins and ceprokins as anticancer drugs (Kavetsky Institute of Experimental Pathology, Oncology and Radiobiology of NASU).
- Bioremediation of soil following oil pipeline accidents (Zabolotny Institute of Microbiology and Virology of NASU).
- Biological monitoring of herbicide pollution of soil (Institute of Cell Biology and Genetic Engineering of NASU).
- Implementation of biotechnological approaches to sewage purification from heavy metal ions (Institute of Colloid Chemistry and Water Chemistry of NASU).
- Implementation of biotechnological approaches to water purification following contamination with oil products (Institute of Bioorganic Chemistry and Oil Chemistry of NASU).
- Cloning of agricultural animals of predetermined gender (Institute of Cattle-breeding of UAAS).
- Creation of recombinant enterovirus strains for development of vaccines against enteroviral diseases of pigs (Institute of Agricultural Microbiology of UAAS).

It should be pointed out that funding shortages and the emigration of qualified biotechnologists has considerably reduced Ukraine's competitiveness in biotechnology.

Major Institutes

- Institute of Cell Biology and Genetic Engineering of NASU, Kyiv. Plant genetic engineering, plant cell culture, phytoremediation, germplasm preservation, biosafety.
- Institute of Molecular Biology and Genetics of NASU, Kyiv. Biotechnology of animal cells, gene therapy, plant cell culture.
- Zabolotny Institute of Microbiology and Virology of NASU, Kyiv. Microbial biosynthesis, manipulations with bacterial DNA, environmental biotechnology.

- Institute of Bioorganic Chemistry and Oil Chemistry of NASU, Kyiv. Environmental biotechnology.

Private Sector Engagement

Until now, private sector involvement in biotechnology has been poor. This can be explained by the fact that the environment for the development of small and middle-sized business in Ukraine has not been favourable, and the internal market of biotechnological products remains underdeveloped.

It is expected that new varieties of agricultural plants obtained by *in vitro* selection and genetic transformation will be commercialized within the next few years. Transnational biotechnological companies (Monsanto, Novartis, AgrEvo, Pioneer, etc.) are in the process of introducing their products into the Ukrainian market. At the time of writing, six transgenic plants are registered in the Ukraine by the Committee on Biosafety. These are insect-resistant potato (a Monsanto product, resistant to the Colorado beetle), glyphosate-resistant corn (Monsanto and DeKalb), glufosinate-resistant corn (AgrEvo), insect-resistant corn (Novartis Seeds), glufosinate-resistant sugar beet (AgrEvo), and glyphosate-resistant sugar beet (Novartis and Monsanto). These plants are currently being considered by the State Committee on Testing and Safety of New Plant Varieties for registration as new varieties. Of these, however, the potato has already been registered as a new variety and will be marketed in 1999.

Constraints

Human resources: Although currently, the number of qualified university graduates working in the field of biotechnology is adequate, the quality of training could be improved. However, the major challenge is the outflow of highly qualified specialists abroad.

State and international funds: It has been widely accepted that financing from the state budget cannot cover the needs of the biotechnology sector. As far as the chances of getting financial support for biotechnological research from abroad are concerned, these are limited to possible international funds, predominantly of charitable character: STCU (Science and Technological Centre of Ukraine), INTAS (International Association for the promotion of cooperation with scientists from the New Independent States of the Former Soviet Union), and CRDF (Civilian Research & Development Foundation).

Legislative policy: The need to develop and adopt flexible new laws that do not limit the development of biotechnology.

International activity: Ukraine maintains contacts with different international organizations involved in biotechnology: OECD, UNIDO, and ICGEB (International Centre for Genetic Engineering and Biotechnology). Nevertheless, in some cases the country cannot take part in international cooperation activities, due to financial constraints.

Future Prospects

The main financial support for biotechnology research in Ukraine remains state contributions to research institutes, with some additional resources granted by the State Fund of Fundamental Research. Thus, the priority areas for research are those in which there has been a tradition of research in the Ukraine, and which can potentially lead to products marketable within the country. These are production of drugs and immunobiological substances, plant genetic engineering, biosensors and bioremediation.

Obviously, the level of finance available for biotechnology in Ukraine is insufficient. It must be increased some 15-25 fold if research is going to have an impact on the economy of the country. Undoubtedly, the creation of technoparks would play an important part in the development of biotechnology, although funds for this purpose are unlikely to be forthcoming from state sources. Perhaps international donor and/or technical aid organizations aiming to improve the investment climate in Ukraine could provide such financial support. Unfortunately, it is premature to claim that there are prospects for stable progress in the development of biotechnology in Ukraine. Thus, it is not possible at present to analyse the impact of biotechnology on meeting social, economic and ecological needs of the country.

Venezuela

Edgar Otaiza-Vásquez and Asdrúbal Arcia M.

National Policy and Regulation

General Policy

Early signs of interest in biotechnology in Venezuela were manifest in the 1970s. However, it was not until 1982 that the former Ministry of Science and Technology (CONICIT) established an *ad hoc* commission to analyse biotechnology research possibilities. This commission proposed agriculture, biomedicine and industry as priority areas for Research and Development (R&D). The Government established the National Commission for Genetic Engineering and Biotechnology (CNIGB, 1985), which proved to be very ineffective as it did not fulfil its main responsibility: the formulation of clear policies and plans for biotechnology development in Venezuela. Other historical aspects are reviewed elsewhere (Hernández *et al.*, 1991).

In 1992, the Venezuelan Government and the Inter-American Development Bank (IDB) signed a four-year agreement known as the CONICIT-IDB Programme (later known as the New Technologies Programme - NTP) to finance research and innovation in five new technology areas, which included biotechnology. The Programme applied for a loan of US $94 million; its biotechnology component assumed almost every activity of the CNIGB unofficially. In 1996, the CNIGB was replaced by the National Commission for Biotechnology, which maintains its advisory character to CONICIT.

Table 4.23. Funding figures (US $ million) of the projects in the biotechnology sub-areas of the New Technologies Programme (NTP).

Sub-area	Number of projects	Amount (US $ millions)	Average amount/ project (US $ millions)
Biomedicine	20	9.06	0.45
Agriculture	9	1.67	0.18
Industry	1	0.20	0.20

Source: Mendible *et al.* (1994).

Table 4.24. Funding figures (US $ million) by NTP to biotechnology research in major academic institutions in Venezuela.

Institution	Biomedicine	Agriculture	Industry
Central University of Venezuela	4.96	0.60	0.20
Simón Bolívar University	0.51	-	-
Los Andes University	0.62	0.42	-
Venezuelan Institute of Scientific Research	0.45	0.18	-
University of Carabobo	1.24	-	-
Others	1.27	0.47	-
TOTAL	9.05	1.67	0.20

Source: Mendible *et al.* (1994); CONICIT (1996).

National and institutional objectives for biotechnology research are either unclear or non-existent in Venezuela, which means an absence of government policies and lack of research priorities in this area. Nevertheless, policy recommendations exist in the form of a document generated by a commission of CONICIT in 1994, whose members were the NTP biotechnology coordinators (Arcia *et al.*, 1995).

Funding

The only reliable information regarding national expenditure and government investment in biotechnology R&D rests with the NTP. The Programme was conceived to promote advanced science and technology (S&T) research, and to develop highly qualified human resources in five areas, including biotechnology. It is a landmark work because of its novel conceptual framework, the adoption of new criteria to evaluate proposals and results in basic or applied scientific research, and the external funding obtained. Economic difficulties have limited its development since 1994. Table 4.23 presents significant figures in three biotechnology sub-areas of the NTP. Projects were classified into: IT (immediate transfer of results to industry), NIT (non-immediate transfer of results to industry) and SPL (shared profit and loss). The Programme considered that state-of-the-art scientific research can be attained only through doctoral and postdoctoral training. It therefore granted 128 biotechnology stipends (93 for biomedicine, 27 for agriculture and eight for industry) towards PhD degrees (61%), highly specialized short courses (22%), postdoctoral training (11%) and MSc degrees (6%) (Mendible *et al.*, 1994). The amount (US $ millions) dedicated to stipends was as follows: 5.43 for biomedicine, 1.61 for agriculture and 0.55 for industry. Short courses and seminars received US $0.70 (J. Mujica, Caracas, 1998, personal communication).

Presently, 14 public universities and three public research institutes are engaged in conducting scientific research in biotechnology. Table 4.24 presents funding in major institutes carrying out most of the biotechnology R&D. Private universities do not demonstrate an interest in such activities. The percentage of gross national product assigned to the S&T sector in Venezuela ranged between 0.31% and 0.49% from 1980 to 1995, less than the 1% recommended by UNESCO as a minimum for developing countries (CONICIT, 1996). For the years 1991 to 1995, the financial resources for S&T increased 80% in 'bolívares' (Venezuelan currency), but inflation devoured it. The situation in 1996 was still worse. CONICIT's budget for 1997 was US $80 million. Between 1991 and 1995, public universities spent a meagre 4% to 6% of their budget on research in science and humanities (US $35-81 million, adjusted to constant prices of 1984) (CONICIT, 1996). Unfortunately, the amount of money allocated by universities to biotechnology research is unknown.

Regulatory Structure

Venezuela as yet has no specific laws on biosafety, despite the fact that Decision 345 of the Andean Pact instructs the five member countries to generate biosafety rules by December 1994. These must cover legal permission for the deliberate release of genetically modified organisms (Kathen, 1996), as well as the development, testing and manufacturing of other biotechnology products. Biosafety has received far less attention in Venezuela than the issues of intellectual property rights (IPRs), and discussion is confined to a limited number of researchers. Public acceptance of biotechnology products is almost unknown. In 1999 an advisory, transdisciplinary Commission of Bioethics and Biosafety was established by CONICIT.

Venezuela has obligations under two international IPR arrangements. The first one is the Trade Related Aspects of Intellectual Property Rights (TRIPS), which is part of the General Agreement on Tariffs and Trade (GATT), signed in 1994. The second is Decision 344 taken by the Commission of the Cartagena Agreement in 1993. Decision 344 establishes a common system for the protection of industrial property in the Andean Pact countries of Bolivia, Colombia, Ecuador, Peru and Venezuela. Another decision of the Andean Pact (Decision 345, October 1993) establishes a common regime for the protection of plant breeders' rights.

Regarding the protection of its biodiversity, Venezuela ratified the Convention on Biological Diversity (CBD) and, as a member of the Andean Pact, must consider its obligations under Decision 391 (July 1996), which regulates access to genetic resources. With the exception of Bolivia, the remaining countries of the Andean Pact are extremely rich in biodiversity, but they do not have either applied research capacity or a biotechnology-based industry that could utilize this wealth. Recently, a bill to protect and rationally utilize Venezuelan biodiversity has been tabled by Parliament.

International Cooperation

Only a few of the many international cooperation agreements in S&T subscribed to by Venezuela have worked properly. Others are more specific to biotechnology, such as the agreement with the International Centre for Genetic Engineering and Biotechnology (ICGEB), the S&T Programme for Development (CYTED) founded by the Spanish Government, and the Biotechnology Programme for Latin America and the Caribbean conducted by the University of the United Nations (UNU). ICGEB provides an integral plan of collaborative research programmes, advisory services and training of human resources in life sciences to its member states, which is being scarcely utilized by Venezuela, despite its position as a Full Member State of the Centre (A. Taylhardat, Caracas, 1998, personal communication). Biotechnology issues are, *inter alia*, the subject of the Cartagena Commission-European Union agreement and UNESCO guidelines. The International Centre for Tropical Agriculture (CIAT, Colombia), the International Potato Centre (CIP, Peru) and the Tropical Agricultural Research and Training Centre (CATIE, Costa Rica) have played an important role in the diffusion of low complexity plant biotechnology strategies in Venezuela. Major Venezuelan universities have signed bilateral agreements with Cuba through its Centre for Genetic Engineering and Biotechnology (CGEB).

In 1990, the Latin-American Economy System and the EU established a Cooperative Programme for Biotechnology closely related with the Bolivar Programme and the Regional Latin American Programme of UNESCO/UNIDO/UNDP. These programmes aim to promote technological innovation and enterprise development in biotechnology. Venezuela is an active member of these programmes. EUREKA, a Venezuelan NGO, has organized several meetings with business-rounds in biotechnology for European and Venezuelan entrepreneurs.

Research, Development and Marketing

Major Projects

Biotechnology research in the country includes work on medicinal and nutraceutical species; pathogen-free cocoa propagation (one of the finest worldwide); pest management of Musaceae; integrated pest management (IPM) in other species; plant extracts and enzymes; transgenic rice and maize resistant to pathogens; prevention and quick diagnosis of plant viral and fungal diseases; virus resistant plants; secondary metabolites; industrial exploitation of tropical fruit species; animal vaccines and components for animal feed and nutrition; bovine embryo transfer; human and veterinary diagnosis kits of viral, bacterial and parasitic tropical diseases; human vaccines; aquaculture (shrimps, soft shell crabs, trout); treatments to enhance nutritional/digestibility properties of forage; enzymatic hydrolysis of lignocellulosic wastes (sugar cane); biomass utilization through microorganisms; cassava for starch production.

The most significant strategies and topics being investigated, according to the analysis of concrete results (Otaiza and Arcia, 1997; Anonymous, 1980-1996; Mendible *et al.*, 1994), are described in the following paragraphs.

Biomedicine

Applied immunology (kits, monoclonal antibodies, Chagas' disease, ferritin); enterokinase production against heart attack; antigen production; drug detection; *Leishmania* vaccine; diagnostics (Enterobacteriaceae, hepatitis, trypanosomiasis); instrumentation for capillary electrophoresis; establishment of a microorganism strains bank.

Agriculture and Industry

IPM in maize, potato, sorghum, sugar cane, Musaceae; biological control of garlic rotting; *Beauveria* for coffee; worm and moth of potato; mites and predators; viral diseases control in citrus, papaya, maize, potato and sweet potato; diagnosis kits for viral diseases; tissue culture of ornamentals (rose, *Heliconia*, *Gerbera*, orchids, *Mussaenda*, *Nephrolepis exaltata*), cocoa, papaya, cassava, potato, rice, garlic, bamboo, strawberry; clonal propagation of pine trees for the timber industry; phosphate stress tolerance in plants;

transgenic papaya resistant to ringspot virus; production of aspartate related amino acids by *Corynebacterium*; biomass utilization; *Rhizobia* inoculants; fruit fly bioassays.

Only NTP-funded projects stipulate the need for technology transfer to industry. Documents of the Ministry of Agriculture make similar stipulations in an attempt to promote biotechnology, in line with the Venezuelan's Government Agricultural Policy up to 1999 (MAC, 1995).

The National Fund for Agricultural Research (FONAIAP) has successfully produced and commercialized some animal vaccines (foot-and-mouth disease) and immunobiologicals (bacteriocins, equine adenitis, *Leptospira* bacteriocin and stomatitis and *Brucella* vaccines) obtained through conventional biotechnology. Many others have been used in industrial production, i.e. vaccines for poultry Newcastle disease, equine encephalomyelitis and bovine rabies and carbuncle. Products for biological control of sugar cane, maize, sorghum, tomato and potato are firmly established in small firms. Shrimp and trout culture is becoming steadily established. A small wine industry is emerging both in academic (through university entrepreneurship) and food and beverage firms, and a production plant for human plasma proteins has been started. Sera against snake venoms are currently produced in one university. CYTED sponsors production and commercialization of diagnostic kits for the human diseases mentioned previously in this report.

Major Institutes

All of the public universities have established Centres for Biotechnology Research. This trend, which tends to disperse expertise, might be attributed to each institution's desire for scientific prestige, researchers' demands for laboratories equipped with the latest devices, and the euphemistic outcomes of renaming small tissue culture laboratories! The critical economic situation of the country suggests the need for concentrating human and financial resources in a few high-quality centres for R&D in biotechnology, which could be strategically located where capacities are best.

Private Sector Engagement

In a broad sense, the Venezuelan private sector is unfamiliar with biotechnology, apparently because of its new features and the traditional slow assimilation of foreign industrial technologies. Industry is not equipped for research and is therefore unable to perform independent biotechnology research to solve problems or to develop new biotechnological products or processes. Local venture capital for biotechnology investment is also scarce. S&T are key resources for economic development today, and the growth of links between public and private sectors is a logical and important requisite. Deep divisions between both sectors are found in Venezuela, overlooking the fact that these are the two ends of the technology innovation chain (Otaiza and Arcia, 1997).

In agrobiotechnology, one fourth of the companies surveyed for their R&D activity are large, consolidated entities that incorporate low complexity biotechnology into their production system. For example, most sugar cane refineries are working with *Diatraea* spp., but only one conducts successful research on sugar cane micropropagation. One notable exception is the establishment of a biotechnology laboratory by a large firm in the food and beverage sector (Polar S.A.). This is of high research capacity, engaging biomedical researchers from public universities. The first result is the patenting of three human diagnosis kits: hepatitis B and papilloma viruses and antigens of *Trypanosoma cruzi* that can be used to diagnose Chagas' disease (Rangel-Aldao, 1996). The firm applied for other patents on beer yeast in the USA.

The remaining companies are small, young, undiversified firms, some with biotechnology strategies, particularly in plant biotechnology; some of them had to re-focus their activities, whilst others disappeared. These small firms and academic institutions of the plant biotechnology area employ almost exclusively strategies and tools of conventional biotechnology with a low complexity level (tissue culture and micropropagation, for example). Viral and fungal protection of horticultural species are followed by biological control of cereal, roots/tuber (potato), one fruit species (banana), sugar cane and sorghum, and resistance to some stress conditions (water and phosphate) and nitrogen fixation. Germplasm banks are rarely mentioned. IPM seems to be the most favoured approach, and perhaps the most consolidated one in the Central-Western region, particularly in small firms (Otaiza and Arcia, 1997). In the animal sub-area a small firm located in Zulia State is the only one commercially successful in bovine embryo transfer. Two small companies profited from two projects of the NTP (IPM and papaya field tests). Multinational enterprises in the biotech-pharmaceutical sector have large-scale interests in Venezuela, but unfortunately conduct their research exclusively in their headquarters abroad. This, of course, hampers possible technology transfer in biomedical research.

Limitations and Constraints

An analysis conducted to determine the status of plant biotechnology in Venezuela (Otaiza and Arcia, 1997) revealed that problems do not differ substantially from those of other S&T sectors in Venezuela, or indeed Latin America as a whole (da Costa, 1995). The complaints of academic researchers included: economic support (scarce funding and low salaries of researchers), training and capacity building (uninformed personnel and difficulties in training them) and infrastructure (shortage of equipment and facilities). Nevertheless, there are other major obstacles that make the transfer of biotechnology results to agriculture and industry impossible, thereby weakening any potential benefit. These include:

- Lack of leadership and an abundance of academically uninformed personnel is common in some universities and in the Ministry of Agriculture (particularly FONAIAP, which works as its executive arm for research).
- Researchers' isolation (individualism) leads to repetitious research, wasting time and money.
- Insufficient control and supervision of projects: Venezuelan researchers frequently conduct indiscriminate research because the academic councils, which are responsible for supplying research funds, have established neither priorities nor policies.
- The inability or unwillingness of researchers and entrepreneurs to communicate effectively with one another is resulting in a paralysing effect on biotechnology as a whole.
- Lack of current biotechnology postgraduate studies in public universities. In a few of them basic biotechnology subjects have been recently incorporated into pharmacy, food technology and plant breeding MSc courses.
- Academic productivity in agrobiotechnology, as measured by the number of publications at the MSc level is low, ranging between 0.6 and 1.5 theses per annum at two of the oldest Venezuelan Agriculture Schools and one Veterinary School. Quick access to databases and how to utilize them is not well mastered by many Venezuelan researchers, i.e. electronic facilities are not widely known despite the existence of the national network (Reacciun of CONICIT).

- Important issues affecting the application of biotechnology such as IPRs, patent protection, bioethics and biosafety are unknown to several academic researchers and some of them consider such items as peripheral to biotechnological activity.
- A proliferation of tiny centres for biotechnology and lack of a national policy for biotechnology.

Future Prospects

As a developing country, Venezuela has to consider both science-based (high-external-input) and the so-called indigenous (low-external-input) biotechnology. For review of these aspects see Bunders *et al.*, 1996. Unfortunately, the social dimension of biotechnology has not yet been incorporated into the postgraduate level curricula at any Venezuelan university. Agrobiotechnology is the most important biotechnological activity of the private sector in Latin America, due to the efforts of some national S&T systems, and to the existence of commercially attractive possibilities which give the region comparative advantages over developed countries. This last fact underscores the strategic importance of identifying such advantages, also existent in biomedicine and industry, and exploiting them to obtain concrete benefits.

Agenda 21 recognized that, although biotechnology cannot provide solutions to all fundamental environmental and developmental problems, it can contribute substantially to sustainable development by improvements in food production and supply, health care, environmental protection and energy (Pilari, 1997). The authors of this report are convinced that the concept of sustainability in biotechnology should not be restricted only to ecological and economic concerns but should also include its social impacts.

Biotechnology in Venezuela is highly dispersed and is not advanced, though other authors have indicated an entirely different status (Jaffé and Trigo, 1993). The situation requires changes, otherwise research will remain a purely intellectual endeavour of academics with little or no benefit to Venezuelan society. Financial resources provided by the Venezuelan Government for S&T have been very scarce traditionally, and the current grave economic crisis has deepened the problem. The main proportion of such resources should be directed toward utilitarian research.

The difficulties accompanying structural adjustment in Venezuela and other Latin American countries make it essential to apply S&T more efficiently in the social sector. There is an urgent need for the development of an integral policy for biotechnology in Venezuela, and this has been long awaited by the scientific community.

Zimbabwe

Joseph Gopo

National Policy and Regulation

General Policy

The government of Zimbabwe has demonstrated its commitment to science and technology by investing in the creation of the Research Council of Zimbabwe (RCZ), which is entrusted with the formulation of science and technology policy, and the establishment of national institutes (such as the University of Zimbabwe, the National University of Science and Technology, and a large infrastructure of technical colleges and institutes of agricultural and medical research). The RCZ commissioned the preparation of a document examining national policy on science and technology in Zimbabwe. The report includes a chapter on national policy on biotechnology, and has been submitted to the government for approval.

The RCZ has also been charged with responsibility for the establishment of the Scientific and Industrial Research and Development Centre (SIRDC) and one of its seven constituent institutes is the Biotechnology Research Institute (BRI). The SIRDC was created to serve as a single centre of excellence for the development, adoption and transfer of technology from the bench top to commercial industrial production.

Regulatory Structure

The Government of the Republic of Zimbabwe has produced and gazetted the necessary regulatory framework on biosafety in the form of Biosafety Regulations/Guidelines. The Biosafety Regulations/Guidelines address the issues of biosafety, biological diversity conservation and regulatory frameworks covering intellectual property rights and patents, as well as issues of access to genetic resources and farmers' rights.

The Biotechnology Research Institute (BRI), in collaboration with the Community Technology Development Association, co-hosted a national consultative workshop on the need to develop intellectual property rights and patents legislation for Zimbabwe (Gopo and Mushita, 1997). Prior to the workshop, a study had been commissioned by SIRDC to assess the usefulness and effectiveness of existing legislation, identify gaps and suggest policy reforms required in terms of harmonizing legislation with signed and ratified international conventions (Mandudzo, 1996; Musa, 1996).

The objectives of this national workshop were:

- To review the existing legislation and mechanisms for IPR protection in Zimbabwe.
- To assess alternatives to current patent protection mechanisms and suggest an appropriate *sui generis* legislation which protects the knowledge, innovations and technologies of local communities and the private and public sectors, with respect to the cultural and socioeconomic interests of Zimbabwe.
- To assess the implications of international trade agreements such as GATT/TRIPS and those of the World Trade Organization (WTO) with respect to IPRs and patents from both the Southern and Northern perspectives.

The results of the study were presented at the national consultative workshop, complemented by position papers given by other stakeholders drawn from the government, NGOs, private companies, research organizations, community-based organizations and universities. This approach was considered essential, as it facilitated participation by all concerned in the process of identifying stakeholders' perceptions and needs with respect to patenting life-forms and traditional knowledge. The consultation process was necessary for the development of rules, regulations, policies and legislation on the management of biodiversity and genetic resources, their exploitation, and their sustainable utilization, including benefit sharing of the results of biotechnology research.

The other important aspect of the study concerned government obligations to sign and ratify international conventions such as GATT/WTO; the Convention on Biological Diversity (CBD) and the International Undertaking on Plant Genetic Resources under the auspices of FAO. The main recommendation arising from the workshop was that a second workshop should be held sometime in 1998 to develop *sui generis* IPR/patent legislation for Zimbabwe, in conformity with GATT/WTO and the CBD articles regarding application of intellectual property rights and patents. It was agreed at the workshop that protection should be provided on verifiable 'indigenous knowledge' that has a demonstrable utility value. In this regard it was agreed that:

- A product eligible for protection should be a combination of extracts from life-forms in a given quantity to effect either a medicinal cure, or resistance to biotic or abiotic stress.
- Technological innovations eligible for protection should have both industrial application and social value.

It was further recognized that it was necessary to define regimes of protection which afford a community a right over its innovation technologies and knowledge. This was considered necessary as the present forms of IPRs and patents do not provide a comprehensive framework of protection in such aspects as farmers' rights, indigenous knowledge and traditional technologies. Specific legal protection regimes (i.e. regulations, policies and legislation) will be suggested for both formal and informal sectors. The legal status of PBR (plant breeders' rights) will be considered in the proposed suitable legislation for plant varieties in line with signed and ratified international agreements, taking account of national socioeconomic and cultural needs.

Research, Development and Marketing

Major Institutes

The University of Zimbabwe

The University of Zimbabwe, Faculties of Science, Agriculture, Medicine and Veterinary Science, and the Departments of Biochemistry, Biological Sciences, Chemistry, Crop Science, Animal Sciences and Pharmacy. The Faculties of Science and Agriculture offer an MSc degree in biotechnology.

The National University of Science and Technology

The National University of Science and Technology in Bulawayo, Faculty of Applied Sciences, Departments of Applied Biochemistry, Applied Biology and Applied Chemistry are involved in some R&D in biotechnology.

Africa University

Africa University, located in Murare City, has a large Faculty of Agriculture which has active research programmes in biotechnology, especially mushroom production.

Solusi University

Solusi University near Bulawayo is a new institution which will be active in biotechnology research.

The Ministry of Agriculture

The Ministry of Agriculture, Department of Research and Specialist Services (DR & SS), Cotton Research Institute, Plant Breeding Institute and other institutes.

Other Institutes

A number of agricultural colleges, including Gwebi College of Agriculture, Esigodini

Agriculture Institute, Mlezu Agriculture Institute, Kushinga-Phikelela Agricultural Institute and Rio-Tinto Agriculture Institute.

The Biotechnology Research Institute (BRI)

The BRI is entrusted with the formulation of a national policy on biotechnology research and its administration, and the prioritization of research activities. The Institute provides scientific and technical know-how and research leadership in all areas and industrial sectors in which biotechnology has an impact. It also plays a leading role in the conservation of genetic resources, providing educational opportunities for human resource development. Finally, the Institute provides leadership in developing legislation on safety issues and intellectual property rights and the commercialization of inventions, and promotes a positive public perception of biotechnology, international relations for research cooperation and biosafety issues.

From a national perspective, the BRI is viewed as a centre of excellence in the transfer of biological, biochemical, biophysical and bioengineering sciences from research to industrial application.

Biotechnology research within the BRI is at present organized into five core research areas as follows:

- The *Basic Biotechnology Programme* is viewed as the central and most important programme aimed at development and capacity building in human resources, and development of infrastructure. This programme will carry out the bulk of basic R&D in biotechnology covering gene mapping and manipulation, DNA analysis, RFLP and RAPD analysis for various medical and legal applications. Also included in the programme are the production of new generations of antibiotics, vaccines, drugs and single-cell proteins as well as the development of cell cultures.
- The plant biotechnology section of the *Agricultural Biotechnology Research Programme* covers plant tissue culture and micropropagation, marker-assisted plant breeding and the development of transgenic plants resistant to pathogens and pests. Development of crop plants with improved and desired traits and provision of biofertilizers and *Rhizobium* inoculums are included as well as reforestation and management of medicinal plants for pharmaceutical purposes. The animal biotechnology section of the programme covers marker-assisted animal breeding for improvement of national beef and dairy herds and small ruminants, artificial insemination and embryo transfer techniques.
- The *Medical Biotechnology Research Programme* (Human and Animal), which includes studies on human genetic diversity, focuses on the production of new generations of antibiotics, anti-allergens, vaccines, monoclonal antibodies, blood derivatives, diagnostic reagents, DNA and RNA probes.
- The *Industrial Biotechnology Research Programme* covers food technology, food processing, quality control and nutritional value of food. The BRI offers services in food analysis, fermentation technology and production of starter cultures. In future it is envisaged that the programme will cover industrial bioprocessing, bioremediation, mineral extraction and sewage treatment.
- The *Environmental Biotechnology Research Programme* provides advisory services on biosafety, drafting of safety guidelines and their implementation, technical assistance on the release of GMOs (contained release and field testing), and risk assessment and management. Services in the areas of conservation of genetic resources and biodiversity are offered and the BRI will assume the role of a watchdog against bioprospecting, biopiracy and intellectual property loss.

During the period 1995-96, the BRI engaged in a number of projects, summarized as follows:

- *Capacity Building* including human resource development and infrastructural development. The BRI also houses and manages the Regional Biosafety Focal Point. The project, which is funded by the Netherlands Directorate-General for International Cooperation (DGIS), aims to build regional capacity in biotechnology, and to establish both an informal focal point for safety in biotechnology and a network for the exchange of information on biotechnology and safety.
- *Research Projects* including the Maize Improvement Research Project supported by DGIS (training of scientists is to be undertaken at CIMMYT, Mexico), and the Cotton Improvement Research Project, a collaborative project between BRI, the Cotton Company of Zimbabwe, DR & SS, the International Service for the Acquisition of Agri-Biotech Applications (ISAAA) and Monsanto Company, USA. Research was also conducted on micropropagation (also funded by DGIS), sorghum (funds sought from the International Sugar Organization through the Common Fund for Commodities) and fermentation technology for indigenous foods (funded by the World Association of Industrial and Technological Research Organizations (WAITRO)).

In future, the BRI hopes to develop closer and more effective links with Zimbabwe's industries, both private and public, as well as with the decision-makers in government, the private sector and the diplomatic and donor communities.

Future Prospects

Forestry

The deforestation that has taken place over recent years needs to be redressed through reforestation programmes. There is a need for more vigorous production and development of nurseries to produce both indigenous plant species and exotic imported ones. Gene technology can be directed towards producing new plant species which are fast-growing, drought tolerant and have improved fibre content for the production of pulp/paper and sawn timber for the construction industry.

Pharmaceuticals and Medicinal Plants

In Zimbabwe there is a rich biological diversity of subtropical and medicinal plant species. This is evidenced by the very strong practice of traditional medicine. The Zimbabwean National Traditional Healers Association possesses a great deal of knowledge on herbal medicines and plants of medicinal value. There is great potential in this area for biotechnology R&D investment to strengthen the pharmaceutical industry.

Pest Control

Crop cultivars of maize, cotton, sorghum, etc. which contain genes for protection against insect pests would be of great benefit to agricultural production in Zimbabwe.

Farming Systems Research

The integration of new technologies into traditional farming systems by involving farmers at all levels is the key to finding the correct balance between food production and population growth, with the aim of reducing the country's reliance on costly imports and food aid programmes.

New Energy Technologies

Biogas production is becoming more prevalent in Zimbabwe. Agricultural and a host of other organic wastes are being utilized. There is still much work to be done to provide an effective package of microorganisms and equipment, and to improve the link between the laboratory and the consumer. Biomass energy is another field that deserves increased attention in Zimbabwe. Plantations consisting of fast-growing shrubs and trees, sugarcane, maize and beetroot for ethanol production can replace inefficient biomass as a fuel.

Animal Husbandry

The BRI is active in the development of genetically engineered vaccines for the prevention and treatment of diseases such as foot-and-mouth disease, rabies, African swine fever, African horse-sickness, etc. A substantial increase in animal supply can be effected by the ability to prevent diseases and death. Bovine trypanosomiasis, one of the most important diseases of cattle in Africa and a major constraint to agricultural and socioecomomic development, deserves particular attention. Besides treatment of disease, genetic manipulation techniques on fertilized eggs are being carried out with the aim of producing animals endowed with greater nutritional value. Improved animal production through embryo transfer, artificial insemination, sire-evaluation and marker-assisted breeding for the improvement of the national beef and dairy herds is being undertaken at the Handerson Research Station, Mazoe through the DR & SS Research Station.

Human Health

The BRI, in collaboration with the Biomedical Research Training Institute and the Faculty of Veterinary Science, is employing biotechnology for the production of several drugs. There is much scope in Zimbabwe for production of vaccines against malaria, trypanosomiasis, schistosomiasis and several other rampant diseases. In its efforts to eradicate smallpox, the World Health Organization (WHO) has already established basic facilities as the first step in the R&D for new vaccines.

Environmental Considerations

It is very clear that for Zimbabwe, any investment in a programme of research that can result in a major reduction of the use of commercial fertilizers as well as a reduction in the use of chemical pesticides through increased use of biological control, will be environmentally friendly. The communal, resource-poor farmer will be able to produce more food for security, with decreased inputs that are harmful to the environment.

Potential for Industrial Growth

Zimbabwe, like other countries of Africa, Asia and Latin America, must embrace and invest in biotechnology research and start to provide funds for both basic and applied research. The developing countries can ignore biotechnology at their own peril and at the risk of missing out on the potential promise. Their industries will not grow and they will miss out on the revenue.

Constraints

In Zimbabwe, R&D investment opportunities are constrained by a number of factors:

- *General economic decline:* General economic decline, symptomatic of most developing countries, results in an inability to invest in R&D in general, and more particularly in biotechnology R&D.
- *Lack of coherent technology policy:* Zimbabwe also has no clear government initiative or support for R&D, and there is no biosafety legislation. The government must commit at least 1% of the GDP to R&D for there to be any significant progress in this area.
- *Inability to exploit genetic resources:* Zimbabwe needs to develop an indigenous capacity to exploit her abundant genetic resources.
- *Lack of private sector investments in biotechnology R&D:* The multinational companies which operate in Zimbabwe invest in R&D in their home countries. The industries which operate in Zimbabwe are externally owned, with their mother companies located in the USA, UK and European countries as well as South Africa.
- *Lack of an indigenous industrial base:* Zimbabwe lacks a strong indigenous industrial base, which results in lack of private sector investment in biotechnology R&D. Programmes in biotechnology R&D depend on the generosity of the donor community. Industrial development and growth must be an outgrowth of national investment.
- *Absence of universally accepted legislation:* In Zimbabwe, as in the majority of developing countries, there is no accepted legislation covering intellectual property rights. The country remains a supplier of biological materials which are converted to value-added products in industrialized countries. As yet, there is no mechanism to ensure mutual benefit-sharing between the owners of the genetic resources and the holders of proprietory technologies.

References Cited in Country Profiles

Academy of Sciences of Cuba (1996) *Annual reports of nationally awarded results, 1993-1996.*

Anonymous (1980-1996) Supplements of the Annual Conventions of the Venezuelan Association for the Advancement of Science-AsoVAC. *Acta Cient. Venezolana*, Vols. 31-47.

Arcia, A., Otaiza-V.E., Rivas, N. and Mendible, J.C. (1995) Documento base del área de Biotecnología, CONICIT-BID, Programa de Nuevas Tecnologías, Caracas, 84 pp.

Ayala, F.J. (1995) Science in Latin America. *Science,* 267, 826-827.

BIOMUNDI Counsellors (1994) *Biociencias en Cuba (Biosciences in Cuba), Directorio 1994-95,* 2nd edn, BIOMUNDI Counsellors, Calle 200 No. 1922 e/ 19 y 21, Playa, Ciudad de La Habana, Cuba. P.O. Box 16015, Habana 11600 (Fax: +53-7-336529; E-mail: direct@biotec.cu).

BIOMUNDI Counsellors (1997) *The Presence of Cuban Publications from the Biomedical Sector in International Databases. Report to the West Havana Scientific Pole. February, 1997.* BIOMUNDI Counsellors, Calle 200 No. 1922 e/ 19 y 21, Playa, Ciudad de La Habana, Cuba.

Bunders, J., Haverkort, B. and Hiemstra, W. (eds) (1996) *Biotechnology; Building on Farmers' Knowledge.* MacMillan Education Ltd., London and Basingstoke, UK, 240 pp.

Cabinet publication (1994) *HI-Tech Companies in Egypt.* Information and Decision Support Centre, Technology Development Programme.

Centre for Genetic Engineering and Biotechnology (CGEB) (1997) *Research and Development Profile of the Divisions.* Havana, Cuba.

CNIGB (Comisión Nacional de Ingeniería Genética y Biotecnología) (1985) *Informe de la Comisión*, Caracas, 126 pp.

CONACYT (1996) *México. Indicadores de actividades científicas y tecnológicas*. Consejo Nacional de Ciencia y Tecnología, México.

CONICIT (1996) *Indicadores de la capacidad de investigación en Ciencia y Tecnología 1995*. La Galera de Artes Gráficas, Caracas, pp. 58-65, 165-170.

Craveiro, A.M. and Guedes T.M.M. (1996) *Infra-estrutura Cientifica e Tecnologica. Infra-estrutura tecnologica: Perfil das Empresas Brasileiras de Biotecnologia*. Workshop Biodiversidade: Perspectivas e Oportunidades Tecnologicas. Campinas, Brazil, 29 April to 4 May, 1996.

CSIR-FRD AMI (1993) Report on Potential Markets for Biotechnology in South Africa, 1993.

Cuban Group for the Study of Recombinant Streptokinase (1996) *National extension of the use of recombinant streptokinase in acute myocardial infarct. Final Report*. Havana, Cuba.

da Costa, L.N. (1995) Future of science in Latin America. *Science* 277, 827-828.

Díaz-Otero S. (1996) *Report by the President of the Science and Technology Agency of the Ministry for Science, Technology and Environment to the XIII General Assembly of the Ibero-American Programme for Science and Technology (CYTED), November, 1996*. Ministry for Science, Technology and Environment, Havana, Cuba.

Directory of Biotechnology in South Africa (1995) FRD Programme Report Series No. 21, Pretoria.

Eastmond, A. *et al.* (1992) Agrobiotechnology in Mexico Faces the Challenge of the Free Trade Agreement. *Advanced Technology Assessment Systems* 9, 107-114. United Nations, New York.

Gopo, J.M. and Mushita, A.T. (1997) Development of a Model IPR-Patent Legislation for Zimbabwe. In: *Proceedings of the National Workshop on the Development of a Model IPR-Patent Legislation in Zimbabwe*. October 9-11, 1996.

Hernández, A.G., Mendible, J.C., Arcia, A., Revel-Chion, L. and Celis, L.G. (1991) Biotechnology Worldwide: Venezuela. In: Coombs, J. and Campbell, P.N. (eds) *Biotechnology Worldwide*. CPS Press, Science House, Newbury, UK, pp. 224-230.

Jaffé, W.R. and Trigo, E.J. (1993) Agrobiotechnology in the developing world. Trends, issues and policy perspectives. In: Tzotzos, G.T. (Ed.) Biotechnology R&D Trends, *Annals of the New York Academy of Sciences* Vol. 700, 111-127.

KARI-NARP (1995) *Phase II Project Preparation Report 1995*. Kenya Agricultural Research Institute (KARI), National Agricultural Research Programme, Nairobi, Kenya.

Kathen, A. D. (1996) The impact of transgenic crops releases on biodiversity in developing countries. *Biotechnology Development Monitor* 28, September, 10-14.

KFAS (1995) *Research Projects Directorate, Annual Report*. Kuwait Foundation for the Advancement of Sciences, Kuwait.

Kirkpatrick, A.F. (1996) Role of USA in shortage of food and medicine in Cuba. *The Lancet*, 348, 1489-1492. (See also the editorial in the same issue: Sanctions on health in Cuba. *The Lancet*, 348, 1461, 1996.)

KISR (1995) *Fourth Strategic Programme Plan 1995-2000*. Kuwait Institute for Scientific Research, Kuwait.

Kuwait (1989) *Plan of Work for the State of Kuwait in Selected Priority Areas in the Field of Biotechnology*. National Committee on Genetic Engineering and Biotechnology, Kuwait.

Kuwait University (1994) *Sponsored Research 1992*. Kuwait University, Kuwait.

Kuwait University (1995) *Research Abstracts Volume VIII (1993)*. Kuwait University, Kuwait.

MAC (Ministry of Agriculture) (1995) *Jornadas de reflexión y diagnóstico sobre la investigación agrícola en Venezuela. Conclusiones*. Compil.: Chicco, C.F. and L. Vásquez, FONAIAP, Maracay, 117 pp.

Mandudzo, D.T. (1996) Policy Paper on a Model IPR-Patent Legislation for Zimbabwe. In: *Proceedings of the National Workshop on the Development of a Model IPR-Patent Legislation in Zimbabwe*. October 9-11, 1996.

Mendible, J.C., Otaiza-V.E. and Rivas, N. (1994) *Informe sobre la situación actual del área de Biotecnología*, Programa de Nuevas Tecnologías (CONICIT-BID), 25 pp.

Ministry for Public Health (1996a) *National Programme for Hepatitis, 1996.* Vice-ministry for Hygiene and Epidemiology, Ministry for Public Health, Havana, Cuba.

Ministry for Public Health (1996b) *Annual Report, 1996.* Ministry for Public Health, Havana, Cuba.

Ministry for Science, Technology and Environment (1996) *Annual Report, 1996.* Havana, Cuba.

Musa, T.M. (1996) Strategies for Sustainability and Use. In: *Proceedings of the National Workshop on the Development of a Model IPR-Patent Legislation in Zimbabwe.* 9-11 October.

Omuse, J.K. (1993) Resource allocation and financing of science and technology. Paper presented at the National Workshop on Science and Technology Management for Development, Kenya.

Otaiza-V.E. and Arcia, M.A. (1997) Plant biotechnology in Venezuela: a myth? A critical evaluation to determine its status and social impact. *Interciencia* 22, 238-246.

PADCT (1995) *Programmea de Apoio ao Desenvolvimento Cientifico e Tecnologico, Assessoria de Programmeas Especiais.* Ministerio da Ciencia e Tecnologia. Brasilia, DF, Brazil.

Pilari, D. (1997) Policy issues related to the transfer and sustainable use of biotechnologies. In: Mulongoy, K.J. (Ed.) *Transboundary Movement of Living Modified Organisms Resulting from Modern Biotechnology: Issues and Opportunities for Policy-Makers.* The International Academy of the Environment, Geneva, Switzerland, pp. 3-11.

Quintero, R. (1994) Retrospectiva de la biotecnología en México. *Biocit Siglo XXI* 3, 8. CIT-UNAM, México.

Rangel-Aldao, R. (1996) South American starter cultures. With a little planning, commercial biotechnology was kick-started in South America. *Nature Biotechnology* 14, 951-952.

Simeûn-Negrìn R.E. (Minister for Science, Technology and Environment) (1996) *Strategy of Science and Technology in Cuba.* Opening Lecture at the Ibero-American Seminar on Modern Tendencies of Management on Science and Technological Innovation (IBERGECYT'96). Havana, Cuba.

Sociedad Mexicana de Biotecnología y Bioingeniería (1996), Biotecnología en México. CD-ROM Edition, Sociedad Mexicana de Biotecnología, México.

Webster, J.R. and Koch, P.A. (1995) South African Plant Biotechnology. In: Altman, D.W and Watanabe, K.N. (eds) *Plant Biotechnology Transfer to Developing Countries.* R.G. Landes Co., Austin, Texas, pp. 70-83.

Zakri, A.H. (1992) *Commercialization of Biotechnology in a Developing Economy.* Penerbit, Bangi, Malaysia.

Appendix I
Directory of Research Centres

Following is a list of research centres carrying out research and development in biotechnology arranged by alphabetical order of country. The list does not claim to be exhaustive or representative of the state of biotechnology research and development in the individual countries. The research centres included are those which replied to the questionnaire circulated to over 500 centres worldwide and for which accurate information was therefore available. The questionnaire is given in Appendix II.

Argentina

CENTRE FOR PHOTOSYNTHETIC AND BIOCHEMICAL STUDIES [CEFOBI]
(CENTRO DE ESTUDIOS FOTOSINTETICOS Y BIOQUIMICOS [CEFOBI])

Director Dr Rubén H. Vallejos

Contact Person(s) Dr Hugo R. Permingeat

Address Suipacha 531
Rosario 2000
Argentina

Tel: +54 41 371955 **Fax:** +54 41 370044

E-mail: rncefobi@alpha.arcride.edu.ar

Year Established 1976 **Total Number of Staff** 33

Research Activities

Genetic engineering of cereals. Plant biotechnology. Photosynthesis.

Research Facilities

Cold rooms. Greenhouse. Major capital equipment includes: Protein purification facility; ultracentrifuges; refrigerated centrifuges; harvesting centrifuge; spectrophotometers; HPLC; 'Phast' system; scintillation counters; 'Beta-Ram' for measuring beta-emission in HPLC eluates; automated PCR system.

Collaborative Projects

Long Ashton Research Station, Bristol, UK: Improvement of the nutritional and processing properties of wheat by genetic engineering. Centro de Investigacion y Desarrollo, CSIC, Barcelona, Spain: Aplicación de genes de defensa cuya expresión es inducida por infección con el hongo *Fusarium*. Joint projects also established with Washington State University, Lincoln, USA, Nebraska University, Lincoln, USA and Osnabruck University, Osnabruck, Germany.

Brazil

BIOTECHNOLOGY GROUP, CHEMISTRY DIVISION, TECHNOLOGICAL RESEARCH INSTITUTE OF SÃO PAULO
(AGRUPAMENTO DE BIOTECNOLOGIA, INSTITUTO DE PESQUISAS TECNOLÓGICAS DO ESTADO DE SÃO PAULO S.A. - IPT)

Director Dr Vera Maria L. Ponçano A. Silva

Contact Person(s) Margaretti Simões Oliveira

Address Av. Prof. Almeida Prado, 532
São Paolo 05508-901
Brazil

Tel: +55 11 2682211 ext. 315 **Fax:** +55 11 8693566

E-mail: krajevic@dce03.ipt.br

World Wide Web: http://www.ipt.br

Year Established 1975 **Total Number of Staff** 29

Research Activities

Biotechnological production of biodegradable polymers from sugarcane. Bioinsecticides and veterinary vaccines. Bioinoculants for agriculture. Development and study of fermentative processes and/or genetic improvement of microorganisms involved in the production of a number of products (vaccines, enzymes, antibiotics, vitamin C). Maintenance of industrial organisms, modelling, automation and control of fermentative processes.

Research Facilities

Microbiology laboratory facilities, containment level P2. Industrial fermentation laboratory containing twelve fermentors from 1 to 100 litre vessel capacity. Major capital equipment includes: fermentors; HPLC; GC; spray-drier; freeze-drier.

Collaborative Projects

Microbiology Institute, Münster, Germany, and Biomedical Sciences Institute ICB-USP, Brazil: Molecular biology of industrial strains of microorganisms. Unicamp, University of Campinas, Brazil: Veterinary vaccines. Copersucar, Brazil: Biodegradable polymers.

Publications

Fourteen publications in peer-reviewed journals over past five years.

Commercial Products

Technology transfer to industry. Bioinoculants. Automation of fermentative process - ethanol production. Biodegradable polymers.

CENTRE OF BIOTECHNOLOGY, BUTANTÃN INSTITUTE
(CENTRO DE BIOTECNOLOGIA, INSTITUTO BUTANTÃN)

Director Dr Ana Maria Moro

Contact Person(s) Dr Ana Maria Moro; Prof Isaias Raw

Address Av. Vital Brazil, 1500
São Paolo 05503-900
Brazil

Tel: +55 11 8143790 **Fax:** +55 11 8151505

E-mail: anammoro@usp.br

Year Established 1985 **Total Number of Staff** 76

Total Annual Budget US$ 500,000

Research Activities

Vaccines for human use. Biopharmaceuticals (surfactants, erythropoietin, monoclonal antibodies). Haemoderivatives.

Research Facilities

Fermentation plant. Clean area for production of animal cell derived proteins. Downstream processing plant. Major capital equipment includes: Biopilot; ÄKTA/FPLC; HPLC; fermentors; hollow fibre bioreactor; spectrophotometer; ultracentrifuge; beta counter.

Collaborative Projects

FIOCRUZ/Instituto Adolfo Lutz, Rio de Janeiro, Brazil: Development of anti-meningitis vaccine B/C.

Publications

Forty-three publications in peer-reviewed journals over past five years.

Commercial Products

Vaccines: DPT, tetanus, diphtheria, rabies. Sera: Antivenoms, antithymocytic monoclonal antibodies, venom enzymes.

COCOA RESEARCH CENTRE [CEPEC]
(CENTRO DE PESQUISAS DO CACAU [CEPEC])

Director Dr Raúl Raná Valle

Contact Person(s) Dr Raúl Rená Valle; Dr João Louis Pereira

Address Km 22 Rod. Ilheus
Bahia
Itabuna
Brazil

Tel: +55 73 2143201 / 3215 **Fax:** +55 73 2143204 / 3215

E-mail: cepec@ceplac.gov.br

Year Established 1957 **Total Number of Staff** 903

Total Annual Budget US$ 7.3 million

Research Activities

Plant pathology. Genetics. Biotechnology. Plant physiology. Agricultural entomology and zoology. Soil science. Post-harvest processes (fermentation, drying, etc.). Agricultural economics. Climatology. Social sciences. Crop diversification programme. Computer science and information technology.

Research Facilities

Specialized laboratories. Greenhouses. Analytical equipment.

Collaborative Projects

Penn State University, USA: Projects in molecular biology. Centre de Coopération International en Recherche Agronomique pour le Développement (CIRAD), Montpellier, France: Projects in molecular biology. EMBRAPA Research Centres, Brazil: Projects in genetics and crop development. Ongoing collaborative projects with various domestic universities in the areas of plant pathology, molecular biology and fermentation technology.

Publications

Numerous publications in peer-reviewed journals over past five years.

Commercial Products

Cocoa by-products; plant material.

INSTITUTO DE TECNOLOGIA DE ALIMENTOS (ITAL)

Director Dr Nelson Jose Beraquet

Contact Person(s) Dr Nelson Jose Beraquet; Dr Airton Vialta; Dr Terezinha de Jesus G. Salva

Address Av. Brasil, P.O. Box 139
Campinas
SP 13.073-001
Brazil

Tel: +55 19 241 5222 ext. 190 **Fax:** +55 19 241 5034

World Wide Web: http://www.ital.org.br

Year Established 1969 **Total Number of Staff** 453

Total Annual Budget US$ 10 million

Research Activities

Food Science and Technology covering the areas: juices, pulps, concentrates and canning of fruits and vegetables; industrial fermentations; post-harvest of fruits and vegetables; baking; beverages; edible oils; dehydrated foods; fish and marine resources; meat and derivatives; dairy products; engineering properties of foods; microbiology; quality control; pesticides residues; sweets and chocolate; packaging; grain storage and technology.

Research Facilities

Twenty-eight laboratories. Twelve processing pilot plants. One fermentation plant.

Collaborative Projects

Centro de Referencia para Lactobacilos (CERELA), Argentina: Evaluation of antimicrobial peptides in pathogens control in foods. Instituto de Investigaciones Biologicas (INSIBIO), Argentina: Evaluation of antimicrobial peptides in pathogens control in foods.

Publications

Twenty-eight publications in peer-reviewed journals in last five years.

Commercial Products

Coconut milk without preservatives. Dried milk flour for food supplement. 'Fish-stick' made of fish by-products. Enriched rice. Enriched refined sugar.

NATIONAL RESEARCH CENTRE FOR GENETIC RESOURCES AND BIOTECHNOLOGY - CENARGEN
(CENTRO NACIONAL DE PESQUISA DE RECURSOS GÉNETICOS E BIOTECNOLOGIA - CENARGEN)

Director Dr Afonso Celso Candeira Valois

Contact Person(s) Dr Afonso Celso Candeira Valois

Address SAIN Parque Rural, Final da W5 Norte
Brasília 70.770-900
Brazil

Tel: +55 61 3403501 **Fax:** +55 61 3403624

E-mail: valois@cenargen.embrapa.br

World Wide Web: http://www.cenargen.embrapa.br
Year Established 1974 **Total Number of Staff** 278
Total Annual Budget US$ 6.5 million

Research Activities

Genetic resources programme: Interchange and quarantine; collection, characterization, conservation, evaluation, documentation and use of germplasm (vegetal, animal and microbial). Biotechnology and biological control programmes: Molecular biology (genetic engineering), cellular biology (cell, organ and tissue cultures), biological control of pests, diseases and weeds, besides molecular characterization of species.

Research Facilities

About forty laboratories (multiple uses). Greenhouses. Animal facilities. Biotery. Experimental farm. Electron microscope.

Publications

One hundred and eighty publications in peer-reviewed journals over past five years.

Commercial Products

Bioinsecticide using *Bacillus sphaericus* as a base for control of the urban mosquito. A high-pressure hand-held helium device for biobalistic mediated gene expression. Embryo freezer for farm use. Software to identify fungus on seeds and treatment recommendations.

Bulgaria

INSTITUTE OF WHEAT AND SUNFLOWER 'DOBROUDJA' (IWS)

Director Prof Dr Peter Ivanov

Contact Person(s) Prof Dr Peter Ivanov

Address near Gen. Toshevo
9520
Bulgaria

Tel: +359 58 27454 **Fax:** +359 57 314448

Year Established 1951 **Total Number of Staff** 284

Total Annual Budget US$ 20.3 million

Research Activities

Wheat breeding. Sunflower breeding. Legume breeding. Genetic engineering. Crop technology research.

Research Facilities

Laboratories. Greenhouses. Experimental fields, machinery. Scientific and technical information bureau.

Collaborative Projects

Institute of Genetic Engineering, Kostinbrod, Bulgaria: Transfer of powdery mildew resistance to bread wheat. Semundo Saatzucht, Germany: New gene pools of B and Rf lines in sunflower. ICGEB, Trieste, Italy: Tagging of alien genes for fungal resistance in stable wheat-*Aegilops* aneuploid lines. Collaborative projects also underway with seed production companies in Bulgaria and following foreign companies. Rustica Prograin Genetiques, France: Genetic transformation in sunflower; Limagrain Genetics: Genotypes adapted to the *in vitro* culture of sunflower; Hilleshög NK: Genetic transformation in sunflower.

Publications

Twenty publications in peer-reviewed journals over past five years.

Commercial Products

Wheat varieties: Pryaspa, Yassen, Slavyanka, Shabla, Dobrotiza, Svilena. Varieties of winter triticale: Belitza, Persenk. Sunflower varieties: Albena, Super Start, Obitel, Viola, Dobrich. Bean cultivars: Astor, Voulkan, Prelom, Abritous. Lentil cultivars: Naslada, Stela, Stanka1, Stanka2. Seed production.

Cameroon Republic

IMMUNOLOGY - BIOTECHNOLOGY LABORATORIES (IBL)

Director Mr Jacob L. Ngu

Contact Person(s) Mr Jacob L. Ngu

Address B.P. 4563, Nlongkak
Yaoundé
Cameroon Republic

Tel: +237 20 3796 **Fax:** +237 20 3796

Year Established 1992 **Total Number of Staff** 10

Total Annual Budget US$ 100,000

Research Activities

Onchocerciasis ('River Blindness'): novel diagnostic tests for onchoceriasis based on detection of specific antigens in tears, urine and dermal fluid. Medicinal plants: screening of plant extracts for the following biological properties: a) regeneration of kidney cells; b) anticancer activity; c) lactogenic activity in cattle. Molecular biology: screening of bacteria from one of the oldest rain forests for new restriction enzymes. Plant extracts with growth-promoting properties useful for micropropagation of tubers.

Research Facilities

Facilities for cloning and expression of genes. Facilities for screening for restriction enzymes produced by bacteria. Tissue culture facilities. Chromatographic techniques. Ultracentrifugation equipment. Immunological studies. Animal house.

Collaborative Projects

New England Biolabs: New restriction enzymes produced by rare bacteria from the 'Korup' forest.

Commercial Products

Onchocerca diagnostic tests. The Centre has developed and field-tested three new diagnostic tests for river blindness - DSIA, TADA, DIA for detecting specific antigens in tears, skin and urine respectively; the specificity of the tests is 100%, and their respective sensitivities are 100%, 97% and 96%. Twenty-two gene clones are being expressed and evaluated for a vaccine for 'river blindness'. Plant extracts with growth-promoting properties useful for micropropagation of tubers have been identified.

China

FRESHWATER FISHERIES RESEARCH CENTRE OF CHINESE ACADEMY OF FISHERY SCIENCES

Address Qitang
Jinagsu Province
Wuxi
China

Tel: +86 510 580 1424 **Fax:** +86 510 580 3304

E-mail: cc.ffrc@public1.wx.js.cn

Year Established 1978 **Total Number of Staff** 224

Total Annual Budget US$ 0.7 million

Research Activities

Application of modern biotechnology in fisheries. Genetic improvement and breeding of aquatic animals. Health management in aquaculture. Protection of aquatic environment. Resource management and aquaculture in inland open waters. Integrated fish farming. Fisheries economics.

Research Facilities

Seven laboratories with various specialized research equipment and instruments (including electronic microscope). Experimental fish farm with six hectares fish ponds of different sizes. Training facilities adequate for various international and domestic training.

Collaborative Projects

Nanjing Geographic and Lake Research Institute, China: Study of aquaculture model on large scale in inland lakes. Yangtze River Fisheries Research Institute, China: External culture of parasites of Japanese eel. Shanghai Fisheries University, China: Selected breeding of *Tilapia.*

Commercial Products

Fish medicines. Water purifying agents and disinfectants. Eggs of *Tilapia nilotica* and *Tilapia aurea* and their hybrids. Eggs of Jian carp (a much improved strain of common carp developed by Centre).

HEBEI ACADEMY OF AGRICULTURAL AND FORESTRY SCIENCES

Contact Person(s) Li Guangmin, President

Address No 598, Western Heping Road
Hebei Province

Shijiazhuang City
China
Tel: +86 311 704 4613 **Fax:** +86 311 704 2490
Year Established 1958 **Total Number of Staff** 1834
Total Annual Budget US$ 2.5 million

Research Activities

Main crops variety breeding and heterosis utilization including gene transfer, tissue culture, germplasm recreation, etc. Farming system and crop cultivation technology for high yield, quality and benefit including plant stress-resistance physiology research and plant growth regulator research. New fertilizers and their application technology. Plant protection including chemical and biological control. Vegetable and fruit variety breeding and cultivation including non-virus seedling culture.

Research Facilities

Plant genetics and variety breeding centre. Plant stress resistance physiology laboratory. Pests chemical resistance monitoring and new pesticide research laboratory. Fruit tree germplasm and variety breeding laboratory. Dryland farming laboratory with GIS. National centre for semi-arid farming engineering and technology. AGRIS sub-centre and information system.

Collaborative Projects

Chinese Academy of Agricultural Sciences, China: Graduate student training. Academies of Beijing, Tianjin, Henan, Shanxi and Inner Mongolia: Information exchange in administration and research progress. Lethbridge Research Centre; Agriculture and Agri-Food Canada: Hebei dryland farming project.

Publications

Centre publishes Acta Agriculturae Boreali-Sionica (quarterly).

Commercial Products

Variety and hybrid seed of main crops including wheat, corn, cotton, millet, soybean and vegetables. Normal and non-virus seedlings of fruit trees such as apple, grape. High effectiveness and low toxicity pesticide. Compound element and micro-element fertilizer. Farm machinery and agro-product and by-product processing machines. All the above are in commercial production.

INSTITUTE FOR MOLECULAR BIOLOGY, NANKAI UNIVERSITY
Address Weijin Road 94
Tianjin 300071
China

Tel: +86 22 350 8419 **Fax:** +86 22 350 1180

Year Established 1980 **Total Number of Staff** 38

Total Annual Budget US$ 150,000 to 200,000

Research Activities

Molecular genetics: structure and function of bacterial plasmids; expression and regulation of eukaryotic genes. Enzyme engineering: synthesis and characterization of abzymes; enzymatic bioconversion in organic media; immobilization and encapsulation of enzymes. Bioactive materials: DNA immunoadsorbent in haemoperfusion; specific absorbents in blood purification; artificial liver support.

Research Facilities

DNA synthesizer. Ultracentrifuge. UV spectrophotometer. Amino acid analyser. FPLC, HPLC, CO_2 incubator.

Collaborative Projects

Hedong Hospital, China: DNA, Immunoadsorbents for SLE therapy in clinical trials. First Central Hospital, China: Artificial liver support. Donau University, Krems, Austria: Artificial liver support. McGill University, Canada: Use of one waste to clean up another waste.

Publications

Publications in the following international journals: Artificial Organs (1 to 2 papers/year); Artificial Cells Blood Substitute and Immobilization Biotechnology; Plasmid; Nucleic Acids Research; Gene.

Commercial Products

Tetrodotoxin production by extraction from ova of puffer fish (large scale). Phenyl-methylene-sulphonyl fluoride (large scale). The Centre is involved in commercial production.

RESEARCH CENTRE FOR ECO-ENVIRONMENT SCIENCES, ACADEMIA SINICA

Address 15, Shuangqing Road
Haidian District
Beijing
China

Tel: 010 629 23549 **Fax:** 010 629 23563

Year Established 1975 **Total Number of Staff** 458

Total Annual Budget US$ 2.4 million

Research Activities

Environmental analysis. Organic synthetic chemistry. Biotechnology. Systems ecology. Pollution ecology. Regional ecology. Atmospheric chemistry. Atmospheric pollution control. Environmental aquatic chemistry. Polymeric membrane technology. Water pollution control. Environmental information and publication. Environmental toxicology. Chemical engineering. Geochemistry. Soil chemistry. Sustainable development and environmental planning. Environmental impact assessment. Industrial waste resources re-use. Pharmaceutical chemistry. Environmental protection industry.

Research Facilities

High resolution X-ray fluorescence spectrometer with double crystals. High frequency plasma emission spectrograph. Fourier transform infrared spectrometer. GC/MS. X-ray diffraction and fluorescence spectrometer. Laser spectrometer. UV/visible light spectrophotometer. Ion chromatometer. Atomic absorption spectrophotometer. Gas chromatograph. HPLC. Thermal energy analyser. Differential thermal analyser. Computer working station.

Collaborative Projects

Daqing Petroleum Administrative Bureau, China: Technology for production of citric acid by submerged fermentation with *Aspergillus niger*. Urban System Consulting GmbH Berlin, Germany: Ecological study for the landuse of cities.

Publications

Recent publications in following international journals: Biomedical and Environmental Sciences; Biological Trace Element Research; Journal of Environmental Science and Health; Journal of Fermentation and Bioengineering; Journal of Toxicology and Enviornmental Health; Geo-ökodynamik.

Commercial Products

Joint production technology and engineering for paper-pulp making. Ultrafiltration membrane and equipment. Technology and equipment for water purification. Inorganic polymer, floccuflapron poly-aluminium chlorine. Reclamation treatment technique for the chrome leather scraps. The technology of citric acid/sodium citrate production with high yield. New Bt pesticide.

SHANGHAI INSTITUTE OF PLANT PHYSIOLOGY, CHINESE ACADEMY OF SCIENCES

Address 300 Fenglin Road
Shanghai
China

Tel: +86 21 6404 2090 **Fax:** +86 21 6404 2385

E-mail: XUZK@fudan.ihep.ac.cn

Year Established 1944 **Total Number of Staff** 423

Total Annual Budget US$ 1.8 million

Research Activities

Photosynthesis. Biological nitrogen fixation. Plant developmental biology. Plant cell biology. Plant environmental biology. Secondary metabolism of plants and microbes. Plant physiology of nutrition transport in plants. Microbial metabolism. Microbial genetics. Bacteriophages. Molecular genetics (plant and microbe) and space life science.

Research Facilities

Major capital equipment includes: Continuous batch fermentor; FPLC system; capillary electrophyorasis; HPLC; ultracentrifuge; protein/peptide sequencer; DNA synthesizer; particle delivery system; automatic DNA sequencer; phytotron.

Collaborative Projects

Shanghai Academy of Agricultural Sciences, China: Study on monoclonal antibody of the deoxycinalenoin (DON) in wheat.

Commercial Products

Free Radical SP88. Soya protein concentrate.

STATE KEY LABORATORY OF BIOCHEMICAL ENGINEERING

Contact Person(s) Prof X. Yusheng

Address P.O. Box 353
Beijing 100080
China

Tel: +86 010 625 61817 **Fax:** +86 010 625 61822

E-mail: biochem@istic.sti.ac.cn

Year Established 1991 **Total Number of Staff** 63

Total Annual Budget US$ 550,000

Research Activities

Biochemical reaction engineering (scale-up of bioreactor). Biochemical separation engineering (separation and purification of bioproducts and equipment). Biochemical control engineering (control of bioprocesses). Biochemical system engineering (evaluation and optimization of biosystems and bioprocesses).

Research Facilities

Major capital equipment includes: Beckman GS-KR centrifuge; Beckman 56-MC centrifuge; Beckman XI-io ultracentrifuge; Beckman HPLC system; Mettler RM-180 rheometer; Nikon AFX-DX labophot system. Nature CO_2 water-jacketed incubator; Pharmacia LKB phase system; Pharmacia BioPilot system; Sonics and material Vibra Cell T.M.

Collaborative Projects

Institute of Botany, Chinese Academy of Sciences (CAS), China: Large-scale reproduction and embryo formation of plant cells in bioreactors. Institute of Oceanology, CAS, China: A bioreactor for agar and its utilization. Shandong University, China: Principles in separation of cells and cell fragments by block technology. Institute of Microbiology, CAS, China: Immobilized enzyme and bioreactor for 6-APA. Nanjing University of Chemical Technology, China: Inorganic media for bioseparation. Purdue University: Fumaric acid production by *Rhizopus* strains.

STATE KEY LABORATORY OF MICROBIAL TECHNOLOGY, SHANDONG UNIVERSITY

Director Prof Yinbe Qu

Contact Person(s) Dr Guanjun Chen

Address Jinan
Shandong 250100
China

Tel: +86 531 8902610 **Fax:** +86 531 8915234

E-mail: lifezds@sdu.edu.cn

Year Established 1995 **Total Number of Staff** 28

Total Annual Budget US$ 200,000

Research Activities

Bioconversion of renewable biomass resource. Molecular biotechnology. New fermentation process and products. Bioprocess technology. Environmental and natural resources biotechnology.

Research Facilities

Pilot fermentation plant. Major capital equipment includes: GC/LC/MS system; six sets of fermentors (2 - 50 litre vessel capacity); UV-3100 UV-VIS-NIR spectrophotometer; HPLC analysis/preparation system.

Publications

Eighteen publications in peer-reviewed journals over past five years.

Commercial Products

Cellulase preparation. Xanthan gum. Bt biocide. Fungal polysaccharides.

YELLOW SEA FISHERIES RESEARCH INSTITUTE

Director Dr Qisheng Tang

Contact Person(s) Dr Qisheng Tang

Address 106 Nanjing Road
Qingdao 2266071
China

Tel: +86 532 582 2941 **Fax:** +86 532 581 1514

Year Established 1947 **Total Number of Staff** 416

Total Annual Budget US$ 1.1 million

Research Activities

Aquaculture. Marine living resources assessment and management. Marine ecosystem. Environment protective measures and control. Aquatic products and their utilization. Technology on marine research.

Research Facilities

'Bei Dou' marine sceintific research vessel. Xiamomaidou marine culture research base.

Collaborative Projects

Institute of Marine Research, Bergen, Norway: 'Bei Dou' research and fisheries management project.

Publications

Institute publishes the journal *Marine Fisheries Research,* in Chinese with English abstracts (twice yearly).

Colombia

CENTRE OF TROPICAL PARASITOLOGY
(CENTRO DE PARASITOLOGIA TROPICAL, UNIVERSIDAD DE LOS ANDES)

Director Prof Felipe Guhl

Contact Person(s) Prof Felipe Guhl

Address Apartado Aereo 4976
Bogotá
Colombia

Tel: +57 1 2867593 **Fax:** +57 1 2841890

E-mail: fguhl@uniandes.edu.co

Year Established 1970 **Total Number of Staff** 20

Total Annual Budget US$ 120,000

Research Activities

Control programmes for Chagas' disease in Colombia. Molecular biology of trypanosomiasis, amoebiasis and toxoplasmosis. Diagnosis of tropical diseases.

Research Facilities

Well-equipped laboratory for molecular biology studies and diagnostic work.

Collaborative Projects

Universidad de Antioquia, Colombia: Control of insect vectors in tropical diseases. Ministry of Health, Colombia: Epidemiological study of Chagas' disease in Colombia. London School of Hygiene and Tropical Medicine, University of Leiden, The Netherlands, University of Barcelona, Spain and University of Mexico: Field detection and characterization of pathogenic *Entamoeba histolytica.* Collaborative work on parasite control also underway with British Petroleum.

Commercial Products

Antigen products for diagnostic use.

INSTITUTE OF ECOLOGICAL GENETICS AND NEO-TROPICAL BIODIVERSITY

Director Prof Dr Hugo F. Hoenigsberg

Contact Person(s) Prof Dr Hugo F. Hoenigsberg

Address Calle 57, No. 7-45

Bogotá
Colombia
Tel: +57 1 2490401 or 3463257
E-mail: hoenigsb@colciencias.gov.co
Year Established 1996 **Total Number of Staff** 15
Total Annual Budget US$ 40,000

Research Activities

Projects to be initiated in population genetics and specialization in plants, rodents and *Drosophila willistoni.* Systematics and taxonomy in tropical forests of the Amazonian basin. Molecular genetics in transgenic *Drosophila.* Molecular biology in inventory of biodiversity.

Research Facilities

Microscopy. Incubators. Electrophoresis equipment. Equipment for basic molecular biology research.

Collaborative Projects

Collaborative projects being initiated with the Universidad del Rosario and the Escuela Colombiana Medicina in Colombia as well as several botanic gardens, research centres, and university departments in Scotland and England.

Publications

Institute publishes International ASIBE Journal, *Evolución Biológica.*

Croatia

DEPARTMENT OF MOLECULAR GENETICS 'RUDER BOSKOVIC'
(ZAVOD ZA MOLEKULARNU GENETIKU INSTITUTA 'RUDER BOSKOVIC')

Director Dr Nikola Ljubesic

Contact Person(s) Dr Dragutin Petranovic

Address P.O. Box 1016
Zagreb 10001
Croatia

Tel: +385 1 4561037 **Fax:** +385 1 425497

World Wide Web: http://nippur.irb.hr

Year Established 1991 **Total Number of Staff** 67

Total Annual Budget US$ 994,500

Research Activities

DNA recombination: regulation and role played in DNA repair and plasmid stability. Genetics, structure, function and evolution of aminoacyl-tRNA synthetases. Phylogenetically conserved genes of potential importance in biotechnology. Gene expression in plants. Structure and function of photosynthetic apparatus. Organization and evolution of eukaryotic genome. Molecular mechanisms of carcinogenesis. Cell response to genotoxic agents.

Research Facilities

Computer centre, animal facilities, radiation sources, central service for analytical chemistry, NMR centre. Electron microscope, phase contrast/fluorescence microscope, image analysers, Coulter counter, spectroscopes, centrifuges, ultracentrifuges, CO_2 incubators.

Collaborative Projects

ICGEB, Trieste: Genes and enzymes involved in the inhibition of DNA recombination; protein engineering of aminoacyl-tRNA synthetases. Yale University, New Haven, Connecticut, USA (Prof D. Söll): Transfer RNA recognition by class II synthetases.

Publications

Ninety-seven publications in peer-reviewed journals over past five years.

Affiliations/Subsidiaries

ICGEB affiliated centre.

INSTITUTE OF IMMUNOLOGY - DEPARTMENT OF RESEARCH AND DEVELOPMENT

Director Dr Vlatko Silobroio

Contact Person(s) Dr Vlatko Silobroio

Address Rockefellerova 2
Zagreb HR-10000
Croatia

Tel: +385 1 430333 **Fax:** +385 1 277278

Year Established 1961 **Total Number of Staff** 65

Total Annual Budget US$ 500,000

Research Activities

Immunology. Vaccine development. Immunochemistry. Biochemistry. Microbiology. Transfusion technology. Design and production of blood products.

Research Facilities

Laboratories well equipped with standard equipment for research in the respective fields mentioned above; cold rooms; animal farm; animal facilities.

Collaborative Projects

Ruder Boskovic Institute, Zagreb, Croatia: Preparation and properties of novel immunomodulators; the role of oncogenes and B-lymphocyte antigens in B-CLL. Institute of Pharmacology and Department of Chemical Sciences, University of Trieste, Trieste, Italy: Preparation and properties of novel immunomodulators.

Publications

Fifty publications in peer-reviewed journals over past five years.

Commercial Products

Small-scale production of radioimmunoassay kits (T_3, T_4, cortisol, total IgE, specific IgE). The Institute is also involved improving control and production of vaccines, interferon and blood products.

Cuba

BETERA LABORATORIES

Director Dr Julio Beltrán Hernández

Contact Person(s) Dr Julio Beltrán Hernández

Address Ave 102 between 31 & 31b, Marianao
Havana
Cuba

Tel: +53 20 0711 to 14 or 20 3086 **Fax:** +53 20 4710

Year Established 1992 **Total Number of Staff** 76

Total Annual Budget US$ 600,000

Research Activities

Research and development of serum and diagnostic cells of blood groups. Research and development of class I histocompatibility reagents. Purification of proteins from human, animal and vegetal sources. Alternative procedures to obtain human plasma. Development of microtechniques for clinical research. Research and development of drugs with antimitotic and antitumoural effects obtained by chemical synthesis. Research and development of chemical products aimed at eliminating dependence on hallucinating drugs.

Research Facilities

Installations and equipment to obtain human plasma by automatized methods. Installations and equipment to obtain purified proteins from human, animal, and vegetal sources using chromatographic methods in medium-scale production. Preservation facilities for blood cells. Equipment for automatized identification of Class I HLA antigens. Microtechniques for application in the clinical laboratory.

Collaborative Projects

National Oceanology Institute, Cuba: Purification of phycobiliproteins. Hermanos Amejeiras Hospital, Cuba: Production of HLA antiserum. Centre of Haemoderivatives, Cuba: Quantification of anti Rh using microELISA methods; Production of human plasma using alternative automized methods. National Centre of Isotopes and Pharmacia Institute: Development of antitumoural and antimitotic product. National Centre of Nuclear Development (CEADEN): Sterilization of biological products by radiation.

Commercial Products

BETECEL (Reactive Red Cells): used to determine blood group in the ABO System and to detect and identify atypical antibodies; longer storage life (four months) than that of similar products on the market (15 days). Class I HLA antiserum: used to determine histocompatibility antigens in cases of organ transplants, diagnosis of autoimmune diseases, etc. Dry plates for HLA typing: histoBET ABC72 and histoBET B27: used in histotyping donors and receptors and in the positive diagnosis of 'anquilosante espondilitis' and rheumatoid arthritis, respectively. Standard, low toxicity and baby rabbit complement: these are the main source of complement for typing and screening HLA class I, II. Polyclonal haemoclassifier serum of ABO system: obtained from human serum with a high titre of agglutinins from donors tested for HIV, HTLVI, HVC, HBsAg. Phycoentrin, phycocianin and alophycocianin: fluorescent proteins purified from seaweed and *Spirulina.* Phytohaemoagglutinin (PHA) and concavalin A (Con A): lectins obtained from *Phaseolus vulgaris* and *Canavalia ensiformis* beans, respectively. Human albumin, fibronectin, transferrin and other plasma proteins. All products produced on a semi-industrial scale (medium-scale) due to the specific characteristics of these products.

CENTRE FOR GENETIC ENGINEERING AND BIOTECHNOLOGY (CGEB)
(CENTRO DE INGENIERIA GENÉTICA Y BIOTECNOLOGÍA [CIGB])

Director Dr Manuel de Jesús Limonta Vidal

Contact Person(s) Dr Luis Herrera Martínez, Dr José de la Fuente García, Dr Marisel Quintana Esquivel

Address Avenida 31 entre 158 y 190, Cubanacán, Playa, P.O. Box 6162
Havana
Cuba

Tel: +53 7 218008, +53 7 216367 **Fax:** +53 7 336008, +53 7 218070

E-mail: despacho@cigb.edu.cu

World Wide Web: http://www.cigb.edu.cu

Year Established 1986 **Total Number of Staff** 350

Research Activities

Pharmaceuticals: growth factors, immune response modifiers, anti-inflammatory drugs, interferons and cytokines, thrombolytic agents, neurological drugs. Vaccines: human and veterinary. New expression systems in microorganisms, mammalian cells and transgenic animals. Plant and animal biotechnology. Transgenic plants: pest resistance and improved traits in crop plants (sugarcane, rice, potato, tomato, sweet potato). Transgenic animals as experimental models; disease resistance and growth promotion in fish and shrimps. Diagnostics: antibody fragments for diagnosis and other purposes. Recombinant and therapeutic antibodies. Industrial and laboratory enzymes (for sugar industry). Protein structure and protein engineering.

Research Facilities

The Centre comprises a network of four research and production facilities: the principal CIGB facility; the Centre for Biological Research (founded in 1982) in Havana; the Centre for Genetic Engineering and Biotechnology of Camaguey; and the Centre for Genetic Engineering and Biotechnology of Sancti Spiritus. The principal CIGB facility contains laboratories (including P3 area), a fermentation plant, a diagnostics production facility, an animal facility (including P3 and P4 areas) and three greenhouses. The Centre for Biological Research contains research and production laboratories. The Centre for Genetic Engineering and Biotechnology of Camaguey, and the Centre for Genetic Engineering and Biotechnology in Sancti Spiritus comprise research laboratories and animal houses. Major capital equipment includes: mass spectrometer; electron microscopes; fermentors (up to 300 litre vessel capacity); computer network; peptide synthesizer; amino acid analyser; oligonucleotide synthesizer; spectrophotometers; centrifuges; beta and gamma counters.

Collaborative Projects

ICGEB Collaborative Research Programme: Transformation of sweet potato for increasing its nutritional value as food and animal feed. Modification of therapeutic monoclonal antibodies by recombinant DNA techniques. Genetic engineering of rice for pest resistance. Mechanism of cell transformation by HPV-16. Generation of therapeutic monoclonal antibodies against endothelial adhesion molecules. Studies of growth control in fish. Transgenesis for manipulating growth in fish. Epitope mapping of a high molecular weight antigen of *Neisseria meningitidis* and the study of its identity among several isolates. Molecular characterization of geminiviruses.

Publications

More than 158 publications in peer-reviewed journals over past five years.
The Centre publishes the quarterly journal *Biotecnología Aplicada*, where many of the research results from the Centre are published (157 articles over past five years).

Affiliations/Subsidiaries

ICGEB affiliated centre.

Commercial Products

Vaccines: recombinant anti-hepatitis B vaccine (for human use), recombinant anti-tick vaccine (for cattle), recombinant anti- *E. coli* (for swine). Pharmaceuticals: natural human leukocyte interferon (IFN) alpha, recombinant human IFN alpha-2b, recombinant human IFN gamma, recombinant human epidermal growth factor, recombinant streptokinase, recombinant human interleukin-2, natural human transfer factor. Diagnostics kits: anti-HIV (1+2), anti-hepatitis C, anti-*Toxoplasma gondii* (for testing in pregnancy). Monoclonal antibodies (49) for diagnosis, laboratory analytical reagents and protein purification. Laboratory reagents: restriction and modification enzymes, vectors and probes. Industrial enzymes (recombinant alpha amylase, dextranase, lactase). Software for biotechnological research and production (BIOCROM for chromatographic processes, BIOSOS for nucleic acid and protein sequence analysis, FERMACS for fermentation processes).

CUBAN RESEARCH INSTITUTE OF SUGARCANE BY-PRODUCTS (ICIDCA)

Director Dr Miriam Klibansky

Contact Person(s) Dr Miriam Klibansky

Address Via Blanca 804 y C. Central, P.O. Box 4006
Havana
Cuba

Tel: +53 7 983005 or 983003 **Fax:** +53 7 338236

E-mail: icidca@ccniai.cu

Year Established 1963 **Total Number of Staff** 150

Total Annual Budget US$ 2.1 million

Research Activities

Industrial microbiology. Food additives. Biofertilizers. Bioplaguicides. Biopolymers. Pollution control.

Research Facilities

Fermentation pilot plant (fermentors of capacity 2 m^3, 1 m^3, 450 l, 50 l and 5 l). Centrifuges. Evaporators. Driers. Ultrafiltration equipment.

Collaborative Projects

INISAV, Cuba: Development of bioplaguicides. INCA, INICA, Cuba: Development of biofertilizers. IFAL, Cuba: Pharmaceuticals. UNAM, Mexico: Solid state fermentation, bioconversion. CIB, Madrid, Spain: Genetic engineering in fungi. Federal University Paraná, UFPR, Brazil: Solid state fermentation. CIATEJ, Mexico: Utilization of sugarcane derivatives.

Publications

Fifty publications in peer-reviewed journals over past five years.

Commercial Products

Aged rum. Resins. Starter cultures (*Lactobacillus*). Mushrooms. *Rhizobium* cultures.

RESEARCH INSTITUTE OF TROPICAL ROOT AND TUBER CROPS, BANANAS, PLANTAINS, POTATOES AND VEGETABLES

Director Dr S.R. Morales

Contact Person(s) Dr S.R. Morales

Address INIVIT Apartado 6, Villa Clara
Santo Domingo 53000
Cuba
Tel: +53 4 2344 or 2103 **Fax:** +53 42 4 2201
E-mail: inivit@uclv.cu *or* inivit@quantum.cu
Year Established 1967 **Total Number of Staff** 365

Research Activities

Plant breeding. Biotechnology. Culture methods. Biological control. Mineral nutrition and fertilizing methods. Plant pathology. Cytogenetics. Entomology. Seed production. Conservation and study of genetic resources. Nematology. Virology.

Research Facilities

Specialized laboratories for following areas of research: chemistry, bromeliad studies, entomology, plant pathology, microbiology, biotechnology. Installation for storage of seeds of tropical root and tuber crops. Representative soils for field experiments.

Collaborative Projects

IBP (Plant Biotechnology Institute), Cuba: *In vitro* optimization of the adaptation phase of tropical roots and tubers, banana and plantain. INCA, Cuba: Use of steroidal and other natural and synthetic phytohormones and use of *in vitro* mutagenesis in the genetic improvement of tropical roots and tubers, banana and plantain. Bioplanta Ciego de Avila, Cuba: Use of *in vitro* mutagenesis for the genetic improvement of yucca. CEADEN, Cuba: Characterization of induced variability. INISAV, Cuba: Diagnostics and phytopathology of viral diseases in tropical roots and tubers, banana and plantain.

Commercial Products

Seeds of tropical root and tuber crops. Tissue culture plants (dasheen, bananas and plantains). Papaya seeds. Biological control products: *Verticillium lecanii, Bacillus thuringiensis, Trichoderma, Beauveria bassiana.*

Estonia

ESTONIAN BIOCENTRE - INSTITUTE OF MOLECULAR AND CELL BIOLOGY
(EESTI BIOKESKUS - MOLECULAAR JA RAKUBIOLOOGIA INSTITUUT)

Director Dr Richard Villems

Contact Person(s) Dr Richard Villems

Address 23 Riia Str.
Tartu EE2400
Estonia

Tel: +372 7 420443 **Fax:** +372 7 420194

E-mail: rvillems@ebc.ee

World Wide Web: http://www.ebc.ee

Year Established 1985 **Total Number of Staff** 100

Total Annual Budget US$ 4.1 million

Research Activities

Gene therapy and DNA vaccination vectors. DNA chip technology; DNA diagnostics. Bioremediation.

Research Facilities

Transgenic animal facility. DNA sequencing facility. Major capital equipment includes: three DNA sequencers; two-laser cell sorter; a series of Braun fermentors; four 'Spincos'; Origin 200 computer.

Collaborative Projects

Collaborative projects with commercial companies Glaxo Wellcome and Therexsys, UK.

Publications

One hundred publications in peer-reviewed journals over past five years.

Commercial Products

DNA vectors. DNA chips. Aromatic hydrocarbon-degrading bacteria.

Hungary

BIOLOGICAL RESEARCH CENTRE (BRC) - HUNGARIAN ACADEMY OF SCIENCES
(MAGYAR TUDOM ÁNYOS AKADÉMIA SZEGEDI BIOLÓGIAI KÖZPONT)

Director Prof Dénes Dudits

Contact Person(s) Dr László Dallmann

Address P.O. Box 521
Szeged H-6701
Hungary

Tel: +36 62 433434 **Fax:** +36 62 433434

E-mail: kulugy@everx.szbk.u-szeged.hu

World Wide Web: http://www.szbk.u-szeged.hu

Year Established 1971 **Total Number of Staff** 487

Total Annual Budget US$ 5.5 million

Research Activities

I. Institute of Biophysics: molecular neurobiology; stress physiology in plants. II. Institute of Biochemistry: molecular genetic studies on brain glutamic acid decarboxylase; anti-tumour effect of cytokines; molecular studies of genes coding for connective tissue proteins; sequence-specific DNA recognition by restriction and modification enzymes; human mitochondrial genetics; synthesis and radioactive labelling of biologically active compounds. III. Institute of Enzymology: molecular mechanisms of learning and memory; structure-function studies on proteases involved in fibrinolysis and tumour metastasis; production of catalytic antibodies (abzymes); computer-aided study of drug-action; molecular basis of the immune response; protein structure research. IV. Institute of Genetics: bacterial genetics; nitrogen fixation; molecular immunology. V. Institute of Plant Biology: molecular stress- and photobiology; RNA processing; *Arabidopsis* molecular genetics.

Research Facilities

Greenhouses, animal facilities, phytotron. 'Opton' electron microscope; 'Excimer' laser, model Lextra 100; spectrometer, model IFS 66; ultramicrotome; HPLC; spectrofluorimeter, model LS 100-base; phosphor imager analyser; laser microscope, model LSM 410.

Collaborative Projects

ICGEB, Trieste: Cell signalling, molecular biology. Max-Planck Institute, Cologne: *Arabidopsis* genetics. CNRS, Paris: Nitrogen fixation. NIBB, Okazaki: Stress adaptation. Max-Planck Institute, Frankfurt: Biological energy conversion. Biocentrum, Basel: Bioelectronics. National Laboratory, Los Alamos: Protein dynamics.

Publications

Number of publications in peer-reviewed journals: 136 (1992); 190 (1993); 145 (1994); 133 (1995); 130 (1996).

Affiliations/Subsidiaries

ICGEB affiliated centre.

Commercial Products

Electrophoretic systems. Small-scale laboratory equipment.

CENTRAL FOOD RESEARCH INSTITUTE [KÉKI] (KÖZPONTI ÉLELMISZERIPARI KUTATÓ INTÉZET [KÉKI])

Director Prof Dr Peter A. Biacs

Contact Person(s) Prof Dr Peter A. Biacs; Dr Mária Váradi; Dr Bálint Czukor

Address Herman Ottó út 15
Budapest H-1022
Hungary

Tel: +36 1 1558991 **Fax:** +36 1 1558928

E-mail: H9740BIA@Ella.Hu

Year Established 1959 **Total Number of Staff** 100

Research Activities

Research and development in food science and technology.
Fermented foods. Biosensors. Enzymatic modification of food proteins.

Research Facilities

Animal facilities. HPLC, GC, SFE.

Publications

Seventeen publications in peer-reviewed journals over past five years.

India

ATUCON RESEARCH AND DEVELOPMENT CENTRE

Director Dr Ratan C. Paliwal

Contact Person(s) Dr Ratan C. Paliwal

Address A-7, Pushpanjali Enclave
New Delhi 110034
India

Tel: +91 11 7014590 **Fax:** +91 11 7014572

Year Established 1973 **Total Number of Staff** 24

Total Annual Budget US$ 500,000

Research Activities

Centre conducts studies on chemicals; agronomy; agrochemistry, including fertilizers, pesticides and insecticides; drugs and pharmaceuticals; dyes and dye intermediates; cosmetics; petroleum products; synthetic resins; food products and processing; soaps and detergents; perfumes; flavours; essential oils and oleoresins; waste and effluent treatment; environmental impact studies of fibre and reinforced plastic; alternative energy sources; computer aided design; biotechnology and biosynthesis; spice and herb processing and extraction; food and food products adhesives.

Collaborative Projects

Collaborative work with Atul Consultants (P) Ltd. on production of enzymes for detergent powders and with Herbo Agro India Ltd. on citric acid production.

Publications

Research work rarely published in journals.

Commercial Products

Centre not involved in commercial production.

BIOTECHNOLOGY DIVISION, REGIONAL RESEARCH LABORATORY, CSIR

Director Dr R.S. Kapil

Contact Person(s) Dr G.N. Qazi

Address Canal Road
Tawi
Jammu 180001
India

Tel: +91 191 549051 or 549084 **Fax:** +91 191 548607

Year Established 1964 **Total Number of Staff** 25

Total Annual Budget US$ 150,000

Research Activities

Microbial biotechnology. Industrial enzymes. Biotransformations. Genetic engineering of microbes of industrial importance. Agricultural biotechnology.

Research Facilities

Fermentation pilot plant. All requisite facilities for genetic manipulation studies.

Collaborative Projects

NATCO Pharma Ltd., Hyderabad, India: Scale-up studies for the production of 2-ketogluconates. University of Dortmund, Germany: Bioprocessing with enzyme membrane reactors. University of Kaiserslautern, Germany: Genetic studies on keto acid-producing bacteria.

Publications

Twenty-four publications in peer-reviewed journals over past five years.

Commercial Products

Resolution of racemic mixtures of drugs. Glucose oxidase. Lipases. Gluconates. Biofertilizers. Biocides.

CENTRAL DRUG RESEARCH INSTITUTE

Director Dr V.P. Kamboj

Contact Person(s) Dr Alka Singh

Address Chatter Manzil, P.O. Box 173
Lucknow 226 001
India

Tel: +91 522 212411 to 18 **Fax:** +91 522 223405

E-mail: root@cscdri.ren.nic.in or root%cdrilk@sirnetd.ernet.in

Year Established 1951 **Total Number of Staff** 939

Total Annual Budget US$ 6.0 million

Research Activities

Development of new drugs, immunodiagnostics and vaccines. Development of contraceptive agents and devices. Basic biomedical studies to understand disease processes. Systematic evaluation of medicinal properties of natural products. Development of technology for drugs, intermediates and biologicals.

Research Facilities

Broad-based biological screening. Problem area based drug development (antifertility, malaria, filariasis, leishmaniasis, fungal infection, cardiovascular and central nervous system disorders, hepatoprotection, immunomodulation, wound healing, anti-ulcer treatment). Regulatory pharmacology, toxicology, pharmacokinetics and clinical trials. Medicinal chemistry, biopolymers and chemical and fermentation technologies. Regional/national facilities in laboratory animal, sophisticated instrumentation and technical information.

Collaborative Projects

The Institute has collaborative programmes underway with national and international research laboratories, medical centres and the pharmaceutical industry.

Publications

Research contributions are published regularly in national and international journals of repute. Those relevant to biotechnology are published in journals covering this area, namely *Immunology, Gene, Vaccine, Immunotherapy of Infectious Diseases, Biotechnology Letters*, etc.

Commercial Products

The Institute does not directly undertake commercial activity. The products (new drugs, diagnostic agents, vaccines) developed are licensed to established industrial houses for commercialization. Significant products and technologies licensed to industry or commercialized are as follows: New drugs marketed: Gugulipid (hypolipidaemic); Centchroman (non-steroidal oral contraceptive); Centbucridine (local anaesthetic); *Bacopa monnieri* standardized extract as herbal remedy for memory improvement. Drug technologies developed and in commercial production: 1-ephedrine; Dextropropoxyphane hydrochloride; Pyrithioxin.

CENTRAL FOOD TECHNOLOGICAL RESEARCH INSTITUTE

Director Dr V. Prakash

Contact Person(s) Dr V. Prakash

Address Cheluvamba Mansion, CFTRI P.O.
Mysore 570 013
India

Tel: +91 821 565760 or 565603 **Fax:** +91 821 520758

E-mail: director@nicfos.ernet.in

Year Established 1950 **Total Number of Staff** 799

Total Annual Budget US$ 3.44 million

Research Activities

Food biotechnology and bioengineering. Biochemistry, nutrition, fermentation, microbiology, plant cell biotechnology. Food safety, quality control. Traditional foods. Milling, baking, confectionery. Human resources development, business development, technology transfer. Post harvest technology, food processing, food machinery design, solid state and submerged fermentation. Mathematical modelling. Downstream processing. By-products, waste utilization. Fermented foods, beverages.

Research Facilities

Fermentors (laboratory and pilot plant scale) for submerged and solid state fermentation. Downstream processing facility. Pilot plant and workshop. Scale-up facility for food and biotechnological processes for commercial application. Fully equipped analytical quality control laboratory. Package testing laboratory. Instrumentation for advanced research in biological sciences and food biotechnology. Animal house facility. Modern abattoir with meat processing equipment. International school of milling technology with equipment for collaboration with industry. Comprehensive library facilities. Printing and reprographics department.

Collaborative Projects

Australian Council for International Agricultural Research, Australia: Development and application of simple test kits for pesticide residues in plant-derived foods. EC, Brussels: Development of aromatic concentrated juices of mango from developing countries for their export markets. National Institute of Health, USA: Lipid peroxides as promoters and dietary factors as antipromoters. SAREC, Sweden: Cultivation of microorganisms and production of lactic acid. Numerous collaborative projects on food technology and biotechnology with research institutes in India and commercial companies in India and Japan.

Publications

Two hundred and fifty publications in international peer-reviewed journals over past five years.

Commercial Products

The Institute is not involved in commercial producion. However, pilot scale production is undertaken for test-marketing. The major thrust is on development of viable and globally competitive technologies.

CENTRAL INSTITUTE OF FISHERIES TECHNOLOGY

Director Dr K. Gopakumar

Contact Person(s) Dr K. Gopakumar

Address Matsyapuri P.O.
Kerala

Cochin 682 029
India
Tel: +91 484 666845 **Fax:** +91 484 668212
E-mail: cift@x400.nicgw.nic.in
Year Established 1957 **Total Number of Staff** 470

Research Activities
Fishing craft and gear technology. Fish processing. Biochemical and nutritional aspects of fish. Microbiology, fermentation and biotechnological aspects of fish. Quality assurance of seafoods. Marine and fish processing engineering. Instrumentation. Extension research.

Research Facilities
The Institute is solely engaged in research in fishing, fish processing and allied aspects.

Collaborative Projects
Fisheries College, Panangad, India: Development of pelleted feed for prawns in aquaculture. FAO and European Commission: Utilization of low value fish. National Research Institute (NRI), UK: Post harvest technology. Collaboration with several commercial companies in production of surgical sutures and chitin/chitosan.

Publications
The Institute publishes the *Fishery Technology Journal* (half yearly).

Commercial Products
Chitin/Chitosan from crustacean shell waste. Surgical sutures from fish gut. Canned, frozen and dried fish products. Speciality and by-products from fish. (The Institute does not undertake commercial production of any of the above products.)

CENTRE FOR CELLULAR AND MOLECULAR BIOLOGY
Director Prof D. Balasubramanian
Contact Person(s) Dr G. Kranti Kumar (Head, Planning and Management)
Address Uppal Road
Hyderabad 500 007
India
Tel: +91 40 672241 to 672250 **Fax:** +91 40 671195
E-mail: kranti@ccmb.globemail.com

Total Number of Staff 531
Total Annual Budget US$ 3.5 million

Research Activities

Biotechnology and biomedicine: Tumour biology, cataract and other eye diseases, DNA typing and fingerprinting, transgenics and aquaculture, applied processes and products. Genetics and evolution: Molecular basis of sex determination, microbial genetics, extremophiles, sequence analysis of nucleic acids and proteins, mathematical modelling of biological processes. Cell biology and development: Mechanism of action of hormones, molecular basis of maturation/capacitation of spermatozoa, stress proteins and response to stress, molecular basis of development. Molecular biology: Signal transduction, transcription, nuclear pores and lamin proteins, host-guest interactions in plants, biotic and abiotic stresses in plants, enzyme structure, function and regulation. Biochemistry and biophysics: Intracellular transport of proteins, membrane active molecules and membrane organization, receptors of neurotransmitters, protein structure, conformation and function.

Research Facilities

Automatic amino acid analysers, gas-phase protein sequencers, HPLC and FPLC units. Semi-automated and manual peptide synthesis. Pharmacia Gene Assembler™. NMR spectroscopy. Electron microscope (JEOL 100 CX) suite. FACS. Confocal laser scanning microscopy with fluorescence imaging. Other major capital equipment includes an atomic absorption spectrometer, fluorescence spectrophotometer, nanosecond fluorimeter, laser densitometer, circular dichroism spectrometer and gene sequencer. Cell culture facilities. Animal house with specialized services such as immunization of mice for monoclonal antibody production, management of mutant mice stocks, induction of tumours in athymic nude mice. Greenhouse for plant molecular biology work. Pisciculure: three recirculating water systems for intensive fish culture and a hatchery. Comprehensive library facilities.

Collaborative Projects

Osaka University Medical School, Osaka, Japan (India-Japan Cooperative Science Programme): Assembly of the nuclear pore complex. Institute Pasteur, Paris, France: Osmotic regulation of *proU* transcription and interaction of nucleoid protein H-NS with DNA in *Escherichia coli:in vitro* studies. Université de Paris-Sud, Orsay, France: Isolation of genes encoding sterol biosynthetic enzymes from *Ascobolus immersus* and their use to test for a novel mechanism of nystatin resistance in fungi. Defence Research & Development Organization, India: Biodegradation of human wastes. Central Silk Board, India: Molecular characterization of silkworm races by DNA fingerprinting. Identification of sex determining genes and molecular characterization of sex- and tissue-specific DNA binding protein(s) in silkworm. Central Zoo Authority, India: Molecular characterization of Indian wild animals by DNA fingerprinting. Semen analysis and cryopreservation of spermatozoa of Indian wild animals. Bangalore Cancer Research

Foundation, Bangalore, India: Stress in breast cancer patients: a psychoimmunological study. L.V. Prasad Eye Institute, Hyderabad, India: Origin, structure and role of pigments in human lenses. Collaborative projects with commercial companies include the following: Dr. Reddy's Research Foundation, India: Screening methodology for anticancer compounds and evaluation of compounds prepared by the company. Shantha Biotechnics Pvt. Ltd., India: Technical advice and laboratory facilities in preparation of hepatitis B antigen and α-interferon. NATCO Pharma Ltd., India: Synthesis of biologically important peptides, calcitonin, octreotide, etc. Torrent Pharmaceuticals Ltd., India: Synthesis of commercially exploitable peptides and screening for antibacterial activity.

Publications

Sixty-five publications in international peer-reviewed journals over past three years.

DIVISION OF BIOCHEMISTRY

Director Dr S.L. Mehta

Contact Person(s) Dr S.L. Mehta

Address Indian Agricultural Research Institute (IARI)
New Delhi 110012
India

Tel: +91 11 5784038 **Fax:** +91 11 5766420

E-mail: guest%bic-iari@dbt.ernet.in

Year Established 1966 **Total Number of Staff** 64

Total Annual Budget US$ 60,000

Research Activities

Biological nitrogen fixation. Development of strains of *Lathyrus sativus* with low content of ODAP (β-N oxalyl L, -α, β-diamino propionic acid). Application of molecular biology to improve 'chapati' and bread-making characteristics of wheat.

Research Facilities

Ultracentrifuge. Growth chamber. PCR equipment. UV spectrophotometer. HPLC. DNA sequencer. Refrigerated high speed centrifuge. Photodyne.

Publications

Journal of Plant Biochemistry and Biotechnology (twice a year).

Commercial Products

Somaclones of *Lathyrus sativus* with very low content of ODAP (β-N oxalyl L, -α, β-diamino propionic acid) have been developed. These are being tested in coordinated field trials throughout India.

INDIAN INSTITUTE OF CHEMICAL BIOLOGY

Director Dr Jyotirmoy Das

Contact Person(s) Dr Anjana Majumder

Address 4 Raja S.C. Mullick Road
Calcutta
India

Tel: +91 33 4733491 **Fax:** +91 33 4730284

E-mail: nandy@x400.nicgw.nic.in

Year Established 1956 **Total Number of Staff** 362

Total Annual Budget US$ 2.3 million

Research Activities

Leishmaniasis. *Vibrio cholerae.* Developmental neurobiology and neurological disorders. Cell biology and immunobiology. Chemistry of bioactive substances.

Research Facilities

Leishmania strain bank (National facility). GC-MS system. NMR installation. Electron microscope. Animal house. FACS automatic cell sorter. Automatic DNA sequencer. Equipment for pharmacological and microbiological analysis.

Collaborative Projects

Numerous collaborative projects with research institutions in India. With the University of Düsseldorf, Germany: Action of neurotoxins on thyroid activity and in parkinsonism.

Publications

One hundred publications in biotechnology in international peer-reviewed journals over past five years.

Affiliations/Subsidiaries

Constituent establishment of CSIR.

Commercial Products

Anti-ulcer drug from Neem bark. Anti-cholera vaccine. (No commercial production.)

INSTITUTE OF MICROBIAL TECHNOLOGY

Director Dr C.M. Gupta

Contact Person(s) Dr C.M. Gupta

Address Sector 39A
Chandigarh 160036
India

Tel: +91 172 690785 or 690132 or **Fax:** +91 172 690585 or 690632

E-mail: cmg@imtech.ernet.in

Year Established 1984 **Total Number of Staff** 154

Total Annual Budget US$ 2 million

Research Activities

Molecular biology and microbial genetics. Fermentation technology. Protein engineering. Immunology including animal cell/tissue culture.

Research Facilities

Scanning and transmission electron microscopes. Ultraspeed and superspeed centrifuges. HPLC/FPLC. Amino-acid analyser. Gas chromatograph. Protein sequencer and protein synthesizer. Fermentation pilot plant (1500 litre capacity) and associated downstream processing equipment. Alpha digital and Microvax II computer systems. Modern facilities to identify, preserve and maintain microorganisms. LN_2 plant. Protein X-ray crystallography facility. P3 laboratory facility. Animal house.

Collaborative Projects

Collaborative projects underway with the Central Scientific Instrument Organization, Chandigarh, India and the Indian Oil Corporation R&D Centre, Faridabad. Collaboration also with the following commercial companies: Vittal Mallya Scientific Research Foundation, Bangalore and IOC R&D Centre, Faridabad.

Publications

Regular publications of research work in international peer-reviewed journals.

NATIONAL RESEARCH CENTRE ON PLANT BIOTECHNOLOGY

Director Dr R.P. Sharma

Contact Person(s) Dr R.P. Sharma

Address Indian Agricultural Research Institute (IARI)
New Delhi 110012
India

Tel: +91 11 5788783 or 5785080 **Fax:** +91 11 5766420

Year Established 1985 **Total Number of Staff** 106

Total Annual Budget US$ 286,000

Research Activities

Increasing co-productivity through exploitation of hybrid vigour. Management of biotic stresses. Management of post-harvest quality in crop plants. Molecular plant breeding. Bioprospecting.

Research Facilities

The Centre is fully equipped with major facilities needed for molecular biology and biotechnology research.

Collaborative Projects

Division of Genetics, IARI, India: Molecular tagging of cooking quality traits in rice. Central Cotton Research Institute, Nagpur, India: Development of Bt transgenic cotton. Central Potato Research Institute, Simla, India: Development of Bt transgenic potato. Rockefeller Foundation: Use of transgenic techniques in rice improvement. The Centre received assistance from UNDP from 1986 to 1991.

Publications

Regular publications in international peer-reviewed national and international journals.

Commercial Products

Mustard variety through tissue culture. Non-sporic related *Bacillus thuringiensis* strains. (The Centre is not involved in commercial production.)

SCHOOL OF BIOTECHNOLOGY, MADURAI KAMARAJ UNIVERSITY

Director Prof K. Dharmalingam

Contact Person(s) Prof K. Dharmalingam

Address Madurai Kamaraj University
Madurai 625 021
India

Tel: +91 452 859280 or 859115 **Fax:** +91 452 859105

E-mail: mkubic@giasmd01.vsnl.net.in

Year Established 1994 **Total Number of Staff** 96

Total Annual Budget US$ 300,000

Research Activities

Antibiotics. Biopesticides (Bt based). Rice transformation. Plant viruses. Fungal molecular biology. Protein crystallization. High resolution graphics.

Research Facilities

Greenhouse. Animal house. Insectarium. Fermentor (100 litre vessel capacity). High resolution graphics installation.

Publications

Twenty-two publications in peer-reviewed journals over past five years.

TROPICAL BOTANIC GARDEN AND RESEARCH INSTITUTE

Director Dr P. Pushpangadan

Contact Person(s) Dr P. Pushpangadan

Address Karimancode, Pacha-Palode P.O.
Kerala
Trivandrum
India

Tel: +91 471 437230 or 431178

Year Established 1979 **Total Number of Staff** 203

Research Activities

Horticulture and garden development. Biotechnology. Orchidology. Microbiology. Ethnomedicine. Phytochemistry. Ethnopharmacology. Conservation biology. Plant systematics and evolutionary science. Ecological economics and environmental planning. Drug development. Studies on edible wild plants. Mushroom cultivation. Essential oils and aroma chemicals extraction. Validation, verification and standardization of herbal drugs. Primary healthcare programme.

Research Facilities

Research facilities for post-graduate studies in areas mentioned above. Training offered in mushroom cultivation, floriculture, landscape gardening, tissue culture, plant-based drug development, essential oils and aroma production, validation, verification and standardization of herbal drugs.

Collaborative Projects

Numerous collaborative efforts with other research centres in India as well as the following organizations abroad: Royal Danish School of Pharmacy, Denmark; University of Singapore, Singapore; World Wide Fund; International Foundation of Science, Sweden and IPRGI, Italy. Collaboration with the companies Arya Vaidya

Pharmacy, Coimbatore, India and Jhawar Chemicals, Bombay, for the commercial production of drugs.

Publications

Occasional reports of work carried out at Institute in the following publications: *Plant Cell Reports,* Springer Verlag, USA; *Plant Cell, Tissue and Organ Culture,* The Netherlands; *Indian Journal of Experimental Biology.*

Commercial Products

No commercial production undertaken, but Institute has developed 18 herbal formulations, a biofertilizer and a mushroom soup, whose production will be undertaken by commercial companies subsequent to technology transfer.

Indonesia

INDONESIAN INSTITUTE OF SCIENCES - R&D CENTRE FOR BIOTECHNOLOGY

Director Dr Made Sri Prana

Contact Person(s) Dr Made Sri Prana

Address Jalan Raya Bogor Km. 46
Bogor
Cibinong 16911
Indonesia

Tel: +62 021 875 4627 **Fax:** +62 021 875 4588/3651

E-mail: kapus@biogtw.biotek.lipi.go.id or prana001@rad.net.id

Year Established 1987 **Total Number of Staff** 150

Total Annual Budget US$ 2.5 million

Research Activities

Animal biotechnology: embryo technology/transfer. Plant biotechnology: rice improvement, tree (forest tree species) improvement. Enzyme technology: inulinase, iso-amylase. Bioactive substances for drug development. Microalgae culture: *Chlorella*, *Spirullina*. Biofertilizers: *Rhizobium.*

Research Facilities

Complete set of facilities for molecular studies: DNA sequencer, PCR equipment, electrophoretic apparatus, ELISA, equipment for RFLP analysis. Facilities for genetic engineering: Particle guns, electrofusion apparatuses, micromanipulator, sterile room. Equipment for chemical analysis: HPLC, GC/MS, GC/IR, atomic absorption spectrophotometer, protein analyser. Fermentors of up to 100 litre capacity. Facilities for aqua/microalga culture.

Collaborative Projects

JSPS and JBA, Japan: Microbiological studies. ASEAN; EC: *Rhizobium*, taro, etc.

Commercial Products

Calf embryos. *In vitro* culture derived seedlings (banana, bamboo, ornamental plants). Amyloglucosidase. Microalgae: biomass of microalgae, fish feeds (pellets).

INTER-UNIVERSITY CENTRE ON BIOTECHNOLOGY - INSTITUTE TEKNOLOGI BANDUNG

Contact Person(s) Dr G.N. Qazi

Address Jalan Ganesya 10
Bandung
Indonesia
Tel: +62 22 250 4256 or 250 4987 **Fax:** +62 22 250 4987
E-mail: banliang@indo.net.id

Year Established 1986 **Total Number of Staff** 61
Total Annual Budget US$ 700,000

Research Activities
Microbiology; molecular biology; biochemistry; fermentation technology; downstream processing; waste water treatment.

Research Facilities
Major capital equipment includes: DNA synthesizer; automatic DNA sequencer; peptide synthesizer; peptide sequencer; GLC-MS; GLC (several); HPLC; ultracentrifuges; spectrophotometers; laminar flow cabinets.

Collaborative Projects
Institute Bogor, Indonesia: Project on 1,3-lipases; inulinase secretion. International Institute of Biotechnology, UK: Environmental biotechnology. CSIRO Division of Tropical Animal Production, Australia: Screw worm fly vaccines.

Publications
The Centre contributes regularly to the scientific journal, *BIOSAINS*, in the Indonesian language.

Commercial Products
The Centre is not geared to commercial production; rather it supports graduate education. Graduate student research in the field of biotechnology is carried out at the Centre.

RESEARCH INSTITUTE FOR FOOD CROP BIOTECHNOLOGY (RIFCB)

Director Dr Djoko S. Damardjati
Contact Person(s) Dr Djoko S. Damardjati
Address Jl. Tentara Pelajar 3A
Bogor 16111
Indonesia
Tel: +62 251 338820, 33440, 337975 **Fax:** +62 251 338820
E-mail: dsdjati@server.indo.net.id or rifcb@indo.net.id

Year Established 1995 **Total Number of Staff** 340
Total Annual Budget US$ 905,375

Research Activities
The major areas of research of RIFCB is plant biotechnology with emphasis on: plant biology systems; microbiology and bioprocesses; physiology; entomology; phytopathology; biochemistry; ecology; tissue and cell culture; molecular biology.

Research Facilities
Laboratories for molecular biology, cellular biology and tissue culture, biochemical and enzymatic studies, microbiology, electron microscopy. Pilot plant for mass production of *Rhizobium* sp. Screen and/or glass house. Field experimental station. Library.

Collaborative Projects
IUC for Biotechnology, Bogor Agriculture University; Research and Development Centre for Biotechnology, Indonesian Institute of Science; Centre for the Assessment and Application of Industrial and Agricultural Biotechnology; Indonesian Sugar Research Institute: Genome project for soybean (under preparation). ACIAR: Improved diagnosis and control of peanut stripe virus. Control of bacterial wilt (*P. solanacearum)* by agricultural biotechnology. Cassava safety, development and evaluation of simple test of the agronomic potential of cassava flour and cassava tubers. The Rockefeller Foundation, USA: Population genetics of BLB and blast pathogen. JICA: After-care of ATA 378. IRRI: Population genetics of important rice pathogen and crop improvement in Indonesia (BLB, Blast, and Tungro). Clemson University: Project on integrated pest management. ABSP, Michigan State University, USA: Regeneration study of Indonesian sweet potatoes (*Ipomoea batatas*).

Publications
The Centre publishes the journal *Buletin Agrobio*.

Commercial Products
Biofertilizer for soybean (Rhizo-plus). Biofosfat. The Centre is also involved in pilot-plant scale preparations.

Iran

BIOCHEMICAL AND BIOENVIRONMENTAL RESEARCH CENTRE (BBRC)

Director Dr M Vossoughi

Contact Person(s) Dr M Vossoughi

Address P.O. Box 11365-6891
Teheran
Iran

Tel: +98 21 6005417 **Fax:** +98 21 6012983

E-mail: vossoughi@alborz.sharif.ac.ir

Year Established 1971 **Total Number of Staff** 20

Total Annual Budget US$ 100,000

Research Activities

Biochemical engineering. Environmental technology.

Research Facilities

Greenhouses and fermentation facilities.

Publications

Fifteen publications in international peer-reviewed journals over past five years.

Commercial Products

Single cell protein production. Organic acids. Amino acids.

Malaysia

CENTRE FOR GENE ANALYSIS AND TECHNOLOGY (CGAT)

Director Assoc Prof Dr Rahmah Mohamed

Contact Person(s) Assoc Prof Dr Rahmah Mohamed

Address Centre for Gene Analysis and Technology
Selangor
Bangi 43600
Malaysia

Tel: +60 3 829 3338

E-mail: ram@pkrisc.cc.ukm.my

World Wide Web: http://www.fsh.ukm.my/cgat/

Year Established 1996 **Total Number of Staff** 45

Research Activities

Molecular biology of tropical pathogens and parasites. Molecular systematics for wildlife and marine organisms.

Research Facilities

Animal facilities. Automated DNA sequencing. 372 ABI automated sequencer. Monoclonal antibodies laboratory. Tissue culture facilities. Electron microscope. Library.

Collaborative Projects

Wildlife Department, Malaysia: DNA fingerprinting of species of mousedeer. Fisheries Research Institute, Malaysia: DNA fingerprinting and molecular systematics of freshwater fish. Veterinary Research Institute, Malaysia: Development of DNA methods for diagnosis of soil pathogens.

Commercial Products

Patents applications.

FOREST RESEARCH INSTITUTE OF MALAYSIA [FRIM]
(INSTITUT PENYELIDIKAN PERHUTANAN MALAYSIA)

Director Dr Abdul Razak Mohd. Ali

Contact Person(s) Dr Abdul Razak Mohd. Ali

Address Kepong
Kuala Lumpur 52109
Malaysia

Tel: +60 3 6342633 **Fax:** +60 3 6367753
E-mail: razak@frim.gov.my
World Wide Web: http://frim.gov.my
Year Established 1929 **Total Number of Staff** 530
Total Annual Budget US$ 8.0 million

Research Activities

Research and development in all areas of forestry and forest products, including conservation and biodiversity of the forest. Tropical forestry.

Research Facilities

Nursery. Herbarium. Xylarium. Various laboratories/workshops for testing of timber and other forestry products including seed technology laboratory. Controlled environment facilities. Cold rooms. Field test areas. Plant tissue culture laboratory. Library specialized in forestry publications. Major capital equipment includes: HPLC; GC/MS and mass spectrometer; scanning electron microscope; sawmill; various woodworking machinery, kiln-drying chambers, fire testing furnace; environmental chambers; dynamic strength testing machines; clean rooms for biotechnology research.

Collaborative Projects

Fletcher-Challenge, New Zealand: Tissue culture of rattan and trees.

Publications

FRIM publishes two internationally peer-reviewed journals: *The Journal of Tropical Forest Science* (quarterly) and *The Journal of Tropical Forest Products* (biannual).

Commercial Products

Activated charcoal from wood. Modular drying system for timber. White charcoal and wood briquette production. Electricity generation from gasification of wood. SEMSET, software for forest management. Protection system for rubberwood. Tissue culture of rattan plants and trees (to be commercialized).

MALAYSIAN AGRICULTURAL RESEARCH AND DEVELOPMENT INSTITUTE (MARDI) - BIOTECHNOLOGY RESEARCH CENTRE

Contact Person(s) Dr Hassan Bin Mat Daud, Head Biotechnology Programme

Address P.O. Box 12301
Kuala Lumpur 50774
Malaysia

Tel: +603 943 7617 **Fax:** +603 948 3411
E-mail: mdhassan@mardi.my
Year Established 1990 **Total Number of Staff** 39
Total Annual Budget US$ 882,000

Research Activities

Genetic engineering of plants. *In vitro* culture of plants. Cryopreservation of plant and animal materials. Bioprocess enhancement. Diagnostics. Biosensor.

Research Facilities

Well-equipped laboratories (molecular and cellular biology, *in vitro* technology, immunology, biochemistry, microbiology, fermentation technology, cryopreservation technology). Transgenic glasshouses and ordinary glass houses. Animal holding facilities (animal house and grazing areas).

Collaborative Projects

Agricultural University of Malaysia: Improvement of orchids through genetic engineering. National University of Malaysia: Molecular manipulation and engineering of rice for resistance to diseases. ILTAB/TSRI (Rockefeller Foundation's Rice Biotechnology Network): Development of transgenic rice conferring resistance to tungro virus. QDPI (ACIAR): Pineapple quality improvement.

Publications

Thirty publications in peer-reviewed journals over the last five years.

Commercial Products

Tissue cultured plants (banana, orchids, mangosteen). Cryopreserved embryos (cattle and sheep). The Institute is not involved in commercial production.

TROP BIO RESEARCH SDN. BHD.

Director Dr M. N. Salleh

Contact Person(s) Dr Francis Ng

Address Technology Park Malaysia, Bukit Jalil
Kuala Lumpur 57000
Malaysia

Tel: +60 3 9682003 **Fax:** +60 3 9682888
E-mail: fng@vcmail.tpm.com.my
Year Established 1996 **Total Number of Staff** 22

Research Activities

Plant micropropagation. Plant molecular biology.

Research Facilities

Tissue-culture laboratory. Molecular biology research facilities. Plant nursery with shadehouse. Experimental farm and gene-bank.

Mexico

INSTITUTE OF BIOTECHNOLOGY OF THE NATIONAL AUTONOMOUS UNIVERSITY OF MEXICO [IBT/UNAM]
(INSTITUTO DE BIOTECNOLOGIA - UNIVERSIDAD NACIONAL AUTONOMA DE MEXICO)

Director Dr Xavier Soberón Mainero

Contact Person(s) Dr Augustín López Munguía

Address Av. Universidad 2001, Col. Chamilpa
Morelos
Cuernavaca 62210
Mexico

Tel: +52 73 291600 or 116265 **Fax:** +52 73 172388 or 172399

E-mail: augustin@ibt.unam.mx

World Wide Web: http://www.ibt.unam.mx

Year Established 1982 **Total Number of Staff** 282

Total Annual Budget US$ 8.0 million

Research Activities

Molecular biology and biochemistry of: bacteria, parasites, viruses, plants; genetics and molecular biology of the microorganism-plant interaction; molecular and cellular biology of animals; activation and regulation of immune response; cellular and molecular neurobiology; structure, function and manipulation of peptides and proteins; development and methodological consolidation in molecular biology; industrial microbiology; fermentation engineering and technology of cellular cultures; product recovery and purification; design of process and control equipment; enzyme engineering and technology; prospective biotechnology; integration and optimization of processes and prototypes; technology development; bioinformatics.

Research Facilities

The Centre has 21 well-equipped laboratories housing 36 research groups, organized into five departments: bioengineering; molecular microbiology; molecular genetic biology of plants; molecular recognition and biostructure; genetic and molecular physiology. Technical services and facilities are provided by the following units: laboratory animal house unit; scaling and pilot plant unit; chemical synthesis of macromolecules unit; microscopy unit; greenhouse unit.

Collaborative Projects

University of Arizona, USA: Charaterization of membrane transport mechanisms conferring salinity tolerance in *Mesembryanthemun crystallinum.* University of California, Berkeley, USA: Signal transduction in the root system of *Zea mays.* Max-Planck Institut für Züchtungsforschung: Functional analysis of Zea proteins. Autonomous Metropolitan University of Mexico: Collaboration in training in submerged fermentation technology. John Innes Centre, Norwich, UK and University of Sevilla, Spain: The *Rhizobium tropici* respiratory chain: isolation of mutants with enhanced respiration and symbiotic nitrogen fixation. Instituto de Cibernetica e Biofisica, CNR: Effect of Mexican scorpion toxins on the potassium channels of nervous membranes. Imperial College of Science, Technology and Medicine, London, UK: Structure and function of the transcriptional activator protein NifA from nitrogen-fixing bacteria. Institute of Molecular and Cellular Biology of Plants, Polytechnic University of Valencia, Spain: A functional approach to salt tolerance. BioEurope S.A.: Isolation, production and characterization of glycosyltransferase enzymes for the production of new carbohydrates. School of Medical Sciences, University of Bristol, UK: Construction of potential rotavirus vaccines and evaluation of their capacity for inducing both humoural and cellular immunity. University of Santiago de Compostela, Catholic University of Valparaiso, Universidade do Minho, University of Vigo, Spain: Development of new processes for the extraction of oils and active products from non-conventional oil seeds and vegetables for the pharmaceutical and food industries. University of Maryland, Baltimore, USA: Regulation of enzymes of the uride biosynthesis in tropical varieties of legumes. University of Arizona, USA: Biosurfactant-enhanced remediation of metal contaminated soils. University of Wisconsin, Madison, USA: Molecular properties of intracellular Ca^{2+} channels of striated muscle. Lineal Acceleration Centre, University of Stanford, USA: Access to the synchrotron laboratory. CIMMYT, Mexico: Provision of crystal-spores of certain Bt strains. Centre of Research and Advanced Studies, National Polytechnic Institute, Mexico: Analytical services.

Publications

Three hundred and sixty-two publications in peer-reviewed journals over past five years.

Commercial Products

The Centre does not supply commercial products but some bioprocesses and process improvements developed at IBT have been taken up by domestic industry.

NITROGEN FIXATION RESEARCH CENTRE
(CENTRO DE INVESTIGACION SOBRE FIJACION DE NITROGENO-UNAM)

Director Dr Georgina Hernández

Contact Person(s) Dr Georgina Hernández

Address Av. Universidad s/n, Col. Chamilpa
Morelos
Cuernavaca 62210
Mexico

Tel: +52 73 139877 **Fax:** +52 73 116710

E-mail: direccio@cifn.unam.mx
World Wide Web: http://www.cifn.unam.mx
Year Established 1980 **Total Number of Staff** 144
Total Annual Budget US$ 1.4 million

Research Activities
Metabolic engineering of *Rhizobium*. Structure, dynamics and manipulation of the genome of nitrogen-fixing microorganisms. Molecular mechanisms of the dynamics of the *Rhizobium* genome. Participation of plasmids in the cellular functions of *Rhizobium*. Structure, origin and evolution of the symbiotic plasmid of *Rhizobium etli*. Evolution and ecology of the chemical communication signals between bacteria and plants. Genetic engineering of legumes. Nitrogen fixation in non-legume plants. Theoretical approaches to the regulation of gene expression.

Research Facilities
Greenhouses. Plant tissue culture rooms. Dark rooms. Library. Informatics section. Major capital equipment includes: Electron microscope; DNA sequencer; microscopes; HPLC.

Collaborative Projects
Instituto de Matemáticas Aplicadas y Sistemas, Universidad Nacional Autónoma de México (UNAM), México: Grammatical models of regulation of gene expression. Facultad de Química, UNAM, México: Production and physiological characterization of transgenic plants modified in the ammonium assimilation pathway. PANGEA, USA: Regulatory information in the encyclopaedia of *Escherichia coli*. Magdeburg University, Germany: Grammatical simulation of dynamics of regulation of gene expression in *E. coli*. Instituto Nacional de Investigaciones Forestales, Agrícolas y Pecuarias, México: The effect of reserve polymers of *Rhizobium* in the symbiosis with beans: implications for biological nitrogen fixation in greenhouse and experimental field determinations.

Publications
Fifty-six publications in peer-reviewed journals over past five years.

Morocco

NATIONAL CENTRE FOR COORDINATION AND PLANNING OF SCIENTIFIC AND TECHNICAL RESEARCH

Director Dr Ahmed Jebli

Contact Person(s) Dr Ahmed Jebli

Address 52, Av. Omar Ibn Khattab, B.P. 8027
Rabat 10102
Morocco

Tel: +212 7 774215 or 772803 or 770987 **Fax:** +212 7 771288

E-mail: cnr@mtds.com

Year Established 1981 **Total Number of Staff** 120

Research Activities

Geophysics. Microelectronics. Remote sensing. Renewable energies.

Research Facilities

National network on seismic risk. Maintenance of scientific instrumentation. CNCPRST edition of Moroccan scientists' work.

Collaborative Projects

Associated laboratories/research networks/international collaborating authorities: United Nations specialized agencies (UNU, UNDRO, CNUCED, IDNDR); International Council of Scientific Unions; International Astronomic Union; Third World Academy of Sciences; African Academy of Sciences; Standing Committee on Scientific and Technological Cooperation (Islamic conference-organization); Federation of Arab Scientific Research Council; Regional African Centre of Technology; European Economic Community; Research Centre for International Development; Islamic Foundation for Science, Technology and Development; Centre National de la Recherche Scientifique (CNRS), France; Deutsche Forschungsgemeinschaft, Germany; Consiglio Nazionale delle Richerce, Italy; Consejo Superior de Investigaciones Cientificas, Spain; Instituto Nationale de Geophysica, Italy; Instituto Nationale de Geografia, Spain.

Publications

Moroccan research units directory. Moroccan scientists directory. Lettre d'information. Moroccan sciences abstract.

Nigeria

NIPSSET - SPACE UNIVERSITY - NDONI - RVS (UNESCO)
SPAAR - FOCAL POINT (WORLD BANK)

Director Prof Ephraim Nkem Okeya

Contact Person(s) S. Naglaa

Address c/o 149 Hospital Road
Abia State
Aba
Nigeria

E-mail: z.franca@cgnet.com

Year Established 1977 **Total Number of Staff** 60

Total Annual Budget US$ 1.9 million

Research Activities

Biotechnology and genetic engineering. Lean beef production (genetically determined). Marine biotechnology (marine biology/chemistry). Release of microorganisms (atmosphere). Biological weapons. Genetic studies: human, animal, plant, viral, bacterial. Drugs. Biodiversity. Microbiology (agriculture/food). Environmental studies. Space research.

Research Facilities

Greenhouses (solar). Ranch laboratory. Fermentors. Energy production. Resins. Phyto-laboratory. Satellite communication. Biotechnology laboratory.

Collaborative Projects

New York Academy of Sciences, USA: Biological weapons. ISNAR, The Netherlands: Transgenic studies. University of Greenwich, UK; Open University, The Netherlands: Mycological studies, specifically mushrooms. University of Illinois, Chicago, USA: Pharmaceuticals (phytosource). Institute Pasteur, Paris, France; Institute of Refrigeration, France; Stockholm University, Sweden: Bacterial power (cryogenic). Harvard University/Agribusiness, USA: Transgenic studies.

Commercial Products

Cassahol (ethyl and methyl-based). Cultures/fermentors (genetic/transgenic). Dust particles monitor.

Pakistan

AGRICULTURAL BIOTECHNOLOGY INSTITUTE

Director Dr Azra Quraishi

Contact Person(s) Dr Azra Quraishi

Address Tissue Culture Laboratory, P.O. Box NIH
Islamabad
Pakistan

Tel: +92 51 240972 **Fax:** +92 51 240909

Year Established 1995 **Total Number of Staff** 21

Total Annual Budget US$ 29,000

Research Activities

Tissue culture: virus elimination and micropropagation in potato, date palm, banana, jojoba and ornamental plants. Genetic engineering: protoplast culture and somatic hybridization in tomato and oil crops.

Research Facilities

Glass houses and a screen house.

Publications

Twenty-five publications in peer-reviewed journals over past five years.

Commercial Products

Pre-basic seed potato from tissue culture provided to two seed marketing companies for further multiplication and distribution on a commercial scale.

NATIONAL INSTITUTE FOR BIOTECHNOLOGY AND GENETIC ENGINEERING (NIBGE)

Director Dr Kauser Abdulla Malik

Contact Person(s) Dr Kauser Abdulla Malik

Address P.O. Box 577, Jhang Road
Faisalabad
Pakistan

Tel: +92 41 651475-479 **Fax:** +92 41 651472

E-mail: director@nibge1.fsb.erum.com.pk

Year Established 1994 **Total Number of Staff** 200

Total Annual Budget US$ 800,000

Research Activities

Agriculture: Production of transgenic cotton plants; tissue culture for production of disease-free vegetatively-grown crops like potato, banana, sugarcane; biofertilizer (nitrogen-fixing microbes and growth hormone-producing bacteria) production for field application. Industry: Microbial conversion of lignocellulose biomass into alcohol; production of biochemicals and energy-rich compounds from agricultural waste; bacterial solubilization of metal from low grade ores; biodesulphurization of coal. Health: Molecular diagnostic methods for early detection of tuberculosis, hepatitis C, typhoid; studies on disease control. Environment: Biotechnological methods for detoxification of industrial wastes from textile and pharmaceutical industries; studies on mutagenic effect of these hazardous effluents; development of biosensors for monitoring pollutants in natural ecosystems.

Research Facilities

Two greenhouses. Four fermentors. Pilot plant for waste effluent treatment. Facility for large scale production of biofertilizer. Major capital equipment includes: UV spectrophotometer; ultramicrotome MT-7000 including knife maker and accessories; automatic DNA analyser; ultracentrifuge; scanner (MAGI Scan); electron microscope (JEO, Japan).

Publications

Forty-nine publications in peer-reviewed journals over past five years.

Affiliations/Subsidiaries

ICGEB affiliated centre.

Commercial Products

Agricultural: Biofertilizer 'BioPower' for cereal crops and legumes. Virus-free seed potato. Turn key basis/consultation on development of tissue culture facilities.

NUCLEAR INSTITUTE FOR FOOD AND AGRICULTURE (NIFA)

Director Dr Mohsin Iqbal

Contact Person(s) Dr Amin M. Hussain

Address P.O. Box 446
Peshawar
Tarnab
Pakistan

Tel: +92 0521 261 683 **Fax:** +92 0521 262 733

E-mail: CTC@shell.portal.com.nifa

Year Established 1982 **Total Number of Staff** 173

Total Annual Budget US$ 470,000

Research Activities

Mutation breeding. Soil science. Entomology. Food science.

Research Facilities

About 20 research laboratories, some of them air-conditioned. About six hectares of experimental land; wire-net house. Major capital equipment includes: atomic absorption spectrophotometer, 15 N analyser, neutron moisture probe, intralyzer, laminar flow bench, amino acid analyser, 'Soxtech', 'Kjeltech', 'Fibertech', electrophoresis equipment, food processing equipment, radiation source (Co-60).

Collaborative Projects

NARC, Islamabad, Pakistan: Conducting National Uniform Yield Trials of wheat, chickpea, lentil and oilseed brassica. NIAB, Faisalabad, Pakistan; AEARC, Tandojam, Pakistan: Coordinated projects on mutation breeding, soil science, entomology and food science. IAEA, Vienna, Austria: Two ongoing coordinated research projects (CRP) on food irradiation and haploidy breeding. Many CRPs completed. IFS, Sweden: Evaluation of nitrification inhibitors for potato and its influence on nitrogen losses from labelled urea under various soil conditions. PPIC (Phosphorus & Potash Institute of Canada) - through PARC: Effect of cumulative application of muriate of potash on crops and soil.

Publications

Twenty-three publications in peer-reviewed journals in last five years.

Commercial Products

Squashes (mango, guava, lemon, orange, etc.). Jams and marmalades. These products are produced on semi-commercial scale.

Peru

INTERNATIONAL POTATO CENTRE (CIP)

Director Dr Ali M. Golmirzaie

Contact Person(s) Dr Ali M. Golmirzaie

Address P.O. Box 1558
Lima 100
Peru

Tel: +51 1 4366920 **Fax:** +51 1 4351570

E-mail: CIP@cgnet.com

Total Number of Staff 63

Total Annual Budget US$ 25.2 million

Research Activities

Production systems. Germplasm management and enhancement with emphasis on application of biotechnology. Disease management. Integrated pest management. Propagation crop management. Post-harvest management marketing.

Research Facilities

Plant tissue culture facilities. Facilities for the following areas of research: genetic engineering for plant transformation; DNA manipulation (sequencing, cloning, characterization by PCR, RFLP analysis); molecular virology; serology for plant pathogen analysis; entomology and nematology; screenhouses and nethouses.

Collaborative Projects

Financial and Economy Ministry, Peru: Training and guidance for the improvement of potato seed production quality in Peru. Plant Genetic Systems, Belgium: The production of transgenic potato clones with resistance to potato tuber moth. Axis Genetics Ltd., England: Evaluation of genetically modified potato for resistance to insects. The Weizmann Institute of Science, Israel: Potato clones with genes for defence against bacterial diseases.

Publications

The Centre contributes regularly to many international and national scientific journals such as: *Euphytica, American Potato Journal, Hortscience, Cell Physiology, Theoretical* and *Applied Genetics.*

Commercial Products

CIP does not carry out large scale (commercial) production. Instead, CIP distributes native and enhanced germplasm worldwide to research centres and national agricultural research systems.

Philippines

INSTITUTE OF BIOTECHNOLOGY AND MOLECULAR BIOLOGY

Director Dr Nina Gloriani-Barzaga

Contact Person(s) Dr Nina Gloriani-Barzaga; Ms Jeane U. Balana

Address National Institute of Health Building, University of the Philippines
Manila
Philippines

Tel: +63 2 5255874 **Fax:** +63 2 5284041 or 5211394

E-mail: nbarzaga@cph.upm.edu.ph

World Wide Web: http://www.upm.edu.ph

Year Established 1995 **Total Number of Staff** 26

Total Annual Budget US$ 200,000 (start-up)

Research Activities

Development of diagnostic reagents. Development of indigenous herbal products for antimicrobial, anticancer and other pharmacological applications.

Collaborative Projects

Monash University, Gippsland, Victoria, Australia: Collaborative project on *Culicinomyces*. US-NAMRU-2 in Jakarta, Indonesia: Project on HIV. Gunma University, Japan: Project on malaria. Queensland Institute of Medical Research, Australia: Project on schistosomiasis.

NATIONAL INSTITUTE OF MOLECULAR BIOLOGY AND BIOTECHNOLOGY (BIOTECH)

Director Dr Reynaldo E. Dela Cruz

Contact Person(s) Dr Reynaldo E. Dela Cruz

Address University of the Philippines
Laguna 4031
Philippines

Tel: +63 94 536 2721 to 2725 **Fax:** +63 94 536 2721

E-mail: red@mudspring.uplb.edu.ph or red@biotech.uplb.edu.ph

Year Established 1979

Total Annual Budget US$ 1.0 million

Research Activities

Agricultural and forest biotechnology. Food and speciality products biotechnology. Biotechnology for health. Environmental biotechnology.

Research Facilities

Pilot plant for semi-commercial scale-up operations. Electron microscope laboratory. Central analytical services laboratory. Microbial culture services laboratory (natural repository).

Publications

Institute publishes the *Philippine Journal of Biotechnology* (twice a year).

Commercial Products

NitroPlus, legume inoculant. Mycogroe, reforestation inoculant. Rastenrella vaccine for domestic animals such as cow and water buffalo. Bio-N, inoculant for rice, corn and sugarcane.

Poland

CENTRE OF MOLECULAR AND MACROMOLECULAR STUDIES [CMMS], POLISH ACADEMY OF SCIENCES
(CENTRUM BADAN MOLEKULARNYCH I MAKROMOLEKULARNYCH, PAN)

Director Prof Marian Mikolajczyk

Contact Person(s) Prof Wojciech J. Stec (Department of Bioorganic Chemistry)

Address Sienkiewicza 112
Lodz 90-363
Poland

Tel: +48 42 819744 **Fax:** +48 42 815483 or 847126

E-mail: wjstec@lodz1.p.lodz.pl

Year Established 1972 **Total Number of Staff** 188

Total Annual Budget US$ 2.67 million

Research Activities

Chemistry, stereochemistry and bioorganic chemistry of organophosphorus compounds in combination with other heteroatoms. Synthetic chemistry of organosulphur and organophosphorus compounds and the reaction mechanisms. Stereospecificity and molecular recognition in biological systems; design of new strategies and methodologies of synthesis of biopolymers including their phosphorothioate, phosphorodithioate and methanephosphonate analogues. Protein engineering. Synthesis of silicon-based polymers and copolymers. Kinetics, thermodynamics and mechanisms of polymerization. Synthesis and application models of biopolymers and bioanalogous polymers. Studies on the relationship between supermolecular structure and mechanical, thermal, optical and electrical properties of polymeric materials.

Research Facilities

Laboratory for analysis of organic compounds and polymers: high resolution NMR laboratory; high resolution mass spectrometry laboratory; FT-IR laboratory; laboratory for elemental analysis; laboratory for polymer properties. Bioorganic chemistry department: laboratory for DNA/RNA synthesis; laboratory for protein analysis; cell culture laboratory; laboratory for structural analysis of nucleic acids and proteins (UV- CD). Major capital equipment includes: FT-NMR 500 MHz, 300 MHz, 200 MHz; mass spectrometer; FT-IR; tensile testing machine; automatic elemental analysator; DNA/RNA synthesizers; 420A amino acid analyser; 477A protein sequencer; circular dichrograph CD-6; HPLC laboratory; capillary electrophoresis; centrifuges and ultracentrifuges.

Collaborative Projects

Chiba Institute of Technology, Department of Industrial Chemistry, Narashino, Chiba, Japan (Prof H. Takaku): Improvement of the oxathiaphospholane method of stereocontrolled synthesis of stereoregular oligo(nucleoside phosphorothioate)s. Columbian University, Department of Medicine, New York, USA (Prof C.A. Stein): P-Stereoregular oligo(nucleoside phosphorothioate)s. Synthesis and biological studies. University of Pittsburgh, Department of Biological Sciences, Pittsburgh, USA (Prof L. Jen-Jacobson): Chiral phosphate probes for protein-DNA interactions. Lynx Therapeutics, Inc., Hayward, USA (Prof G. Zon): Stereopure antisense analogues of oligonucleotides. The Katholieke Universiteit Leuven, Rega Institut, Leuven, Belgium (Prof P. Herdevijn); Stereocontrolled synthesis of Heptadeca(nucleoside phosphorothioate)s complementary to Ha-Ras mRNA. Medical University of Lodz, Department of Biophysics (Prof C. Cierniewski): Antisense Inhibition of PAI-1.

Publications

Three hundred and eighty-seven publications in international peer-reviewed journals over past five years.

Affiliations/Subsidiaries

ICGEB affiliated centre.

Commercial Products

Macdafen (Ifosfamide, anticancer drug). Tamoxifen (anticancer drug). Chalk-filled polypropylene.

DEPARTMENT OF ANIMAL REPRODUCTION, NATIONAL RESEARCH INSTITUTE OF ANIMAL PRODUCTION
(INSTYTUT ZOOTECHNIKI, ZAKLAD FIZJOLOGII ROZRODU ZWIERZAT)

Director Prof Jedrzej Krupinski (Director of Institute)

Contact Person(s) Prof Zdzislaw Smorag (Head, Department of Animal Reproduction)

Address Balice
Krakow 32-033
Poland

Tel: +48 12 855711 Ext.: 309 **Fax:** +48 12 856733

E-mail: zsmorag@izoo.krakpw.pl

Year Established 1950 **Total Number of Staff** 23

Total Annual Budget US$ 300,000

Research Activities

Development of effective methods of *in vitro* production of cattle embryos. Rabbit and cattle embryo cloning, embryo splitting and electrofusion. Rabbit embryo vitrification and establishment of effective methods of vitrification of cattle embryos produced by *in vivo* and *in vitro* methods. Genotype selection using laboratory methods and transgenic animals; work concerns rabbit, pig and cattle, in all three of which transgenic animals have been produced. Molecular method of embryo sexing using PCR technique. Experiments on semen separation into 'male' and 'female' enriched fractions using flow cytometry.

Research Facilities

Animal facility and livestock. Laboratory facilities: laboratory for *in vitro* fertilization and culture; surgery room; molecular biology laboratory; semen and embryo bank. Major capital equipment includes: flow cytometer; apparatus for embryo micromanipulation; equipment for ovum pick-up in cattle.

Collaborative Projects

Edison Biotechnology Institute, Ohio University, USA: Transgenesis in rabbit, pig and cattle. Harvard University Medical School, USA: Transgenesis in cattle. INRA, Jouy-en-Josas, France: Transgenesis in cattle. Forschungsinstitut für die Biologie Landwirtschaftlicher Nutztiere, Fortpflanzungsbiologie, Dummerstorf, Rostock, Germany: *In vitro* production of bovine embryos; embryo micromanipulation. Tohoku National Agricultural Experimental Station, Japan: Cloning of mammalian embryos; *in vitro* culture of bovine embryos.

Publications

Fifteen publications in international peer-reviewed journals over past five years.

DEPARTMENT OF BIOCHEMISTRY AND BIOTECHNOLOGY,
AGRICULTURAL UNIVERSITY OF POZNAN
(KATEDRA BIOCHEMII I BIOTECHNOLOGII AR POZNAN)

Director Prof Witold Walerych

Contact Person(s) Prof Ryszard Slomski

Address ul. Wolynska 35
Poznan 60-637
Poland

Tel: +48 61 487202 **Fax:** +48 61 487146

Total Number of Staff 24

Research Activities

Mechanisms of transcription in higher plants. Structure and expression of plant genes. Molecular studies of genes involved in nitrogen fixation. Metabolism and function of dinucleotide oligophosphates. Uptake and transport of cobalt and other microelements in bacteria. Characteristics of plant viruses. Analysis of polymorphic DNA fragments in plants and animals.

Research Facilities

Major capital equipment includes: Automatic DNA sequencer; water purification system; chromatography systems.

Collaborative Projects

INRA, Versailles, France: Transgenic plants. Soil Science Department, University of Minnesota, USA: Soybean nodulation specificity. Department of Biochemistry, University of Texas, USA: Metabolism of dinucleotides.

Publications

Nine publications in peer-reviewed journals over past five years.

DEPARTMENT OF BIOTECHNOLOGY AND FOOD MICROBIOLOGY, AGRICULTURAL UNIVERSITY OF POZNAN

Director Prof Dr Wiodzimierz Grajek

Contact Person(s) Prof Dr Wiodzimierz Grajek

Address ul. Mazobriecko 48
Poznan 60-623
Poland

Tel: +48 61 487354 **Fax:** +48 61 487146

Year Established 1991 **Total Number of Staff** 20

Research Activities

Industrial microbiology: organic acid fermentation; vitamin biosynthesis; denitrification processes; antibacterial substances. Animal tissue culture: insect cell cultures; baculovirus expression system. Plant cell cultures: somatic embryogenesis; secondary metabolites; cryopreservation.

Research Facilities

Pilot plant for fermentation process. Installation for animal tissue cultures. Major capital equipment includes: Industrial bioreactor 1,500 litre capacity; industrial bioreactor 150 litre capacity; bioreactor for animal cell culture 75 litre capacity; one bioreactor of 30 litre capacity and six bioreactors of five litre capacity; ultracentrifuge; industrial centrifuge; spray drier; ultrafiltration apparatus; membrane bioreactor; gas chromatograph; HPLC; industrial shakers.

Collaborative Projects

University of Gdansk, Poland: Production of foreign protein by insect cell/baculovirus expression system. Agricultural University of Warsaw, Poland: Production of somatic seeds of cucumber. Adam Mickievicz University, Poland:

Production of foreign protein by the system insect cells/baculovirus. Medical University of Poznan, Poland: Secondary metabolites of *Gentina* sp. INRA, Nantes, France and University of Sofia, Bulgaria: Production of bacteriocin and starter cultures for food industry.

Publications

Eleven publications in international peer-reviewed journals over past five years.

Commercial Products

Bacterial production of vitamin B_{12}. Water denitrification system. Bacteriocin production by LAB.

DEPARTMENT OF FOOD TECHNOLOGY AND STORAGE
(KATEDRA TECHNOLOGII PRZEMYSLU ROLNO-SPOZYWCZEGG I PRZECHOWALNICTWA)

Director Prof Dr Zdzislaw Targonski

Contact Person(s) Prof Dr Zdzislaw Targonski

Address Agricultural University, Skromna St. 8
Lublin 20-950
Poland

Tel: +48 81 5254770 **Fax:** +48 81 53533549

Year Established 1953 **Total Number of Staff** 21

Total Annual Budget US$ 30,000

Research Activities

Biosynthesis and induction of cellulolytic, xylanolytic and lytic enzymes using imperfect fungi, *Trichoderma reesei, Auerobasidium pullolans* and others. Bioconversion of ligno-celluloses and their utilization for the production of ethanol and xylitol. Production of organic acids and amino acids, e.g. fumaric acid and lysine. Application and production of inulinases from Jerusalem artichoke. Cell immobilization and its use for the production of enzymes, ethanol and organic acids. Production of fodder's yeast, seleno-protein, seleno-cysteine and seleno-methionine enriched yeast biomasses and yeast biomass with elevated content of glutathione.

Research Facilities

Laboratory fermentation plant equipped with bioreactors and accessory installations for biotechnological processes and downstream processing. Analytical instruments including chromatography, electrophoresis, spectrofluorometer, spectrophotometer, UV-VIS and other related instruments. Major capital equipment includes: Bioreactor, Biostat C; Bioflow III (New Brunswick Scientific); HPLC.

Publications

Eleven publications in international peer-reviewed journals over past five years.

Commercial Products

Pectic enzymes for maceration of vegetables. Seleno-protein enriched yeast biomass.

DEPARTMENT OF PLANT GENETICS, BREEDING AND BIOTECHNOLOGY, WARSAW AGRICULTURAL UNIVERSITY
(KATEDRA GENETYKI, HODOWLI I BIOTECHNOLOGII ROSLIN SZKOLA GLOWNA GOSPODARSTWA WIEJSKIEGO W WARSAWIE)

Director Assoc Prof Zbigniew Przybecki

Contact Person(s) Prof Dr Stefan Malepszy

Address Nowoursynowska 166
Warsaw 02-787
Poland

Tel: +48 22 430982 **Fax:** +48 22 430982

World Wide Web: http://alpha.sggw.waw.pl/info/ogr/kghibr

Year Established 1921 **Total Number of Staff** 16

Total Annual Budget US$ 200,000

Research Activities

Somatic embryogenesis (SE) and its use in plant propagation (cucumber, rye). Genes involved in SE (cucumber): their isolation and characterization. Transgenic cucumber thaumathin and rye. Molecular characterization of genes involved in cucumber sex determination.

Research Facilities

Laboratory equipped for plant biotechnology (molecular biology, tissue culture). Experimental field (including computer-controlled greenhouse). Major capital equipment includes: microprojectile; fluorescence microscope; flow cytometer; bioreactors.

Collaborative Projects

Department of Physics, Agricultural University, Szczecin, Poland: Application of luminescence in determination of quality of *in vitro* regenerated plants. Institute of Chemical Engineering, Warsaw Technical University, Poland: Optimization of some physical parameters not damaging plant embryogenic tissue bioreagents. Institute of Vegetables, Skierniewice, Poland: Biochemical identification of different self-incompatible genes in cabbage haploids. Max-Planck Institute for Plant Breeding, Cologne, Germany: Molecular characterization of genes involved in cucumber somatic embryogenesis.

Publications

Fifteen publications in international peer-reviewed journals over past five years.

FACULTY OF BIOTECHNOLOGY, UNIVERSITY OF GDANSK AND MEDICAL UNIVERSITY OF GDANSK
(MIEDZYUCZELNIANY WYDZIAL BIOTECHNOLOGII UNIWERSYTETU GDANSKIEGO I AKADEMII MEDYCZNEJ W GDANSKU)

Director Prof Dr Wieslaw Makarewicz

Contact Person(s) Prof Dr Wieslaw Makarewicz; Prof Dr Anna J. Podhajska

Address Marii Sklodowskiej-Curie 3B
Gdansk 80-210
Poland

Tel: +48 58 461205 **Fax:** +48 58 461205

E-mail: wmakar@amedec amg.gda.pl

World Wide Web: http://www.biotech.univ.gda.pl

Year Established 1993 **Total Number of Staff** 56

Research Activities

Class IIS restriction endonucleases. Phytopathology; pectinolytic enzymes and role of *Erwinia bacteries* as pathogens for potatoes. Heat shock proteins (HSP) as molecular chaperones. Auto regulation of HSP gene expression and substrate specificity of heat shock proteins. Mutagenic interaction of chemical substances with nucleic acids. Viral gene expression in baculovirus system. Mechanism of cytotoxic effect of TNF and the effect of TNF on cell proliferation. Enzymes of purine metabolism in mammalian tissues. Structure and regulatory properties of phosphoprotein phosphatase 2A. Regulation and function of phospholipase C isoenzymes.

Research Facilities

Major capital equipment includes: Protein sequencer; DNA sequencer; oligonucleotide synthesizer; fermentor; micro calorimeter; 'stop-flow' spectrometer; Perkin-Elmer spectrophotometer.

Collaborative Projects

Department of Cell and Molecular Biology, Lund University, Sweden: Identification of protein phosphatase/s acting *in vivo* on phosphodiesterase type 3B. Department of Neurophysiology, University of Arizona, USA: Angiotensin regulation of gene expression. Friedrich Miescher Institute, Basel, Switzerland: Identification of *in vivo* substrates of type 2A protein phosphatase. Institut National des Sciences Appliques, Lyon, France: Regulation of the genes important in pathogenicity of *Erwinia.* Scottish Crop Research Institute, Dundee, Scotland: Diagnostics of local species of bacteria (from Poland) from the genus *Erwinia.* University of Geneva, Switzerland: Heat shock protein. Regulation of gene expression. De Montfort University, Leicester, UK; Institute for Animal Sciences and Health, The Netherlands; Institute of Virology, Slovak Republic: Construction of herpes virus vaccines against helminth infections in calves and sheep. McArdle Laboratory for Cancer Research, Madison, USA: Class IIS restriction modification system.

Publications

Forty-four publications in peer-reviewed journals over past five years.

Commercial Products

Purified restriction endonucleases. Purified viral and plasmid DNA. DNA size markers. Purified methylotransferases and other enzymes.

FACULTY OF CHEMISTRY AND PROCESS ENGINEERING, WARSAW UNIVERSTIY OF TECHNOLOGY
(WYDZIAL INZYNIERII CHEMICZNEJ I PROCESOWEJ POLITECHNIKI WARSAWSKIEJ)

Director Prof Roman Gawronski

Contact Person(s) Prof Ryszard Pohorecki

Address ul. Warynskiego 1
Warsaw 00-645
Poland

Tel: +48 22 6606314 **Fax:** +48 22 251440

E-mail: pohorecki@ichip.pw.edu.pl

World Wide Web: http://www.ichip.pw.edu.pl

Year Established 1970 **Total Number of Staff** 86

Total Annual Budget US$ 1.85 million

Research Activities

Approximately one-third of the research activity is devoted to bioprocess engineering in the following areas: bioreactor engineering, aeration, mixing, influence of hydrodynamic stress on living cells. Solid state fermentation (SSF): enzyme production in SSF, heat and mass transfer in SSF reactors. Baker's yeast production: high-effective aeration systems for yeast cultivation. Enzyme separation and purification by chromatography and precipitation. Chitosan production by fungi. Biological oxidation of sulphur. Bioaccumulation of metals. Remainder of research activity in chemical engineering.

Research Facilities

Chemical engineering pilot plant laboratory. Major capital equipment includes: seven laboratory bioreactors ranging from 2 dm^3 to 15 dm^3; preparative chromatography system (BioPilot, Pharmacia); HPLC; laser Doppler eneometer; microfiltration equipment.

Collaborative Projects

Institute of Microbial Technology, Chandighar, India: Bioprocess conversion of tannin to gallic acid. Department of Biotechnology, Regional Research Laboratory CSIR, Jammu, Tawi, India: Application of solid state fermentation to enzyme production and waste utilization.

Publications

Forty-five publications in international peer-reviewed journals over past five years.

INSTITUTE OF AGRICULTURAL AND FOOD BIOTECHNOLOGY

Director Prof Dr Roman A. Grzybowski

Contact Person(s) Prof Dr Jerzy Czuba

Address Rakowiecka 36
Warsaw 02-532
Poland

Tel: +48 22 6063655 **Fax:** +48 22 490426

E-mail: ibiopros@warman.com.pl

Year Established 1950 **Total Number of Staff** 209

Research Activities

The Institute comprises six departments, including laboratories. Research is divided into four main areas: Biotechnology: fermentation processes (beer, wine, spirits, vinegar), yeasts, organic acids, microbial preparation, improvement of microbial strains. Culture collection of microorganisms of industrial use. Technology of fruit and vegetable products. Food analysis (physical, chemical, physicochemical, microbiological and sensory methods).

Research Facilities

Pilot plant biotechnology section equipped with: fermentors of 5, 30, 150, 300 and 15,000 litre vessel capacity; centrifuges; filters; ultrafiltration to UF 1/2 Alfa-Laval; thin-layer evaporator CT-1B Alfa-Laval; spray drier. Major capital equipment includes: Atomic absorption spectrometer; HPLC; GC; GC 5890 series II, Hewlett Packard; laboratory fermentors; UV-VIS spectrophotometer.

Collaborative Projects

IUPAC, National Institute of Public Health and Environmental Protection, Bilthoven, The Netherlands: Critical analysis of problems related to beryllium content in foods. International Plant-Analytical Exchange (IPE) Programme, Wageningen Agricultural University, The Netherlands: The IPE Programme provides analytical laboratories with a tool for evaluation of the accuracy and precision of analytical methods. Zabolotny Institute of Microbiology and Virology, Ukraine: Production and application of bioproducts of plant origin; studies on the technology of production of microbial enzymes and their application; culture collection management using the modified programme KKP. University of Agriculture, Department of Agricultural Chemistry, Tokyo, Japan: Conservation of culture collections and application of various preservation methods for industrial strains; stabilization of biotechnological traits; new methods of taxonomy. National Food Research Institute, Tsukuba, Japan: Improvement of properties of new strains of baker's yeast by biotechnological methods.

Publications

Eighteen publications in peer-reviewed journals over past five years.

INSTITUTE OF BIOCHEMISTRY AND BIOPHYSICS [IBB], POLISH ACADEMY OF SCIENCES
(INSTYTUT BIOCHEMII I BIOFIZYKI POLSKIEJ AKADEMII NAUK)

Director Prof Dr Wlodzimierz Zagorski-Ostoja

Contact Person(s) Dr Piotr Ceglowski

Address Pawinskiego 5a
Warsaw 02-106
Poland

Tel: +48 39 121623 **Fax:** +48 39 121623

E-mail: secretari@eaibbrain.ibb.waw.pl

World Wide Web: http://www.ibb.waw.pl

Year Established 1957 **Total Number of Staff** 252

Total Annual Budget US$ 3.7 million

Research Activities

Gene sequencing and mapping. Regulation of gene expression at DNA and RNA level. Protein biosynthesis and post-translational modification. Mutagenesis and DNA repair. Messenger RNA sequencing. Metabolism of nucleic acids. Protein kinases. Regulation of enzyme activity. Protein structure-function studies. Computer modelling of peptides and proteins. NMR studies of proteins and peptides. Antiviral and anticancer nucleotides. Polyprenoid structure and function. Nucleus-mitochondria interaction. Molecular plant virology. Plant transformation. Molecular basis of plant and fungi biotechnology. *Arabidopsis* morphogenesis.

Research Facilities

All facilities necessary for molecular biology research including: DNA sequencing and synthesis; protein purification and analysis; climate cameras and greenhouses. Major capital equipment includes: Nuclear magnetic resonance 500 MHz; ultracentrifuges; ABI 310 sequencer; Alf sequencer; SGI work station 'Challenge'.

Collaborative Projects

Collaborative projects underway with following institutes: Max-Planck Institute for Molecular Genetics, Berlin, Germany; Department of Medical and Physiological Chemistry, Uppsala University, Uppsala, Sweden; CNR, Dept. of Pharmaceutical Sciences, University of Padua, Padua, Italy; Institute of Terpenoid Chemistry, Institute of Chemistry, Moldavia Academy of Sciences, Moldavia; State Scientific Centre Russia - NPO 'VECTOR', Koltsovo, Novosibirsk Region, Russian Federation; Polish-French Centre for Plant Biotechnology (organizationally located at IBB-PAS by common agreement between PAS, KBN and the French CNRS).

Publications

Fifty-eight publications in international peer-reviewed journals over past five years.

INSTITUTE OF BIOORGANIC CHEMISTRY OF POLISH ACADEMY OF SCIENCES
(INSTYTUT CHEMII BIOORGANICZNEJ, PAN)

Director Prof Dr Andrzej B. Legocki

Contact Person(s) Prof Dr Andrzej B. Legocki

Address ul. Noskowskiego 12/14
Poznan 61-704
Poland

Tel: +48 61 528503 **Fax:** +48 61 520532

E-mail: ibch@ibch.poznan.pl

World Wide Web: http://www.ibch.poznan.pl

Year Established 1988 **Total Number of Staff** 177

Total Annual Budget US$ 1.75 million

Research Activities

The Institute represents an interdisciplinary unit with its major scientific interests related to bioorganic chemistry, stereochemistry, molecular biology of nucleic acids, gene organization and function, genetics of higher plants, human molecular genetics, X-ray analysis of proteins and nucleic acids, bioinformatics and phytochemistry.

Research Facilities

Greenhouse. NMR laboratory. X-ray laboratory. Major capital equipment includes: DNA sequencer; DNA synthesizer; NMR spectrometer; IR and Raman spectrometer; silicon graphics work station; X-ray diffractometer.

Collaborative Projects

Institut des Sciences Vegetales CNRS, Versailles, France: Organ-specific expression of lupin plant genes during symbiotic nitrogen fixation. Sektion Polymer, Universität Ulm, Germany: Synthetic oligonucleotide libraries and their application for chemoselection of biomolecules and their analogues. Max-Planck-Institut für Molekular Genetik, Berlin, Germany. Structure and function of α-sarcin domain as tested by antisense DNA probes. Howard Hughes Medical Institute, USA: Crystallographic studies of antitumour amidohydrolases. Institut des Sciences Vegetales, CNRS, Gif-sur-Yvette, France: Five projects on characterization of plant genes and their expression during the effective symbiosis of *Lupinus luteus - Bradyrhizobium* sp. system.

Publications

One hundred and twenty-two publications in international peer-reviewed journals over past five years.

INSTITUTE OF HUMAN GENETICS, POLISH ACADEMY OF SCIENCES (ZAKLAD GENETYKI CZLOWIEKA POLSKA AKADEMIA NAUK)

Director Prof Jerzy Nowak

Contact Person(s) Prof Jerzy Nowak

Address Strzeszynska 32
Poznan 60-479
Poland

Tel: +48 61 233011 **Fax:** +48 61 233235

Year Established 1974 **Total Number of Staff** 56

Total Annual Budget US$ 634,000

Research Activities

Function of nucleic acids. Molecular and clinical genetics. Immunogenetics. Mutagenesis. Reproductive biology.

Research Facilities

Well equipped laboratories for molecular biology and cytogenetics research. Isotope facilities. Animal facilities. Major capital equipment includes: DNA sequencer (ALFExpress); Cytoscan; micromanipulator; DNA synthesizer; gamma and beta scintillation counters.

Collaborative Projects

Division of Experimental Medicine, Harvard University, Boston, USA: Tuberoces sclerosis. Department of Surgery, University of Glasgow, UK: Biology of reproduction. Whitehead Institute, Cambridge, USA: Pathology associated with chromosome Y abnormalities. Abteilung für Medizinische Genetik, Universität Ulm, Germany: Chromosome microaberrations in renal carcinoma. Centre of Nutrition and Toxicology, Karolinska Institute, Sweden: Molecular analysis of suppressor genes in cancer. Medical Academy, Poznan, Poland: Extensive collaboration on various areas. Universitätsklinikum Rudolf Virchow, Freie Universität Berlin, Germany: Gene rearrangements in leukaemias.

Publications

Twenty-one publications in international peer-reviewed journals over past five years.

INSTITUTE OF MICROBIOLOGY AND BIOTECHNOLOGY [UMCS]
(INSTYTUT MIKROBIOLOGII I BIOTECHNOLOGII, UMCS)

Director Prof Dr Nikodem Grankowski

Contact Person(s) Prof Dr Nikodem Grankowski

Address ul. Akademica 19
Lublin 20-033
Poland

Tel: +48 81 5375950 **Fax:** +48 81 5333669

Year Established 1970 **Total Number of Staff** 67

Total Annual Budget US$ 145,000

Research Activities

Genetic and immunochemical studies of surface polysaccharides of *Rhizobium*. Pathogenicity of *Sarcobium*, enzymatic degradation of endotoxins. Enzymatic modification (phosphorylation) of the translational apparatus, kinase structure, function and biological activity. Fungal biosynthesis of industrial enzymes (amylases, pectinases, cellulases, dextranase), enzymatic conversion of cellulose biomass. Microbial detoxification of polluted soils, bacterial and fungal vaccines for improving cereal crops. Interferon production by animals and human organisms. Immunomodulation.

Research Facilities

Major capital equipment includes: HP 5980 gas chromatograph with HP 5971 mass spectrometer; Beckman L565 ultracentrifuge; Beckman Optima ultracentrifuge; Beckman LS5000 liquid scintillation counter; bioreactor Celligen Plus; freeze-drier, Labconco Mo 77535 61; fermentor, Brown Biostat 2.

Collaborative Projects

Informal cooperation with the following institutions: Centre for Disease Control, Atlanta, USA; Public Health Laboratory Service, Leeds, UK; Forschungszentrum Borstel, Borstel/Hamburg, Germany; Department of Applied Chemistry and Microbiology, University of Helsinki, Finland; Department of Applied Cell and Molecular Biology, Umea University, Sweden; Centre for Bioremediation and Detoxification, Pennsylvania State University Park, USA; Friedrich-Miescher Institute, Basel, Switzerland; Odensee University, Odensee, Denmark; Centre of Molecular Biology 'Severo Ochea', Madrid, Spain.

Publications

Thirty-eight publications in international peer-reviewed journals over past five years.

INSTITUTE OF MOLECULAR BIOLOGY, JAGIELLONIAN UNIVERSITY (UNIWERSYTET JAGIELLONSKI, INSTYTUT BIOLOGII MOLEKULARNEJ)

Director Prof Tadeusz Sarna

Contact Person(s) Dr Adam Dubin

Address Mickiewicza Ave. 3
Krakow 31-120
Poland

Tel: +48 12 341305 **Fax:** +48 12 336907

E-mail: ADM@mol.uj.edu.pl/ DUBIN@mol.uj.edu.pl

World Wide Web: http://www.mol.uj.edu.pl

Year Established 1970 **Total Number of Staff** 148

Total Annual Budget US$ 1.82 million

Research Activities

Biochemistry: Structure, biosynthesis and biological function of proteins and peptides; structure and dynamics of proteins; function of the photosynthetic apparatus and mechanism of photoinduced chloroplast movement in the cell. Biophysics: Structure and dynamics of macromolecules and membranes; photobiophysics of melanin pigments; design and development of radiospectroscopic instrumentation. Cell biology: Regulation of cell growth and differentiation in tissue culture. Microbiology and immunology: Immunochemistry and genetics of bacteria; regulation of the immune response; interaction of bacteria and immune cells. Molecular genetics: Viral pathogenesis; virus-host cell interaction; effect of viral infection on the synthesis of acute phase proteins and cytokines.

Research Facilities

Greenhouse. Animal facilities. Confocal microscopy laboratory. Protein chemistry laboratory. Major capital equipment includes: HPLC systems; FPLC; PROCISE -

Protein Sequencing System Model 491; liquid scintillation counter 1211; confocal microscopy BioRad MRC 1024 with silicon graphics, Indy; pulse EPR spectrometer; EPR spectrometer, CW - Bruker; Pulse Nd: YAG Laser, Surelight II; optical parametric oscillator magic prism; silicon graphics Indigo 2; spectropolarimeter, JASCO 710; stopped-flow fluorimeter; live-time spectrofluorimeter, ISS K2; ultracentrifuge, Beckman L7; dual-wavelength spectrophotometer AMINCO D-2000.

Collaborative Projects

Medical College of Wisconsin, USA: Biophysics of melanins, oxygen transport, ESR equipment. University of Illinois at Urbana, Champaign, USA: Multifrequency EPR, oxidation processes in biological systems. University of Wisconsin, USA: EPR-imaging. Department of Biochemistry, University of Georgia, USA: Properties of proteinases and their inhibitors. Fakultät für Biologie, Lehrstuhl Zellphysiologie, Universität Bielefeld, Germany: Studies of α-TQ in biomembranes. Laboratory of Photobiology, University of Liege, Belgium: Greening process in plants. Institut für Allgemeine Botanik, Universität Jena, Germany: Involvement of phytochrome in intracellular movements. Universität Berlin, Humboldt, Germany: Modulation of murine hair cycling. University of California, Los Angeles, USA: Structure and dynamics of biomolecules. Collaborative projects also underway with 43 other centres.

Publications

Fifty-seven publications in international peer-reviewed journals over past five years.

INSTITUTE OF ORGANIC CHEMISTRY, BIOCHEMISTRY AND BIOTECHNOLOGY, TECHNICAL UNIVERSITY OF WROCLAW (INSTYTUTU CHEMII ORGANICZNEJ, BIOCHEMII I BIOTECHNOLOGII POLITECHNIKI WROCLAWSKIEJ)

Director Prof Jacek Mlochowski

Contact Person(s) Prof Andrzej Zabza

Address Wybrzeze Wyspianskiego 27

Wroclaw 50-370

Poland

Tel: +48 71 224064 **Fax:** +48 71 224064

E-mail: zabaz@kchfl.ch.pwr.wroc.pl

Year Established 1968 **Total Number of Staff** 62

Total Annual Budget US$ 500,000

Research Activities

Studies on insect pathogenic fungi of the genus *Zoophthora:* growth conditions, secondary metabolites, biotransformations. Biotransformations of organic compounds using microorganisms and enzymes as biocatalysts: biotransformation of terpenoids, organophosphonates and aromatic compounds having anticancer properties. The use of molecular biology for evaluation of the role of ecdysone binding protein. Biodegradation of organophosphonic compounds. Rational design of biologically active substances.

Research Facilities

Major capital equipment includes: NMR, 300 MHz Brucker; FT-IR; GC-MS system, Hewlett Packard; HPLC, Beckman Gold Line; X-ray apparatus, Kuma.

Publications

Fifty-two publications in international peer-reviewed journals over past five years.

INSTITUTE OF PLANT GENETICS, POLISH ACADEMY OF SCIENCE (INSTYTUT GENETYKI ROSLIN PAN)

Director Prof Dr Jerzy Chelkowski

Contact Person(s) Prof Dr Jerzy Chelkowski

Address Strzeszynska 34
Poznan 60-479
Poland

Tel: +48 61 233511 **Fax:** +48 61 233671

E-mail: office@igr.poznan.pl

World Wide Web: http://www.igr.poznan.pl

Year Established 1961 **Total Number of Staff** 116

Research Activities

Genetics of higher plants: distant crossing, chromosome mapping, interaction of stress factors with crop plants, gene expression, protein polymorphism.

Research Facilities

Greenhouses. Experimental farm. *In vitro* culture laboratory. DNA analysis laboratory. Micro-malting system. Major capital equipment includes: HPLC; capillary electrophoresis equipment; DNA sequencer; scintillation counter.

Collaborative Projects

Cornell University, USA: Generation of genetic bases for lupin breeding. Horticultural Research International, UK: Disease management of turnip mosaic virus on oil seed rape. IAEA/FAO: Mutagenesis for interspecific hybrids between *Cuphea* and *Brassica.*

Publications

Ninety-eight publications in international peer-reviewed journals over past five years.

INSTITUTE OF TECHNICAL BIOCHEMISTRY
(INSTYTUT BIOCHEMII TECHNICZNEJ)

Director Prof Edward Galas

Contact Person(s) Prof Edward Galas; Prof Stanislaw Bielecki

Address ul. Stefanowskiego 4/10
Lodz 90-924
Poland

Tel: +48 42 367704 or 313442 **Fax:** +48 42 313402

E-mail: stanb@lodz1.p.lodz.pl deanbiof@lodz1.p.lodz.pl

Year Established 1970 **Total Number of Staff** 56

Total Annual Budget US$ 650,000

Research Activities

Applied biocatalysis: Microbial enzymes, their characteristics, technology of production and ways of application, biocatalysis in organic solvents, enzymes of psychrophiles, immobilized enzymes and microorganisms. Genetic engineering: Construction of recombinant strains. Analysis of nucleotide sequences of genes, preparation of genomic libraries of selected microorganisms and their characteristics. X-ray structural analysis of peptides, their derivatives and other biologically active substances including new medicines. Environmental biotechnology: Selection of microorganisms which degrade pollutants. Technology of fruit and vegetables. Vitamins and enzymatic hydrolysates of proteins. Biosynthesis of microbial polysaccharides (e.g. cellulose) and enzymatic synthesis of new oligosaccharides.

Research Facilities

Fermentation pilot plant with several fermentors enabling development of technologies for production of commercial enzymes. Major capital equipment includes: gas chromatography; several HPLC systems (Beckman); FPLC; HPAEC (Dionex); fermentor 150 litre capacity (Chemap); thin layer evaporator (Alfa-Laval).

Collaborative Projects

Department of Bioscience and Biotechnology, University of Strathclyde, UK: Genetic engineering - isolation and characterization of genes encoding microbial enzymes. Institute of Biochemistry, Russian Academy of Sciences, Russian Federation: Applied enzymology, environmental biotechnology. Institute of Food Biotechnology, Wageningen Agricultural University, The Netherlands: Applied biocatalysis. Institute of Biotechnology, University of Graz, Austria: Applied biocatalysis. Chair of Chemical Enzymology, University of Lomonosow, Moscow, Russian Federation: Enzymology.

Publications

Fifteen publications in international peer-reviewed journals over past five years.

Commercial Products

Bacterial α-amylases. Bacterial cellulose.

LUDWIK HIRSZFELD INSTITUTE OF IMMUNOLOGY AND EXPERIMENTAL THERAPY
(INSTYTUT IMMUNOLOGII I TERAPII DOSWIADCZALNEJ PAN, IM. LUDWIKA HIRSZFELDA)

Director Prof Marian Mordarski

Contact Person(s) Prof Elwira Lisowska; Dr Hubert Krotkiewski

Address Rudolf Weigl St. 12
Wroclaw 53-114
Poland

Tel: +48 71 732274 **Fax:** +48 71 679111

E-mail: krotkiew@immuno.pan.wroc.pl

Year Established 1952 **Total Number of Staff** 210

Total Annual Budget US$ 1.65 million

Research Activities

Cellular immunology; tumour immunology; experimental anticancer therapy; microbiology; immunogenetics; clinical immunology including bone marrow transplantation; virology; glycoimmunochemistry of microbial and tissue antigens; bacteriophage therapy.

Research Facilities

Facilities for molecular biology research. Tissue culture bank of animal and human cell lines (frozen). Breeding centre for inbred, transgenic and 'knockout' mouse strains. Major capital equipment includes: Two FACS installations (BIAcore); chromatograph (Dionex); GLC/MS; two HPLC installations.

Collaborative Projects

Centre of Molecular Biophysics, CNRS and University of Orléans, Orléans, France: Interactions between lectins and glycoconjugates, their role in homing of metastasing cells, therapeutic possibilities. University of Lille, Laboratory of Biological Chemistry, Villeneve D'Ascq, France: Immunotropic properties of lactoferrin.

Structural changes of surface glycoconjugates on cancer cells in comparison with their tumourigenic and invasive properties. Max-Planck-Institut für Molekular Genetik, Berlin, Germany: Initiation of DNA replication in *Streptomyces*; interaction of dnaA protein with oriC region. Institut of Microbiology Academica Sinica, Beijing, China: Molecular biology of *Actinomycetes*; construction of diagnostic probes and gene technology. Bartholin Instituttet, Copenhagen Kommunehospital, Copenhagen, Denmark: Studies on the role of the host factors in tumour progression and in the expression of the metastatic potential. Centre des Études Nucleaires de Grenoble, France: Structural studies of polysaccharide and glycolipid antigens of bacterial origin; Structural studies of glycoconjugate antigens from pathogenic bacteria. Central Region de Transfusion Sanguine de Nantes, France: Monoclonal antibodies against erythrocyte antigens. Forschungsinstitut für Experimentelle Biologie und Medizin, Borstel, Germany: Natürliche Killerzellen und durch Lymphokine aktivierte Killerzellen: Induktion und Regulation *in vitro*. Institute of Immunobiology of Freiburg University, Freiburg, Germany: Carcinoembryonic antigen (CEA) and related glycoproteins. Zielinsky Institute of Organic Chemistry, Moscow, Russian Federation: Structural studies on *Hafnia alvei* 0-antigens; chemical and serological studies on *Citrobacter* 0-antigens and EAC. University of Agricultural Sciences, Department of Chemistry, Uppsala, Sweden: Structural and serological studies on bacterial antigens. University Hospital, Department of Clinical Chemistry, Linkoping, Sweden: Glycosylation alterations in pathological states.

Publications

One hundred and seventy publications in international peer-reviewed journals over past five years.

MICROBIOLOGY AND VIROLOGY CENTRE, POLISH ACADEMY OF SCIENCES
(CENTRUM MIKROBIOLOGII I WIRUSOLOGII PAN W LOOZI)

Director Prof Leon Sedlaczek

Contact Person(s) Dr Wieslaw Kaca

Address ul. Lodowa 106
Lodz 93-232
Poland

Tel: +48 42 492263 **Fax:** +48 42 491633

E-mail: office@cmiwpan.lodz

Year Established 1988 **Total Number of Staff** 31

Total Annual Budget US$ 430,000

Research Activities

Genetics and physiology of the genus *Mycobacterium*: mobile elements, molecular basis of epidemiology and diagnostics; permeability of *M. vaccae* cell envelopes to sterols; genetic determination of steroid-transforming enzymes. Molecular mechanisms of haemostatic processes. Cytokines in antiviral immunity. Immunochemistry of Gram-negative bacteria. Cellular components of the immune system. Chemistry of nucleic acids. Modified nucleotides for diagnostics and therapeutics.

Research Facilities

DNA sequencer. Beckman ultracentrifuge.

Collaborative Projects

Centre for Thrombosis and Vascular Biology, The Cleveland Clinic Foundation, Cleveland, Ohio, USA: The Role of divalent cations in regulation of B3 integrin activity. University of Surrey, School of Biological Sciences, Guildford, UK: Identification of tuberculosis and MOTT infections isolated in Poland and the UK, using RFLP analysis.

Publications

Thirty-six publications in international peer-reviewed journals over past five years.

NATIONAL VETERINARY RESEARCH INSTITUTE
(PANSTWOWY INSTYTUT WETERYNARYJNY)

Director Prof Dr Marian Truszczynski

Contact Person(s) Prof Dr Jan Zmudzinski

Address Al. Partyzantow 57
Pulawy 24-100
Poland

Tel: +48 81 863051 **Fax:** +48 81 862595

E-mail: ekozska@esterka.piwet.pulawy.pl

World Wide Web: http://www.piwet.pulawy.pl

Year Established 1945 **Total Number of Staff** 362

Total Annual Budget US$ 7.77 million

Research Activities

Diagnosis and prophylaxis of infectious and parasitic diseases of animals. Pathophysiology of reproduction and mammary glands. Toxicology and hygiene of food of animal origin. Prophylaxis against zoonoses.

Research Facilities

Animal facilities. Major capital equipment includes: gas and liquid chromatographs; atomic absorption spectrophotometers; electron microscope; liquid scintillation counter.

Collaborative Projects

Institute for Animal Health, UK: Evaluation of recombinant hog cholera protein as a vaccine and antigen in diagnostic tests using baculovirus expression system. Animal and Plant Health Inspection Service (APHIS), USA: Development of national programme of pseudorabies eradication. Agriculture Research Service, USA: Comparison of serological relationship between US and European porcine respiratory and reproductive syndrome virus strains by means of serological tests.

Publications

Thirty-eight publications in international peer-reviewed journals over past five years.

Commercial Products

Vaccines against the following diseases: rabbit haemorrhagic disease; swine infectious pleuropneumonia; swine atrophic rhinitis; rotaviral infections in swine; parvovirosis in foxes and minks. Antigen for serological diagnosis of bovine leukosis.

PLANT BREEDING AND ACCLIMATIZATION INSTITUTE
(INSTYTUT HODOWLI I AKLIMATYZACJI ROSLIN)

Director Prof Dr Henryk J. Czembor

Contact Person(s) Prof Dr A. Aniol; Dr J. Zimny

Address Radzikow
Hlonie 05-870
Poland

Tel: +48 22 725 2611 **Fax:** +48 22 724 4714

E-mail: aaniol@ihar.edu.pl

Year Established 1951 **Total Number of Staff** 535

Research Activities

Development and utilization of biological potential of crop species for plant breeding. Seed science and technology. Collection and maintenance of genetic resources.

Research Facilities

Greenhouses (approx. 5000 m^2). Growth chambers (seven walk-in, five small).

Major capital equipment includes: Two HPLC installations; five seed processing plants.

Collaborative Projects

Scottish Crop Research Centre: Introduction of a barley cultivar mixture. Cost Action 817: Studies on air-borne pathogens in 12 European countries. Eureka EU 1322; Hungary, Greece, Turkey and Poland: Selection of cereals for resistance to environmental stress. Royal Veterinary and Agriculture University, Copenhagen, Denmark: Improvement of nutritive value of rape seed. USDA-ARS: Septora diseases in triticale and wheat.

Publications

Fourteen publications in international peer-reviewed journals over past five years.

Commercial Products

Seven novel cereal cultivars of wheat, rye, barley and triticale, registered in 1996. One cultivar of maize, fava bean, grass and soybean, registered in 1996.

THE NENCKI INSTITUTE OF EXPERIMENTAL BIOLOGY
(INSTYTUT BIOLOGEE DOSWIADCZALNEJ IM. M. NENCKIEGO)

Director Prof Maciej J. Nalecz

Contact Person(s) Dr Jolanta Skangiel-Kramska

Address Pasteur 3
Warsaw 02-093
Poland

Tel: +48 2 6598571 **Fax:** +48 2 225342

E-mail: sekr@nencki.gov.pl

World Wide Web: http:www.NENCKI.GOV.PL

Year Established 1918 **Total Number of Staff** 308

Total Annual Budget US$ 3.76 million

Research Activities

Molecular and cellular neurobiology. Biochemistry. Cell biology. Neurophysiology. Experimental psychology.

Research Facilities

Animal house. Tissue culture unit. Isotopic laboratory. Data processing laboratory. Major capital equipment includes: Confocal laser scanning microscope; three scanning electron microscopes; three ultracentrifuges; two image analysers; a 'Magical'; two patch clamp recording installation; a motility assay installation; a spectrofluorometer.

Collaborative Projects

The Institute of Bioorganic Chemistry, Byelorussian Academy of Sciences (BAS), Minsk, Byelorussia: Biochemistry, biological membranes, signal transduction. Institute of Cytology, Russian Academy of Sciences (RAS), St. Petersburg, Russian Federation: Structure-function relationships in actin. Sechenov Institute of Evolutionary Physiology-Biochemistry RAS, St. Petersburg, Russian Federation: Bioenergetic processes in mitochondria and cells of lower vertebrates. Institute of Theoretical-Experimental Biophysics, RAS, Pushchino, Russian Federation: Energy regulation of the cell. Institute of Mental Health of Russian Academy of Sciences, Tomsk, Russian Federation: Cell biology and molecular biology. Westfälische Wilhems-Universität, Münster, Germany: Biochemistry, biological membrane signal transduction. University of Bern, Switzerland: Biochemistry, biological membrane signal transduction. University of Linkoping, Linkoping, Sweden: Neurophysiological mechanisms of vision. University of Rherkov, Ukraine: Energy metabolism of the cell. Open University, Milton Keynes, UK: Neuronal mechanisms of plasma changes in the vertebrate brain: learning and memory. University of Ibadan, Nigeria: Bioenergetics; molecular biology; cell biology; neurosciences; biotechnology.

Publications

Sixty-five publications in international peer-reviewed journals over past five years.

Romania

INSTITUTE OF BIOTECHNOLOGY, BIOTEHNOS, S.A.

Director Prof Dr. Ioan Mânzatu

Contact Person(s) Dipl Eng Cornel Dumitrescu

Address 18, Dumbrava Rosie St.
Bucharest 70254
Romania

Tel: +40 1 2102015 **Fax:** +40 1 2109705

E-mail: dcornel@cbb.bth.ro

World Wide Web: http://www.bth.ro

Year Established 1993 **Total Number of Staff** 270

Total Annual Budget US$ 1.8 million

Research Activities

Biotechnology: product, basic and applied technologies, through-processing units, measuring and control devices, automation and computer control. Products from fine chemical and biochemical synthesis and enzymatic biosynthesis. Drug and cosmetic products. Activated and structured aqueous solutions. Environmental protection: research studies, feasibility projects covering measuring devices, automation and computer control of the process units for environmental protection and recovery of re-usable waste products. Artificial intelligence and robotics. Instrumental analysis: pattern recognition systems. Software products and applications.

Research Facilities

Homogenizing and tube-filling line for pharmaceutical creams; full vial-filling line; production plant for sterile purified water; production plant for water for injectables; vial-testing machine; clean steam generator; digester; dielectric spectroscopy measuring chain; rotating evaporator; HPLC; flow cytometer.

Collaborative Projects

Institute for Electronic Structure and Lasers, Foundation for Research and Technology - Hellas, Crete, Greece.

Publications

Twenty-four publications in peer-reviewed journals over past five years.

Affiliations/Subsidiaries

ICGEB affiliated centre.

Commercial Products

Range of creams and skin products.

RESEARCH INSTITUTE OF CHEMISTRY

Director Dr Sever Serban

Contact Person(s) Dr Sever Serban

Address Sp. Independentei 202
Bucharest R-77208
Romania

Tel: +40 1 6153299 **Fax:** +40 1 3213492

Year Established 1981 **Total Number of Staff** 22

Total Annual Budget US$ 160,000

Research Activities

Collection and characterization of microorganisms. Improvement of microbial biosynthetic potential by mutagenesis and protoplast fusion. Production of animal and vegetable extracts for pharmacological, chemical and agricultural use. Production of enzymes by biosynthesis and extraction processes. Elaboration of experimental models for use of biocatalysis in low-water biphasic systems. Bioconversion of gaseous substrates using immobilized enzymes. Biosensors for automatic analysis in biotechnological processes. Bioprocessing of vegetable waste using basidiomycetes. Enzymatic hydrolysis of lignocellulosic materials. Complex biotreatment of lignocellulosic residues. Artificial intelligence (neural network, genetic algorithms, fuzzy and expert systems in bioprocess control).

Research Facilities

Microbiology laboratory and microorganism collection. Biochemistry laboratory: small-scale installation for recovery and disruption of cells and apparatus for physico-chemical and biochemical analysis. Bioreactors. Small-scale pilot plant for biosynthesis and bioconversion. Bioinformatics laboratory.

Collaborative Projects

Institute for Biology, Bucharest, Romania: Mutant strains for cellulase production. Polytechnic University, Bucharest, Romania: Artificial intelligence and neural networks. Cantacuzino Institute, Bucharest, Romania: Animal cell cultivation. Chemical and Pharmaceutical Research Institute, Bucharest, Romania: Enzyme production. University of Bucharest, Romania: Joint training for students. Collaborative work underway with following companies: Romprocef S.A.: Pulp whitening by microbiological methods; Cerpi S.A.: Proteolytic enzymes; Bioing S.A.: Bioengineering, improved bioreactors.

Commercial Products

Enzymes: horseradish peroxidase R2; alcohol oxidase from *Hansenula polymorpha* in solution; proteolytic and amylolytic enzymes; cellulolytic enzymes. Natural extracts from: *Calendula officinalis; Flores tiliae.* Software for bioprocess control. Bioreactors.

Saudi Arabia

GENETIC ENGINEERING AND BIOTECHNOLOGY PROGRAMME,
RESEARCH INSTITUTE OF NATURAL RESOURCES AND ENVIRONMENT,
KING ABDULAZIZ CITY FOR SCIENCE AND TECHNOLOGY (KACST)

Director Dr Mohammed Al-Suwaiyel (Vice President, Research Institute)

Contact Person(s) Dr Mohammed Al-Suwaiyel

Address P.O. Box 6086
Riyadh 11442
Saudi Arabia

Tel: +966 1 4813612 **Fax:** +966 1 4813611

Year Established 1996 **Total Number of Staff** 4

Research Activities

Plant molecular genetics. Industrial biotechnology, bioremediation.

Collaborative Projects

King Faisal Specialist Hospital and Research Centre, Riyadh, Saudi Arabia: Study of glutelin and prolamin proteins and corresponding genes in Al-Hassawi rice.
Petroleum Energy Centre, Tokyo, Japan: Biorefining process for oil desulphurization.

Singapore

WORLD HEALTH ORGANIZATION COLLABORATIVE CENTRE FOR RESEARCH AND TRAINING IN HUMAN GENETICS

Address National University of Singapore
Singapore 119074
Singapore

Tel: 77 24420 or 772 4112 **Fax:** 77 97486

E-mail: paesec@nus.sg

Year Established 1978 **Total Number of Staff** 9

Research Activities

Alpha and beta thalassaemia. Chromosome disorders. Duchenne muscular dystrophy.

Research Facilities

Training of personnel in Singapore and South-East Asia.

Slovakia

INSTITUTE OF PLANT GENETICS, SLOVAK ACADEMY OF SCIENCES

Director Dr Anna Pretova

Contact Person(s) Dr Anna Pretova

Address Akademická 2, P.O. Box 39A
Nitra 950 07
Slovakia

Tel: +42 87 36659 **Fax:** +42 87 36660

E-mail: pretova@savba.savba.sk

Year Established 1990 **Total Number of Staff** 45

Total Annual Budget US$ 200,000

Research Activities

The research is oriented towards plant genetics, particularly molecular and cellular. Plant biotechnology including all tissue culture techniques and techniques of recombination DNA (plant transformation). Cell totipotency. Studies of genetic variability using RFLP methods. Population genetics.

Research Facilities

Tissue culture room. Thermocycler. Facilities and equipment for tissue culture work (laminar-flow box, etc.) High-quality standard greenhouse (automatically regulated).

Collaborative Projects

Potato Research Institute, Velká Lomnica, Slovak Republic: Improvement of the potato genome using new biotechnological approaches. Zeainvent, Slovak Republic: Production of maize haploids via gametic embryogenesis. CPRO-DLO, Wageningen, The Netherlands: Gene expression. Agricultural University, Wageningen, The Netherlands: Several projects. University of Georgia, Athens, USA: Cytological characteristics of somatic embryogenesis. Institute collaborates with the company Arbor Senec in area of biotechnology expertise.

Commercial Products

Large-scale production of plants. Production of forest trees (*Abies, Pinus, Quercus*) via somatic embryogenesis and organogenesis, and some crops (maize, flax and potatoes) via somatic and gametic embryogenesis and organogenesis.

Syria

ATOMIC ENERGY COMMISSION OF SYRIA (AECS)

Director Prof Ibrahim Othman

Contact Person(s) Prof Ibrahim Othman

Address P.O. Box 6091
Damascus
Syria

Tel: +963 11 6111926/7 **Fax:** +963 11 6112289

Year Established 1976 **Total Number of Staff** 26

Research Activities

Application of sterile male technique and genetic sexing in the control and/or irradication of insect pests; use of pheromones and insect hormone analogues in the disruption of insect behaviour and physiology; use of immunoassays to detect pesticides in the environment; mutation breeding for disease resistance and quality in major crops; plant cell and tissue culture; storage protein electrophoresis and isozyme analysis.

Publications

Eleven publications in international peer-reviewed journals over past five years.

Affiliations/Subsidiaries

ICGEB affiliated centre.

THE INTERNATIONAL CENTRE FOR AGRICULTURAL RESEARCH IN THE DRY AREAS (ICARDA)

Director Dr Adel El-Beltagy

Contact Person(s) Dr Mohan C. Saxena (Research Coordinator), Dr F. Weigand

Address P.O. Box 5466
Aleppo
Syria

Tel: +963 21 23477 **Fax:** +963 21 21490

E-mail: ICARDA@cgnet.com

Year Established 1977 **Total Number of Staff** 500

Total Annual Budget US$ 20 million

Research Activities

Plant breeding to improve barley, lentil, chickpea, fava bean, durum wheat and bread wheat tolerance to biotic and abiotic stresses (biotechnology supported research). Conservation and management of natural resources of land, water and agricultural biodiversity (biotechnology supported research). Production system development and management including crop-livestock integration and management of small ruminants, with major emphasis on improved on-farm water use efficiency. Socioeconomics, public policy and public management research.

Research Facilities

Research farms in northern Syria (three) and Lebanon (two) to cover various agroecologies and range of rainfall. Laboratories for pathology, entomology, microbiology, biotechnology, crop quality, virology, soil fertility, plant physiology, GIS. Controlled growth facilities: plastic tunnels, greenhouses, growth chambers. Germplasm collection of chickpea, lentil, fava bean, barley, bread wheat and durum wheat. Recommendations and advice on improved management of natural resources. Trained manpower in the region of West Asia and North Africa.

Collaborative Projects

Strong collaborations with national programmes in West Asia and North Africa as well as with universities and national research institutions in Australia, Canada, France, Germany, Italy, Spain, UK and USA.

Publications

ICARDA produces its corporate annual report and special report of the germplasm programme in which all work carried out in biotechnology research at the Centre is reported. The scientists at ICARDA contribute research articles to international journals dealing with biotechnology.

Affiliations/Subsidiaries

ICARDA is supported by the CGIAR and carries out cooperative research with more than twenty countries in the West Asia and North Africa region.

Commercial Products

New plant varieties (chickpea, lentil, fava bean, barley, bread wheat, durum wheat, vetches and tathyrus). (No large-scale commercial production.)

Thailand

KASETSART AGRICULTURAL AND AGRO-INDUSTRIAL PRODUCT IMPROVEMENT INSTITUTE

Director Assoc Prof Vichai Haruthaithanasan

Contact Person(s) Assist Prof Dr Klanarong Sriroth

Address Kasetsart University
Bangkok 10900
Thailand

Tel: +66 2 9405634 or 9428604 **Fax:** +66 2 9405634

E-mail: aapkrs@ku.ac.th

World Wide Web: http://www.Std.ku.ac.th/~aapkrs

Year Established 1991 **Total Number of Staff** 46

Total Annual Budget US$ 1.0 million

Research Activities

Research and development on the following products: pulp paper/board; textiles; starch (tapioca and others) and sugar.

Research Facilities

Major capital equipment includes: High performance size exclusion chromatography (HPSEC); high performance thin layer chromatography (HPTLC); plarimeter, Schmidt & Haensch No. 26490; protein analyser; differential scanning calorimeter (DSC); rapid visco analyser (RVA) Model-4; visco amylograph (Barbender); pulping machine.

Collaborative Projects

National Institute of Bioscience and Human Technology (NIBH), Japan: Biodegradable plastic from cassava starch. Wageningen Agricultural University, The Netherlands: Cassava breeding development. University of Massachusetts, Amhurst, USA: Physiological and biological properties of cassava starch. Asian Institute of Technology (AIT): Cassava starch utilization. Centre de Coopération International en Recherche Agronomique pour le Développement (CIRAD), France: Textile technology. Japan International Cooperation Agency (JICA), Japan: Research project for increased utilization of forestry and agricultural plant materials in Thailand.

Publications

Five publications in peer-reviewed journals over past five years.

Yugoslavia

INSTITUTE OF MOLECULAR GENETICS AND GENETIC ENGINEERING
(INSTITUT ZA MOLEKARNU GENETIKU I GENETICKO INZENJERSTVO)

Director Dr Zvezdana Popovic

Contact Person(s) Prof Vladimir Glisin

Address Vojvode Stepe 283/3, P.O. Box 794
Belgrade 11000
Yugoslavia

Tel: +381 11 491198 **Fax:** +381 11 492397

E-mail: beogene@eunet.yu

Year Established 1986 **Total Number of Staff** 72

Total Annual Budget US$ 660,000

Research Activities

Penicillin amidases of Gram negative bacteria and yeast; synthesis of high value proteins in yeast. DNA replication and mutagenesis in prokaryotes. Molecular genetics of pathogenic *Streptococci.* Molecular genetics of lactic acid bacteria (LAB); collection of natural isolates of LAB from home-made cheeses. Cloning and characterization of human *sox* genes. DNA replication in eukaryotes; DNA-protein interactions; application of recombinant DNA technology in medicine (molecular diagnostics of genetic diseases). Molecular basis of anaemia; gene therapy-retroviral vectors; genetics of atopy. Characterization of buckwheat seed storage proteins and cloning of corresponding genes.

Research Facilities

Fermentor; cold room; sterile room; ultracentrifuges; laminar flow hoods; scintillation counter; 'Sorval' centrifuges; 'Olympus' microscope.

Collaborative Projects

ICGEB, Trieste: Molecular and prenatal diagnosis of genetic diseases, specifically cystic fibrosis. Investigation of post-translational modifications relevant for obtaining high yields of bacterial heterodimeric penicillin amidases expressed in yeast. Facultad de Veterinaria, Codroba, Spain: Expression of limonate dehydrogenase and limonin penicillin amidases from *Rhodococcus fascians* in yeast.

Publications

Forty-seven publications in international peer-reviewed journals over last five years.

Affiliations/Subsidiaries

ICGEB affiliated centre.

Commercial Products

Penicillin G amidase; human beta-interferon; starter cultures for production of cheeses of specific geographical origin.

Zimbabwe

TOBACCO RESEARCH BOARD

Director L.T.V. Cousins

Contact Person(s) L.T.V. Cousins

Address P.O. Box 1909
Harare
Zimbabwe

Tel: +263 4 575289 **Fax:** +263 4 575288

Year Established 1950 **Total Number of Staff** 102

Total Annual Budget US$ 4.0 million

Research Activities

Tobacco production. Crop protection and breeding.

Research Facilities

Well equipped laboratories and land for field testing.

Collaborative Projects

Scientific and Industrial Research and Development Centre (SIRDC), Zimbabwe: Production of sweet potato planting material. Rhône-Poulenc Agrochimie: Use of bromoxynil resistance as a marker gene in tobacco. Max-Planck Institute, Germany: Production of fungal resistance in tobacco.

Commercial Products

Trichoderma marzianum, a parasitic fungus for control of fungal diseases. Manufactured and marketed in conjunction with local agrochemical company.

Appendix II
Questionnaire

Introduction

The following questionnaire was sent to more than 500 research centres which had registered research activities in biotechnology in international research directories or had been recommended by those responding initially to the questionnaire.

The Questionnaire

Section I - *Your Research Centre*

Identification

1. Official name of your research centre:
2. English name of your research centre:
3. Year of foundation:
4. Director:
5. Contact person(s):
6. Address:
7. Telephone: 8. Telefax: 9. E-mail:
10. Relation (if any) with government organization:

Organization

11. Number of senior scientists:
12. Number of post-doctoral assistants:
13. Number of research students:
14. Number of technicians:
15. Number of administrative personnel:
16 Total number of staff:

Fields of research

17. Major areas of research in which your centre is actively involved:
18. Publications of the centre in international peer-reviewed journals (last five years only):
    Name of Journal/Newsletter Number of papers
19. Commercial products developed by your centre:

Collaborative projects

20. Collaborative projects with other laboratories:

Name of collaborating centre	Project description

21. Research grants obtained in collaborative projects:

Category of agency*	Name of granting agency	Project description	Amount of funds per year (US$)	Percentage of total budget of centre

* *Category of granting agency:*
A - Government
B - Commercial company
C - International agency
D - Overseas governments (including bilateral programmes)

Budget

22. Total annual budget of your research centre? (Please give a figure in US$):

23. Sources of funds (Please, specify sources and percentage of the total budget):

Category	Source	Percentage of total budget
Governments		
Bilateral programmes		
Multilateral programmes		
International research foundations		
Competitive grants		
Other		

Research facilities

24. Major research facilities or installations offered by the centre (e.g. greenhouses, animal facilities, fermentation plants, etc.):

25. Major capital equipment (list only that of value exceeding US$ 30,000):

Information

26. Access to electronic networks:

Internet: No Yes Others: No Yes; please specify:

27. Has your centre a Web site? No Yes; please specify the URL:

28. Library facilities:

Number of subscription to international journals of life sciences:
Access to on-line bibliographic databases: No Yes; please specify:
Access to on-line sequence databases: No Yes; please specify:
Access to other databases: No Yes; please specify:
Availability of electronic media: Microfilms CD-Roms
Other; please specify:

Intellectual Property Rights

29. Does your research centre own or use any patent or license concerning genetically modified organisms (GMOs) or transgenic products?

No Yes Details:

Biosafety

30. Has your research centre a biosafety committee for monitoring work involving genetically modified organisms (GMOs)?

No Yes Contact:

Section II - *Your Country...*

Other biotechnology organizations

31. Major domestic *research centres*:

Name and address	Area(s) of specialization				Comments
	Agriculture	Environment	Health	Other	

32. *Private enterprises* actively involved in biotechnology:

Name and address	Area(s) of specialization				Comments
	Agriculture	Environment	Health	Other	

33. Committees, organizations, institutes or companies whose activities address the issue of *marketing and distribution* of biotechnology-derived products:

Name and address	Area(s) of specialization				Comments
	Agriculture	Environment	Health	Other	

Regulatory environment

34. Does any legislation exist in your country for the *Protection of Intellectual Property* (IPRs) in biotechnology?

No Yes If yes, please specify:

35. Has your country laws, regulations or guidelines concerning biosafety?

For biotechnology in general	No	Yes	Specify:
For contained use of GMOs	No	Yes	Specify:
For field trials and commercialization of GMOs or GMO-derived products	No	Yes	Specify:

36. Has your country a national committee for monitoring work involving genetically modified organisms (GMOs) or recombinant DNA (rDNA)?

No Yes Official name:

Contact person:

37. To which government department or ministry does this Committee report?

38. Does the Committee issue guidelines? Yes No

39. Has the Committee statutory power? Yes No

40. Is compliance mandatory or voluntary? Please specify:

41. Which issues does the Committee address?

Issues	Yes	No	Comments
Genetic manipulation			
Containment			
Release of transgenic organisms			
Biodiversity			

General Comments:

Appendix III
International Agricultural Biotechnology Initiatives

Introduction

Following is a full overview of the various international initiatives referred to in Chapter Two, organized by the following five categories:

- International research programmes for plant biotechnology.
- International research programmes for livestock biotechnology.
- International or regional biotechnology networks for specific crops or regions.
- Special programmes initiated and coordinated by bilateral or multilateral donor organizations.
- Organizations providing advisory services on policy and research management issues.

Name of initiative (host institution)	Priorities	Agricultural focus (crop/livestock)	Region/country focus
CROP BIOTECHNOLOGY PROGRAMMES			
Agricultural Biotechnology for Sustainable Productivity, ABSP (Michigan State University, USA)	• Genetic engineering of crops for pest/disease resistance • Development of micropropagation systems • Integration of biotechnology within a general agriculture and business framework	• Maize • Potato • Coffee • Sweet potato • Horticultural crops	• Indonesia • Egypt • Kenya
Bean/Cowpea Collaborative Research Support Programme, B/C CRSP (various US universities)	• Control of pests and diseases • Increase crop yields • Increase nutritional quality	• Bean • Cowpea	• International
Centre for the Application of Molecular Biology to International Agriculture, CAMBIA	• Novel biotechnologies and methods for agricultural innovation • Genetic markers and diagnostics • Apomixis	• Rice • Cassava • Bean • Agroforestry	• International
CATIE - Biotechnology Research Unit (Centro Agronomico Tropical de Investigacion y Ensenanza, Costa Rica)	• Enhance regional programme capabilities • Genetic improvement of tropical crops	• Banana/plantains • Coffee • Cocoa • Roots and tubers	• Latin America and the Caribbean
CIAT - Biotechnology Research Unit (International Centre for Tropical Agriculture, Colombia)	• Increasing the efficiency of CIAT strategic research • Institutional development in biotechnology	• Cassava • Common bean • Rice • Tropical forages	• International
CIMMYT - Applied Biotechnology Centre (International Centre for Maize and Wheat Improvement, Mexico)	• Enhanced resistance to pests and diseases • Enhanced stress tolerance	• Maize • Wheat	• International

CIP - Applied Biotechnology Programme (International Potato Centre, Peru)	• Reduce dependence on costly toxic chemical pesticides for potato and sweet potato production • Host-plant resistance in potato	• Potato	• International
CIRAD - Plant Breeding Division (Centre de Coopération International en Recherche Agronomique pour le Développement, France)	• Develop genetically improved crops	• Cotton • Rice • Sorghum • Tropical perennials • Tropical fruits • Forestry	• International
Feathery Mottle Virus Resistant Sweet Potato for African Farmers (Agency for International Development, USA)	• Human resource development • Production of virus-resistant, African varieties of sweet potato • Enhance capacity in biosafety regulation of transgenic crop plants • Export of transgenic sweet potato to Africa for field testing • Technology transfer	• Sweet potato	• Kenya
ICGEB - Plant Biology Programme (International Centre for Genetic Engineering and Biotechnology, Italy/India)	• Capacity building • Genetically improved rice	• Rice	• International (ICGEB member countries)
ICRISAT - Molecular and Cellular Biology Programme (International Crops Research Institute for the Semi-Arid Tropics, India)	• Support and complement conventional crop improvement programmes at ICRISAT	• Sorghum • Pearl millet • Groundnut • Chickpea • Pigeonpea	• International

Name of initiative (host institution)	Priorities	Agricultural focus (crop/livestock)	Region/country focus
IIRSDA - Plant Biotechnology Programme (Institut international de recherche scientifique pour le développment en Afrique, Côte d'Ivoire)	• Conservation and characterization of yam germplasm • Micropropagation and genetic improvement of yam and other crops	• Yam • African eggplant	• Africa
IITA - Biotechnology Research Unit (International Institute for Tropical Agriculture, Nigeria)	• Tackle recalcitrant problems in crop improvement • Enhance national research capabilities	• Cowpea • Yam • Cassava • Banana/plantain	• Africa
International Laboratory for Tropical Agricultural Biotechnology, ILTAB (Scripps Research Institute, USA)	• Genetically engineered food crops with virus resistance	• Rice • Cassava • Tomato • Sugarcane	• International
International Programme on Rice Biotechnology (Rockefeller Foundation, USA)	• Rice genetic improvement • Capacity building	• Rice	• International
International Service for the Acquisition of Agri-biotech Applications, ISAAA (Cornell University, USA)	• Transfer and delivery of appropriate biotechnology applications to developing countries and the building of partnerships between institutions in the South and the private sector in the North, and by strengthening South-South collaboration	• Vegetables • Fruits • Field crops (e.g. cotton) • Cereals • Forestry	• International
Overseas Development Administration - Plant Sciences Research Programme (University of Wales, UK)	• Genetically improved crops	• Cereals • Roots and tubers • Legumes • Oilseeds • Fruit and vegetables • Fibres	• International

Philippine-German Coconut Tissue Culture Project (Albay Research Centre, Philippines)	• Micropropagation of coconut	• Coconut	• Philippines
Regional Programme of Biotechnology for Latin America and the Caribbean (several UN organizations)	• Collaborative research projects • Training	• Maize • Potato • Sugarcane	• Latin America and the Caribbean
Research on the Date Palm and the Arid Land Farming Systems (Estacion Phoenix, Spain)	• *In vitro* propagation • Biological control technology • Date palm farming systems	• Date palm	• Africa • Asia
LIVESTOCK BIOTECHNOLOGY PROGRAMMES			
CIRAD - Animal Production Division (Centre de Coopération International en Recherche Agronomique pour le Développement, France)	• Development of heat-stable vaccines through genetic engineering • Improved diagnostic tests • Determination of genetic resistance to diseases	• Cowdriosis • Dermatophilosis • Rinderpest • Peste des petits ruminants • Mycoplasmosis • Trypanosomiasis	• International
ILRI - Tick-Borne Diseases Programme (International Livestock Research Institute, Kenya)	• Novel vaccines • Improve current control methods	• Theileriosis • Cowdriosis • Anaplasmosis • Babesiosis	• International
ILRI - Trypanosomiasis Programme (International Livestock Research Institute, Kenya)	• Improve diagnosis and parasite characterization • Novel vaccines • Breeding for genetic resistance	• Trypanosomiasis	• International

Name of initiative (host institution)	Priorities	Agricultural focus (crop/livestock)	Region/country focus
International Laboratory of Molecular Biology for Tropical Disease Agents, ILMB (University of California, USA)	• Live recombinant virus vaccines for animal diseases • Technology transfer	• Rinderpest • Bovine virus diarrhoea • Equine influenza • Peste des petits ruminants • Foot and mouth disease • Vesicular stomatitis virus	• International
International Programme on Vectors and Vector-borne Diseases (University of Florida, USA)	• Development and commercialization of improved vaccines and diagnostic tests	• Heartwater	• SADC countries • Caribbean
Small Ruminant Collaborative Research Support Programme - Animal Health Component (Washington State University, USA)	• Improve the efficiency of milk and meat production from small ruminants • Virus-vectored vaccines for sheep and goats	• Heartwater • Contagious caprine pleuropneumonia • Nairobi sheep disease	• Kenya • Indonesia • Bolivia
CROP/LIVESTOCK PROGRAMMES			
ICIPE - Biotechnology Research Unit (International Centre of Insect Physiology and Ecology, Kenya)	• Biological control of pests (plant protection) and vectors • Development of anti-tick vaccines • Development of diagnostics tools	• Maize • Sorghum • Cowpea • Cattle	• Africa

Indo-Swiss Collaboration in Biotechnology, ISCB (Federal Institute of Technology, Switzerland)	• Research capacity building • Human resource development • Development, production and commercialization of specific biotechnology products • Partnerships between research groups (public and private sector)	• Foot and mouth disease • Contagious caprine pleuropneumonia • Plant biopesticides	• India
NETWORKS			
African Biosciences Network - Sub-Network for Biotechnology, ABN-BIOTECHNET (University of Nigeria, Nigeria)	• Genetically improved crops and farm animals • Disease control through new vaccines • capacity building		• Africa
Asia Network for Small-Scale Agricultural Biotechnologies, ANSAB	• Plant tissue culture • Biopesticides • Biofertilizers • Mushroom technology	• Potato • Kapok tree • Rice • Mushroom	• Asia
Asian Rice Biotechnology Network, ARBN (International Rice Research Institute, The Philippines)	• DNA fingerprinting of pests and pathogens • Low-cost marker-aided selection • Transgenic rice	• Rice	• Asia
Phaseolus Bean Advanced Biotechnology Research Network, BARN (International Centre for Tropical Agriculture, Colombia)	• Constraint identification • Technology transfer • Information exchange	• Beans	• International

Name of initiative (host institution)	Priorities	Agricultural focus (crop/livestock)	Region/country focus
Cassava Biotechnology Network, CBN (International Centre for Tropical Agriculture, Colombia)	• Stimulate cassava biotechnology research on priority topics • Integrate priorities of small-scale farmers, processors, and consumers in cassava biotechnology research planning • Information exchange	• Cassava	• International
Technical Cooperation Network on Plant Biotechnology, REDBIO (Food and Agriculture Organization of the United Nations, Regional Office for Latin America and the Caribbean, Chile)	• Generation, transfer and application of plant biotechnology • National and regional policies • Information exchange	• Vegetables • Roots and tubers • Cereals	• Latin America and the Caribbean
DONOR AGENCIES			
Australian Centre for International Agricultural Research, ACIAR	• Use biotechnology wherever appropriate as a research tool within any of ACIAR's projects		• International
DGIS Special Programme Biotechnology and Development Cooperation (Ministry of Foreign Affairs, The Netherlands)	• Improve developing-country access to biotechnology, with special emphasis on small-scale producers and women • Technical cooperation • International collaboration and coordination	• 'Orphan' commodities • Cassava	• Colombia • India • Kenya • Zimbabwe
FAO/AGP Programme on Plant Biotechnology (Food and Agriculture Organization of the United Nations, Italy)	• Information dissemination and cooperation • Advisory services • Capacity building • Promote research, technology transfer and adoption	• Rice • Roots and tubers • Horticulture • Industrial crops	• International

GTZ Biotechnology in Plant Production (Agency for Technical Collaboration, Germany)	• Development of micropropagation systems with diagnostic and pathogen elimination • Training and capacity building • Integration of biotechnology within BMZ/GTZ supported projects	• Potato • Cassava • Yam • Date palm • Coconut • Banana	• Africa • Asia
Swedish Agency for Research Collaboration with the Developing Countries, SAREC	• Plant and forestry genetics • Diagnostics and vaccines in veterinary medicine • Environment • Biosafety • Policy research		• Africa • Asia
United Nations Development Programme (UNDP)	• Productive and sustainable agriculture	• Food crops • Cash crops • Livestock	• International
United Nations Educational, Scientific, and Cultural Organization (UNESCO) – Biotechnology Action Council	• Human resource development		• International
World Bank	• Invest in biotechnology as a contribution to economic development in World Bank member countries		• International
POLICY/MANAGEMENT PROGRAMMES			
Biotechnology Advisory Commission, BAC (Stockholm Environment Institute, Sweden)	• Provide independent, impartial advice on biosafety development and implementation to developing countries • Biosafety capacity building		• International

Name of initiative (host institution)	Priorities	Agricultural focus (crop/livestock)	Region/country focus
Canada-Latin America Initiative on Biotechnology and Sustainable Development, CamBioTec (Centre for Technological Innovation, National Autonomous University, Mexico)	• Identify opportunities for biotechnology research and applications by tracking technological trends and carrying out priority-setting exercises • Strengthen public policies in biotechnology • Promote improved management of innovations • Foster partnerships between Canadians and Latin Americans		• Latin America
Intermediary Biotechnology Service, IBS (International Service for National Agricultural Research, The Netherlands)	• Biotechnology research programme management and policy formulation • Country reviews • Identify international programme expertise		• International
Support to Agricultural Biotechnology Policies (Interamerican Institute for Cooperation in Agriculture, Costa Rica)	• Biosafety, IPR • Industry development		• Latin America and the Caribbean

Appendix IV
Acronyms

ABN-BIOTECHNET	African Biosciences Network - Sub-Network for Biotechnology (University of Nigeria, Nigeria)
ABRABI	Brazilian Association of Biotechnological Enterprises
ABSP	Agricultural Biotechnology for Sustainable Productivity (Michigan State University, USA)
ACIAR	Australian Centre for International Agricultural Research
AECI	African Explosives and Chemical Industry
AFRAND	African Foundation for Research and Development
AGERI	Agricultural Genetic Engineering Research Institute (Egypt)
ÄKTA/FPLC	Advanced state of the art lab-scale protein purification system produced by Amersham Pharmacia Biotech
ANSAB	Asia Network for Small-Scale Agricultural Biotechnologies
ARC	Agricultural Research Council
ARBN	Asian Rice Biotechnology Network (IRRI, The Philippines)
ASEAN	Association of South East Asian Nations (Thailand)
ASIBE	Institute of Ecological Genetics and Neotropical Biodiversity (Colombia)
AUPELF	Agence Francophone pour l'Enseignement et la Recherche (Morocco)
B/C CRSP	Bean/Cowpea Collaborative Research Support Programme (various US universities)
BAC	Biotechnology Advisory Commission (Stockholm Environment Institute)
BAK	Bioindustry Association of Korea
BARN	*Phaseolus* bean Advanced Biotechnology Research Network (CIAT, Colombia)
BARN	Bean Advanced Research Network
BCC	Biotechnology Cooperative Centre (Malaysia)
B-CLL	B-cell chronic lymphocytic leukaemia
BID	Banque Islamique pour le Développement
BID	Banco Interamericano de Desarrollo - Interamerican Development Bank (IDB) (cited in Brazil country profile)
BINAS	Biosafety Information Network and Advisory Service (UNIDO)
BIOTEC	National Centre for Genetic Engineering and Biotechnology (Thailand)

BIOTECH	National Intitute of Molecular Biology and Biotechnology (Philippines)
BIRD	International Bank for Reconstruction and Development
BRI	Biotechnology Research Institute (Zimbabwe)
CAMBIA	Centre for the Application of Molecular Biology to International Agriculture
CamBioTec	Canada-Latin America Initiative on Biotechnology for Sustainable Development
CAS	Chinese Academy of Sciences
CATIE	Centro Agronómico Tropical de Investigación y Enseñanza (Tropical Agricultural Research and Training Centre) (Costa Rica)
CBD	Convention on Biological Diversity (Rio Summit, June 1992)
CBN	Cassava Biotechnology Network (CIAT, Colombia)
CBPP	Contagious bovine pleuropneumonia
CCPP	Contagious caprine pleuropneumonia
CEADEN	National Centre of Nuclear Development (Cuba)
CENARGEN	Centro Nacional de Recursos Genéticos e Biotecnologia – National Research Centre for Genetic Resources and Biotechnology (Brazil)
CFC/BIRD	Common Fund for Commodities/International Bank for Reconstruction and Development
CGAT	Centre for Gene Analysis and Technology (Malaysia)
CGEB	Centre for Genetic Engineering and Biotechnology (Cuba, see also CIGB)
CGIAR	Consultative Group on International Agricultural Research (Washington, DC, USA)
CIAT	Centro Internacional de Agricultura Tropical - International Centre for Tropical Agriculture (Colombia)
CIGB	Centro de Ingeniera Genética y Biotecnología (Cuba, see also CGEB)
CIMMYT	Centro Internacional de Mejoramiento de Maíz y Trigo - International Centre for Maize and Wheat Improvement (Mexico)
CINVESTAV	Centro de Investigaciones y Estudios Avanzados - Centre for Advanced Studies (Mexico)
CIP	Centro Internacional de la Papa - International Potato Centre (Peru)
CIRAD	Centre de Coopération International en Recherche Agronomique pour le Développement (France)
CITMA	Ministry for Science, Technology and Environment (Cuba)
CNPq	Brazilian Council for Scientific Research
CNR	Consiglio Nazionale delle Richerce (Italy)
CNRS	Centre National de la Recherche Scientifique (France)
COHRED	Commission on Health Research for Development
COLCIENCIAS	Colombian Institute for Science and Technology Development
CONACYT	National Science and Technology Council (Mexico)
CONICIT	Ministry of Science and Technology (Venezuela)
CONICYT	National Commission of Science and Technology (Chile)
CORPOICA	Colombian Agricultural and Livestock Research Corporation
CNCBD	China National Centre for Biotechnology Development
CNIGB	National Commission for Genetic Engineering and Biotechnology (Venezuela)

CNRF	Centre National de la Recherche Forestière (Morocco)
CPRO-DLO	Centrum voor Plantenveredelingsen Reproductie Onderzoek – Dienst Landbouwkunding Onderzoek (The Netherlands)
CRDB	Central Research and Development Institute for Biotechnology (Indonesia)
CRIFC	Central Research Institute for Food Crops (Indonesia)
CSIR	Council of Scientific and Industrial Research (South Africa)
CYTED	Science and Technology Programme for Development (cited in Venezuela country profile)
DBT	Department of Biotechnology (India)
DENR	Department of Environment and Natural Resources (Philippines)
DFG	Deutsche Forschungsgemeinschaft (Germany)
DFID	Department for International Development (UK)
DGIS	Directorate-General for International Cooperation (Ministry of Foreign Affairs, The Netherlands)
DOST	Department of Science and Technology (Philippines)
EAC	Erythrocyte with antibody and complement
EC	European Community
ECOWAS	Economic Community of West African States
EDF	European Development Fund (EU)
ELISA	Enzyme-linked immunoassay
EMBRAPA	Empresa Brasileira de Pesquisa Agropecuária – Brazilian Agricultural Research Enterprise (Brazil)
EO	Executive order
ESIQIE	Escuela Superior de Ingenieria Quimica e Industrias Extractivas (Mexico)
ESR	Electron spin resonance
EU	European Union
FACS	Fluorescence activated cell sorter
FADIB	Foundation for African Development through International Biotechnology
FAO	Food and Agriculture Organization of the United Nations
FARM	Farmer Centred Agricultural Resource Management (Initiative of UNDP, UNIDO and FAO)
FDA	Food and Drug Administration (USA)
FIDETEC	Fideicomiso de Desarrollo Technologico (Mexico)
FIOCRUZ	Fundação Oswaldo Cruz - Oswaldo Cruz Foundation (Brazil)
FIS	Fondation Internationale pour la Science (Morocco)
FMD	Foot-and-mouth disease
FMST	Federal Ministry of Science and Technology (Nigeria)
FONAIAP	National Fund for Agricultural Research (Venezuela)
FPLC	Fast protein liquid chromatography
FRIM	Forest Research Institute of Malaysia
FT-NMR	Fourier transform NMR
GATT	General Agreement on Tariffs and Trade

GC	Gas chromatograph(y)
GIS	Geographic Information System
GDP	Gross domestic product
GEBRI	Genetic Engineering and Biotechnology Research Institute (Egypt)
GLC	Gas liquid chromatograph(y)
GMAC	Genetic Modification Advisory Committee (Malaysia)
GMM	Genetically modified microorganism
GMO	Genetically modified organism
GNP	Gross national product
GTZ	Agency for Technical Collaboration (Germany)
HAV	Hepatitis A virus
HBV	Hepatitis B virus
HCV	Hepatitis C virus
HIV	Human immunodeficiency virus
HLA-DR	Histocompatibility antigens encoded by the DR locus of the major histocompatibility gene complex
HPAEC	High pH anion exchange chromatography
HPLC	High pressure liquid chromatography
HPV	Human papillomavirus
HSP	Heat shock protein
IAEA	International Atomic Energy Association
IARC	International Agricultural Research Centres
IAV	Institut Agronomique et Veterinaire (Morocco)
IBS	Intermediary Biotechnology Service (ISNAR)
ICARDA	International Centre for Agricultural Research in the Dry Areas (Syria)
ICBiotech	International Centre for Biotechnology (Osaka University)
ICGEB	International Centre for Genetic Engineering and Biotechnology
ICIDCA	Institúto Cubano de Investigaciones de los Derivados de la Caña de Azucar - Cuban Research Institute of Sugar Cane By-Products
ICIPE	International Centre of Insect Physiology and Ecology (Kenya)
ICRISAT	International Crops Research Institute for the Semi-Arid Tropics (India)
IDB	Inter-American Development Bank (cited in Venezuela country profile)
IDB	International Development Bank (cited in Brazil country profile)
IDNDR	International Decade for Natural Disaster Reduction (United Nations)
IDRC	International Development Research Centre (Canada)
IFAD	International Foundation for African Development (Morocco)
IFAD	International Fund for Agricultural Development (United Nations)
IFN	Interferon
IGE	Institute of Genetic Engineering (Bulgaria)
IITA	International Institute of Tropical Agriculture (Nigeria)
ILMB	International Laboratory of Molecular Biology for Tropical Disease Agents (University of California, USA)
ILRI	International Livestock Research Institute (Kenya)
ILTAB	International Laboratory for Tropical Agricultural Biotechnology (Scripps Research Institute, USA)

INCA	Institúto de Ciencia Agricola - Institute of Agricultural Sciences (Cuba)
INCO	Cooperations with Third Countries and International Organizations (EU)
INCO-DC	Scientific and Technological Cooperations with Developing Countries (EU)
INIA	Institúto Nacional de Investigaciones Agronómicas (cited in Chile country profile)
INRA	Institut National de la Recherche Agronomique – National Institute of Agronomic Research (Morocco)
IOC	International Oceanographic Commission
IPB	Institute of Plant Breeding (Philippines)
IPB Net	International Plant Biotech Network
IPM	Integrated pest management
IPR	Intellectual property rights
IPRGI	International Plant Genetic Resources Institute (Italy)
IR	Infrared
IRPA	Intensified Research in Priority Areas (Malaysia)
IRRI	International Rice Research Institute (Philippines)
ISAAA	International Service for the Acquisition of Agri-biotech Applications (Cornell University, USA)
ISCB	Indo-Swiss Collaboration in Biotechnology (Federal Institute of Technology, Switzerland)
ISNAR	International Service for National Agricultural Research (The Hague, The Netherlands)
ITDI	Industrial Technology Development Institute (Philippines)
IUPAC	National Institute of Public Health and Environmental Protection
JBA	Japan Bioindustry Association
JICA	Japanese International Cooperation Agency
KARI	Kenya Agricultural Research Institute
KBN	Komitet Badan Naukowych (Poland)
KBRA	Korean Biotechnology Research Association
KEMRI	Kenya Medical Research Institute
KFAS	Kuwait Foundation for the Advancement of Science
KISR	Kuwait Institute for Scientific Research
LAB	Lactic acid bacteria
LC	Liquid chromatograph(y)
LN_2	Liquid nitrogen
Mab	Monoclonal antibody
MAFF	Ministry of Agriculture, Forestry and Fisheries (Korea)
MARDI	Malaysian Agricultural Research and Development Institute
MIRCEN	Microbiological Resources Centre (Thailand, UNEP, UNESCO)
MOSTE	Ministry of Science, Technology and the En*vironment (Malaysia)*
MOTT	Mycobacteria other than *Mycobacterium tuberculosis*
MPKSN	National Council for Scientific Research and Development

MOTT	Mycobacteria other than *Mycobacterium tuberculosis*
MPKSN	National Council for Scientific Research and Development (Malaysia)
MRC	Medical Research Council
MS	Mass spectrometer
NAFIN	National Financial Institution/Nacional Financiera (Mexico)
NAFTA	North Atlantic Free Trade Association
NARS	National agricultural research systems
NASENI	National Agency for Science and Engineering Infrastructure (Nigeria)
NASU	National Academy of Sciences of Ukraine
NATO	North Atlantic Treaty Organization
NBC	National Biosafety Committee
NBD	National Biotechnology Directorate (Malaysia)
NCB	National Committee for Biotechnology (Chile)
NCBP	National Committee on Biosafety of the Philippines
NCEMB	National Centre of Excellence in Molecular Biology (Pakistan)
NCST	National Council for Science and Technology (Kenya)
NDPA	National Development Planning Agency (Indonesia)
NEDO	New Energy Development Organization (Malaysia)
NGO	Non-governmental organization
NIBGE	National Institute of Biotechnology and Genetic Engineering (Pakistan)
NIH	National Institute of Health (Philippines)
NIR	Near infrared
NMR	Nuclear magnetic resonance
NNIB	Network of National Institutes of Biotechnology (Philippines)
NRC	National Research Centre
NSTDA	National Science and Technology Development Agency (Thailand)
NTP	New Technologies Programme (cited in Venezuela country profile)
OAU	Organization for African Unity
OECD	Organisation for Economic Co-operation and Development
PADCT	National Programme for Scientific and Technological Development (Brazil)
PAHO	Pan American Health Organization
PAS	Polish Academy of Sciences
PBR	Plant breeders' rights
PCASTRD	Philippine Council for Advanced Science and Technology, Research and Development
PCHRD	Philippine Council for Health Research and Development
PCR	Polymerase chain reaction
PORIM	Palm Oil Research Institute of Malaysia
PPR	Peste des petits ruminants
PSRP	Plant Sciences Research Programme
R&D	Research and development

RAPD	Random Amplified Polymorphic DNA
RAS	Russian Academy of Sciences
RCZ	Research Council of Zimbabwe
REDBIO	Technical Cooperation Network on Plant Biotechnology (FAO, Regional Office for Latin America and the Caribbean, Chile)
RIFCB	Research Institute for Food Crop Biotechnology (Indonesia)
RFLP	Restriction fragment length polymorphism
RHAE	Programa de Recursos Humanas em Áreas Estratégicas – Manpower Development Programme for Priority Research Areas
RPB	Regional Programme of Biotechnology (UNDP, UNIDO)
RRIM	Rubber Research Institute of Malaysia
S&T	Science and technology
SAARC	South Asian Association for Regional Cooperation
SAGENE	South African Committee for Genetic Experimentation
SAREC	Swedish Agency for Research Collaboration with Developing Countries
SEBRAE	Brazilian Support Service for Micro and Small Enterprise
SET	Science, Engineering and Technology
SIRDC	Scientific and Industrial Research and Development Centre (Zimbabwe)
SFE	Stopped flow electrophoresis
SNPA	National System of Agricultural Research (Brazil)
SSTCC	State Science and Technology Commission of China
STDB	Science and Technology Development Project (Thailand)
TDR	Special Programme for Research and Training in Tropical Diseases (UNDP, World Bank, WHO)
TNF	Tumour necrosis factor
TRIPS	Trade Related Aspects of Intellectual Property Rights
TWAS	Third World Academy of Sciences
UAAS	Ukrainian Academy of Agricultural Sciences
UN	United Nations
UNAM	National Autonomous University of Mexico
UNCED	United Nations Conference on Environment and Development
UNDP	United Nations Development Programme
UNDRO	United Nations Disaster Relief Office
UNEP	United Nations Environment Programme
UNESCO	United Nations Educational, Scientific and Cultural Organization
UNFPA	United Nations Fund for Population Activities
UNIDO	United Nations Industrial Development Organization
UNU	United Nations University (Japan)
UPOV	Union Internationale por la Protection des Obtentions Végétales - Union for the Protection of New Plant Varieties
URA	Uruguay Round of Agreements 1994
USAID	US Agency for International Development
USDA	United States Department of Agriculture

UV	Ultraviolet
VSV	Vesicular stomatitis virus
WANA	West Asia and North Africa
WAITRO	World Association of Industrial and Technological Research Organizations
WARDA	West Africa Rice Development Association
WHO	World Health Organization
WIPO	World Intellectual Property Organization (UN)
WTO	World Trade Organization